Food Related Enzymes

Food Related Enzymes

John R. Whitaker, *Editor*

A symposium sponsored by
the Division of Agricultural
and Food Chemistry at the
166th Meeting of the
American Chemical Society,
Chicago, Ill.,
Aug. 28–29, 1973.

ADVANCES IN CHEMISTRY SERIES **136**

AMERICAN CHEMICAL SOCIETY

WASHINGTON, D. C.   1974

Library of Congress CIP Data

Food related enzymes.

(Advances in chemistry series; 136)

Includes bibliographical references and index.

1. Fermentation—Congresses. 2. Enzymes—Congresses.
I. Whitaker, John R., ed. II. American Chemical Society. Division of Agricultural and Food Chemistry. III.
American Chemical Society. IV. Series. [DNLM: 1. Enzymes—Congresses. 2. Food-processing industry—Congresses. QU135 A512f]

QD1.A355 no. 136 [TP501] 540'.8s [547'.758] 74-
20861
ISBN 0-8412-0209-5   ADCSAJ 136 1-365 (1974)

UNIVERSITY OF PLYMOUTH

PLYMOUTH LIBRARY

Tel: (0752) 232323
This book is subject to recall if required by another reader
Books may be renewed by phone
CHARGES WILL BE MADE FOR OVERDUE BOOKS

Advances in Chemistry Series

Robert F. Gould, *Editor*

FOREWORD

ADVANCES IN CHEMISTRY SERIES was founded in 1949 by the American Chemical Society as an outlet for symposia and collections of data in special areas of topical interest that could not be accommodated in the Society's journals. It provides a medium for symposia that would otherwise be fragmented, their papers distributed among several journals or not published at all. Papers are refereed critically according to ACS editorial standards and receive the careful attention and processing characteristic of ACS publications. Papers published in ADVANCES IN CHEMISTRY SERIES are original contributions not published elsewhere in whole or major part and include reports of research as well as reviews since symposia may embrace both types of presentation.

CONTENTS

PREFACE

This book is the outgrowth of a 1½-day symposium that brought together those with basic and applied interests in enzymes for a mutual exchange of ideas. We hope this collection will further communication between the two areas.

One might well ask, what is a food-related enzyme? Is it an endogenous enzyme involved in the growth and maturation of plant or animal raw material? Is it an enzyme involved in developing the flavor, odor, texture, and color characteristics of a given food material? Is it an enzyme involved in adverse color, texture, or flavor changes in foods prior to processing? Is it an enzyme used to determine adequate heat treatment of foods? Is it an enzyme involved in digestion and assimilation of ingested foods? Or is it an exogenous enzyme added during processing or cooking to induce some desirable change such as starch hydrolysis, meat tenderization, or beer chillproofing? Is it an enzyme used to analyze for a constituent of a food or useful in treatment of waste from the food industry? Each of us will have our own views on what it is.

Given a broad definition of what a food-related enzyme is, how were the topics chosen for this monograph? Our major criteria were: 1. Is this enzyme, or group of enzymes, important or potentially important to food science and nutrition and the agricultural sciences in general? 2. Is there sufficient basic information available about this enzyme, or group of enzymes, to permit a meaningful discussion at the molecular level? 3. Will updating the present knowledge on this enzyme, or group of enzymes, be useful to the agricultural and food sciences? 4. Will presentation of fundamental knowledge concerning these enzymes assist in speeding the discovery of similar knowledge on other enzymes and the application of all enzymes in the agricultural and food sciences?

Use of enzymes to effect changes in foods during processing, to correct metabolic disorders in nutrition, and to determine constituents of agricultural and food products is increasing rapidly, particularly in relation to application as immobilized enzyme systems. For maximum efficiency and success in using these catalysts, we must have as complete an understanding of their basic properties as possible. Only through cross stimulation between researchers in the basic and the applied areas of enzymology can maximum success be achieved.

Davis, Calif.   JOHN R. WHITAKER
January 1974

ix

Enzymes in Foods—for Better or Worse

CHARLES I. BECK and DON SCOTT

Searle Biochemics, Arlington Heights, Ill. 60005

This review discusses enzymes that are indigenous to or added to foods and that alter the perception of those foods in both desired and undesired ways. Few enzymes have been commercialized for processing foods because the costs associated with optimizing manufacturing procedures and establishing product safety are often too high considering the market potential. Commercial enzyme preparations must have performance in relation to cost, properties compatible with use conditions, shelf stability, and controllable activity. Present enzyme preparations are predominantly used for a single activity, but there is interest in using enzyme systems for foods and food ingredient modifications. Changes in the nature of foods being marketed (e.g., processed vs. fresh) present many new problems that may be resolved with enzymes. A short annotated bibliography is included.

In a typical week in 1973, the number of abstracts referring to enzymes listed in *Biological Abstracts* was over 60; less than 10% of that number appeared in *Food Science and Technology Abstracts* as actually or potentially significant in food. We estimate the enzyme industry to be about $25 million domestically and $100 million worldwide. Of this, food applications are about $15 million domestically and $45 million worldwide (in 1973).

A book of this size can not encompass such a field effectively, but what it covers, it covers thoroughly. It provides both new and review information and an in-depth focus on the protein chemistry of selected enzymes that are commercially significant to the food industry.

This chapter is a review of significant enzymes which naturally occur in or are added to foods. Where the facts permit, we have fantasized the future, but clearly differentiated "what is" from "what will be."

Specifically this chapter:

    1. Relates the importance of specified enzymes to food quality

    2. Distinguishes between the enzymes naturally present in foods and those which are added

    3. Identifies properties and requirements which contribute to the commercial success or failure of a new enzyme

    4. Explains the genuine basis for optimism despite the retarded rate of commercial implementation of known materials and technologies

    5. Indicates how basic information, such as that discussed in the chapters to follow, has been, and will continue to be, important in developing commercial enzymes

    6. Provides a guide to selected related literature, particularly in the area of commercial application

    The title of this chapter implies a marriage which is real. We are married to enzymes in the sense that all living tissue develops, subsists,

Table I. Mixed

Use	Specific Enzymes

Candying fruit	*Trametes* preparations
	  lipase
	  protease
	  cellulase
	  hemicellulase
	  glucanase
	  mannase
	  pectinase (and more)

Desugaring eggs	glucose oxidase catalase

Brewing	sprouted barley
	  α-amylase
	  β-amylase

Egg white whipping	pancreatin
	  lipase
	  protease

Fruit juice clarification	polygalacturonase polymethylesterase pectic transeliminase hemicellulase

Cell separating
  Peeling fruit–mandarin oranges
  Vegetable baby food—carrots
  Starch separating—potatoes
  Remove seed coat—soybeans
	Trichoderma viride preparations
	  cellulase
	  pectinase
	  hemicellulase

and then breaks down through the action of enzymes. They are the change agents of all natural biological systems. Genetic factors (in the chemical form of DNA and RNA) may determine direction and developmental fate through their influence on the enzymes, but it is the enzymes which actually do the work.

Scientists have a penchant for identifying enzymes and postulating why they exist within a given system. In some cases there is a basis, in fact or theory, for such postulation, *e.g.*, the function of an enzyme in the evolution of that organism. More generally, we just acknowledge that the enzyme is there and then to learn how to live with it or how to use it. Although the organism survived the evolutionary process while retaining or developing the adaptive or constitutive ability to produce a particular enzyme, that activity was not developed for our specific benefit.

We learn to live with our environment. In some cases we adapt to enzymes, such as polyphenoloxidases which turn fruit brown, and in

Enzyme Systems

Specific Action	*References*
Enzymes make the fruit porous so that conc. sucrose solutions can penetrate faster without osmotic fruit shrinkage.	(3)
glucose + O_2 ⇌ gluconic acid + H_2O_2 $2H_2O_2 \rightarrow 2H_2O + O_2$ Catalase causes both oxygen atoms to be used. Residual peroxide would damage the fat.	(4)
The combined activities break down starch to maltose and then to glucose to rapidly provide a substrate for the yeast.	historic
Lipase hydrolyzes the fat which would interfere with egg white whipping. Protease generates peptones which may enhance whipping properties.	industrial use
Depolymerization splits ester linkage. Deesterification forms unsaturated monomers of galacturonic acid. With pectin degraded, hemicellulose is exposed.	(5)
Cellulosic walls in plant tissue are held together by pectic layers. The preparation separates cells by hydrolyzing those layers.	(6)

others we exploit enzymes which can be useful to us; whether in enzyme laundering, meat tenderization, or cheese curd development. In adapting, we can select and genetically improve on the status quo. We can select fruit which is low in polyphenoloxidases or select organisms to maximize our yield of a desired enzyme activity. By studying enzymes, we discover activators, inhibitors, and stabilizers. By studying the desired reaction, requirements for materials or conditions can be determined which permit control, optimization, or limitation of the reaction rate, degree, or course.

If one looks at the myriad enzymes in nature, the opportunities for development and innovative application should seem endless. Yet nature keeps thwarting us. Even where the desired reaction can be enzyme catalyzed and the enzyme can be produced, other factors often prevent utilization. Cellulase commercialization is one of the classic examples of unfulfilled application. If cellulase were available with value that exceeded its cost, the residues from wood pulp, spent paper, and vegetables (such as bagasse) could be used as sugars or as fermentation substrates for manufacturing other products. Real progress has been made in research in this area (1), but the products are not yet truly commercial. This is enigmatic since the entire plant kingdom develops, subsists, and dies around us, and yet we are not buried in cellulose. At present, only nature can afford this enzyme.

Enzyme Systems

A great deal is known about enzyme production (from microorganisms), purification, and isolation. However, enzyme activity and efficiency is often not just intramolecular architecture (amino acid sequence and spacial configuration) but also intermolecular organization.

The word organization can be used in a teamwork sense; assembling the necessary abilities to get the job done. The lysosomes (2), where a predominance of catabolic activities are assembled in one subcellular organelle, are an example of this kind of organization. In food-related enzyme systems, enzyme mixtures are often necessary to carry out a given process. Purification might result in the loss of performance in a given application. Examples of mixed enzyme systems are given in Table I.

The word organization is also significant in terms of engineering design. Here the mitochondrion (7) is a perfect example of engineering organization. In this system, the suborganelle structure can be used to explain the efficiency of electron transport, oxidative phosphorylation, and other highly efficient mitochondrial processes. Thus, using the lysosomes to emphasize the need to collect the required enzyme activities in one place and using the mitochondrion to emphasize the need to arrange those activities efficiently, biochemists and biochemical engineers may be able to work in an unlimited number of fields. Techniques for immobilizing enzymes, such as on membranes, provide for localization and possibly sequential activity arrangement. A specific exotic example of biochemical ingenuity was discussed by Rubin, Piez, and Katchalsky (8), who proposed a "motor" based on the contraction of proteins when treated with salt. Whole microbial cells represent sophisticated, naturally occurring immobilized enzyme systems. The subject of using entire organisms will not be covered in this book since it is more closely related to microbiology and fermentation than enzymology. Both fields are significant, but each deserves individual attention.

Enzymes Indigenous to Foods

Food-related enzymes have already been identified as both those enzymes which are added to food and those which occur naturally in food. In discussing enzymes indigenous to food systems, much narrowing and selection is necessary. If we truly dealt with all enzymes indigenous to food systems, we would be dealing with virtually all enzymes. Of the nearly 2000 enzymes now identified, most of them are widespread. Enzymes associated with digestion, synthesis, transport, and respiration function similarly with similar enzymes in widely different organisms. The enzymes (in their specific architecture) and the pathways may vary slightly, but the analogies and the general concepts of metabolism are carried out by enzymes which are operationally the same.

Food-related enzymes indigenous to the food system are enzymes whose characteristics set them apart because of their individual impact on aspects of the food system that we can perceive. Thus, we are dealing here with only a few enzymes; enzymes whose nature distinguishes them for us.

Members of the genus *Allium* (onion, garlic, leeks, etc.) provide a powerful example of an enzyme which distinguishes itself. Though there are hundreds of detectable enzymes occurring naturally in foods, it is difficult to find many examples where an individual enzyme is predominately responsible for a phenomenon. Enzymes and enzyme systems are responsible for the very identity of the food product. It is the enzyme composition that determines the flavor attributes and the textural differences in firmness, as well as the general appearance and color of food items. Table II provides examples of cases where particular changes in food can be primarily ascribed to a specific enzyme or enzyme system.

In addition to those mentioned in Table II, there is a group of enzymes in foods which lower the nutritive content. This group includes: thiaminase which acts on thiamine in flesh food, ascorbic acid oxidase in

Table II.   Enzymes

Enzyme	Food System	Impact
Alliinase	onion, garlic, leek, etc.	Enzyme and flavor precursor react on tissue damage (cutting) to release the active flavors.
Polyphenoloxidase	fruits and vegetables	Aerobic darkening which is associated with and accelerated by tissue damage
Amylases (α and β)	germinated barley (malt)	Converts starchy endosperm to sugars fermentable by yeast to make beer
	sweet potato	Converts potato starch to sugar
Amylase	potato	Catalyzes the equilibrium between potato starch and sugar
Phosphatase Nucleosidase system ATPase Myokinase Deaminase	meat, fish, poultry	Inosinic acid is associated with high quality flesh foods. It is converted to inosine by phosphatase, then to hypoxanthine by nucleosidase enzymes. Inosinic acid is developed from ATP by the other enzymes listed.
Esterases	fruits	Responsible for developing ester flavors during ripening
Lipases	cheese	Act on milk lipids to help generate characteristic cheese flavors
	butter	Lipolyzes butterfat to give rancid flavor, the perception of which may be pleasant or unpleasant depending on degree and individual preference
Peroxidase/ Catalase	vegetables	Causes off flavors on storage
Protease	wheat flour	Degrades gluten and thereby reduces loaf volume and increases cell size
Cathepsins	meat	Autolytic changes in tissue resulting in natural tenderization without visible changes in outer membrane of muscle fiber

Indigenous to Food Systems

Adaptation	*Reference*
To maximize flavor, cut direct into food just prior to use. As a flavorant, freeze prior to pulverizing, lyophilize, and add water to the powder to activate enzyme and produce the flavor.	*(9)*
In prunes this is associated with a typical condition, "norm." Fruits which are processed are protected by SO_2, ascorbic acid, citric acid, or other means of reducing exposure to oxygen while the enzyme remains active. Heat processing such as blanching or retorting destroy the polyphenoloxidase.	*(10)*
The malting process of barley increases the amylase content to hydrolyze the starch from both the barley and adjunctive cereals.	historic
A preheat step activates the enzyme to sweeten the finished product, sweet potato flakes.	*(11)*
Controlled storage at an optimum temperature shifts the equilibrium in favor of starch accumulation rather than glucose, thus reducing browning in frying.	*(12)*
Cold storage of flesh foods can maximize fresh taste by keeping inosinic acid from destruction. Also humane slaughtering of rested animals maximizes the initial ATP level and thereby raises the level of inosinic acid.	*(13)*
Flavor and texture of fruits are factors in choosing harvest, storage, and processing conditions.	*(17)*
Unpasturized milk is used for cheese making and the resultant cheese is aged to insure the death of potentially pathogenic organisms.	*(10)*
Heat process to inactivate these along with others	industrial use
Proteases are inactivated by oxidizing agents such as bromates.	*(18)*
Beef is aged at 4°C and is tenderized. A process involving surface irradiation controls microbial growth and permits somewhat higher temperatures to accelerate tenderization.	*(14)*

Table II.

Enzyme	Food System	Impact
Glycolytic enzyme system	meat	Glycogen is converted to lactic acid which lowers pH from just over 7 to about 5.4 when there were good reserves of glycogen. Attaining this pH is important to obtain the desirable color and keep the meat quality. Where the animal has undergone severe struggling the final pH can be as high as 6.6, and the meat is dark, dry, close-textured and more susceptible to microbial spoilage.
Reductase	marine fish	Enzymes from indigenous bacteria reduce trimethylamine oxide to trimethylamine (NADH dependant) coupled with the oxidation of lactic acid to acetic acid and CO_2.
Myrosinase (Thioglucosidases)	*Brassica* mustard rape-seed horseradish root	Converts thioglucosides to isothiocyanates and sugars. The isothiocyanates are responsible for flavor but are toxic. They are more potent flavorants than toxicants.
Invertase (Bee)	honey	Bees natully make invert sugar.
Lipase	cocoa powder from expeller cake (or milk powder)	Hydrolyzes fats to give an off-flavor to finished milk chocolate products
	In egg white powder from trace content of yolk and from residual pancreatin used to remove fat	Hydrolyzes fats (especially lauric fat) to give off-flavors to nougat creams in candy.
Amylase	wheat flour used in licorice	Breaks down starch and reduces tensile strength of extruded sheet product

Continued

Adaptation	*Reference*
Slaughter is carried out on a rested, well fed animal with a minimum of stress to that animal.	(*15*)
Fish can be stored alive or chilled, but commercial practices vary.	(*16*)
Rape-seed processed for protein can be pre-blanched and leached.	(*19*)
We build hives and raise bees to manufacture this sweetener for us.	historic
The cocoa powders are heated to 107°C in fat for 5 minutes.	(*20*)
The albumin-syrup mixture is heated to 71°C for 20 minutes.	(*21*)
Steam-treated flour, in which amylase activity has been destroyed, is used.	industrial use

fruit and vegetables which oxidizes vitamin C, lipoxygenase in vegetables which oxidizes vitamin A and essential fatty acids, and phosphatase which hydrolyzes pyridoxal phosphates.

Indigenous food enzymes can also serve as processing and/or freshness indicators. Phosphatase activity in milk is a measure of heat processing (pasteurization). The absence or presence of amylase activity in eggs indicates whether or not they were pasteurized. Peroxidase activities in vegetables such as peas, corn, green beans, and turnips can indicate the extent of their heat processing. Enzyme activities have also been considered as thaw indicators in frozen food storage.

Many enzyme effects are combined effects, and the adaptations that we make are combined adaptations. For example, the conversion of starch in a potato to sugar may cause browning when the potato is fried as a chip. This is controlled through storage conditions which minimize the free glucose content. The conditions chosen, such as temperature, humidity, and gaseous composition, will control sugar content but will also control the texture of the potato and other respiratory processes. Similarly, when vegetables or fruits are blanched to control the activity of polyphenoloxidases or peroxidases, the treatment destroys other enzymes as well. There are many industrial processes treating one or more enzymes which have become so commonplace that the enzyme basis for their use is no longer apparent; what is apparent is the quality of the finished product.

Another source of enzyme changes in food stuffs is related to enzyme systems indigenous to the microorganisms which are indigenous to the foods. They are too complex to deal with here. In the preceeding section, we have tried to indicate those enzymes whose impact is understood and to identify our adaptations to improve a food product or prevent its deterioration.

Enzymes Added to Foods

There is a sizable volume of literature which describes, and in some cases promotes, the use of enzymes in foods. Table III contains enzymes and applications considered currently significant to both the food and enzyme industries.

Characteristics of Commercially Successful Enzymes

Economic Considerations. The most important consideration in determining whether an enzyme will be a commercial success is the economics of its use—will the cost of using the enzyme be at least equal to the value of the changes rendered through its use? This principle has interesting ramifications that tend to horrify the classical enzyme chemist.

The enzymologist ordinarily studies an enzyme under conditions of optimum activity for that enzyme. He has essentially complete freedom to set conditions of pH, temperature, time, enzyme concentration, substrate concentration, nature and amount of activators, inactivators, or cofactors, and to disregard cost.

The commercial food enzymologist must accept the conditions that exist in the food product to be treated. In most cases he can only modify these to a small degree, if at all. For microbiological reasons, he often must work at temperatures that are far from ideal for enzyme action. Substrate concentrations are those that exist in complex foods. Most of all, he is not concerned with whether the operating conditions approach the optimum for the enzyme, but rather, he is concerned with the cost of making the changes that are possible under the given conditions.

The ideal conditions for a new enzyme are those that exist in current practice, requiring minimal if any changes in capital equipment or process technology. In some cases, the reactions catalyzed are so novel (*e.g.*, glucose isomerase) that the installation of completely new processing equipment and conditions is justified, and in these cases, attempts at approaching optimal conditions for the enzyme are definitely taken into account.

Many other factors enter into the economics of enzyme use. One of the most difficult problems with a new enzyme is that the production scale must be large enough to lower the price per unit of enzyme activity. However, the use cannot further expand until the price has already come down. Bcause of the comparatively small size of any single enzyme application, most enzymes require several applications to be economical in manufacture.

The cost of technical services rendered by the enzyme manufacturer, particularly in the beginning, also tends to be prohibitive. The limitations on organisms (*see* Safety, below) are restrictive in an economic sense as well. In general, conditions do not encourage new enzyme and new application research by the enzyme manufacturer in the U.S. Traditionally very low portions of gross sales are spent on this type of research. Most expenditures are for extending uses and improving enzymes already in the company's list. The interest and activity in immobilized enzymes has introduced another factor relative to economics—recoverability of the enzyme for reuse.

Economic considerations are not always simple to determine. Consider, for example, the deteriorative changes that take place in egg white solids on storage. Egg white solids produced by drying the whites immediately after separation from the yolks contain substantial amounts (3.5%) of glucose. Over a period of months, solubility loss and color darkening are caused by the interaction of the aldehyde group of the glucose and

Table III.   Commercial Enzymes

Food	*Enzyme(s)*
Beer	amylase (malt)
	papain (papaya) also other plant and microbial proteases
	β-glucanase (various microbial)
Bread	α-amylase (fungal and others)
	protease (fungal and others)
	lipoxygenase (soybean)
Cheese	rennet (calf stomach)
	pepsin, chymotrypsin, microbial proteases esterases-oral (throat) lipases (pregastric of kid) catalase (fungal)
Confections	invertase (yeast)
Confection (sugar or chocolate coated)	pectinase
Corn syrup (high-maltose)	amylase (fungal)

and Their Applications to Foods

Action	*Utility*
Cleaves starch	Permits fermentation by converting starch to fermentable sugar
Cleaves high molecular weight protein in beer	Prevents chill haze in beer (a complex which includes protein, carbohydrates and tannins)
Cleaves β-glucans	Depending on type, amount, and activity of malt, β-glucans can accumulate in beer processing. Treatment of them with β-glucanase prevents slowing of lautering and blinding during fine filtration. The enzyme is predominately applied in Europe on an as-needed basis.
Starch degradation	Increases loaf volume of bread and improves crumb score and softness.
Degrades wheat gluten	Slackens the dough, increases loaf volume of bread, and improves the grain and texture (especially for bucky flour that makes rubbery dough)
Coupled air oxidation of carotene and unsaturated fatty acids	Bleaches natural pigments in flour to yield a whiter bread
Renders casein precipitable by calcium	Curdles milk in cheese formation
Renders casein precipitable by calcium	Rennet extender
Cleaves fats	Develops characteristic flavors in cheeses, especially Italian cheeses
Converts H_2O_2 to H_2O and O_2	H_2O_2 is used to destroy pathogens in milk, and the residual H_2O_2 is removed with catalase to permit growth of a pure culture.
Cleaves sucrose to invert sugar (equilibrium limiting)	Permits manufacture of liquid center candies by allowing handling of sucrose in semi-solid form for enrobing (*e.g.*, with chocolate) and subsequent liquification of sucrose
Cleaves pectins	As with sucrose and invertase, but in this system original firmness is maintained with pectin. With age, the center liquifies and will be clearer than the creamy invert sugar centers.
Cleaves starch oligosaccharides predominately to maltose	Takes liquified starch products to a maltose-based syrup

Table III.

Food	Enzyme(s)
Corn syrup	amylase (bacterial)
Corn syrup (glucose)	glucoamylase (fungal)
Corn syrup (containing fructose)	glucose isomerase (varied microbial)
Egg albumen	catalase
Eggs	glucose oxidase and catalase (fungal)
Fruit juices	pectinase preparations (fungal) a very mixed enzyme system
Ice milk, etc.	lactase (yeast)
Meat	papain (papaya) and other proteases
Protein hydrolyzates	proteases, bacterial and other proteases
Soft drinks	glucose oxidase and catalase (fungal)

the amino group of the protein in the Maillard reaction. This reaction can be prevented by one of the following:

1. Removing the glucose aldehyde group by conversion of glucose to gluconic acid with the glucose oxidase/catalyase enzyme system. It also can be fermented out through natural fermentation or through controlled fermentation by *Aerobacter aerogenes* or *Saccharomyces cerevisiae*

2. Adding an acidulant prior to drying, removing the precipitated mucin portion, which amounts to 15% of the total protein, drying the remaining portion, and blending it with a dry alkaline material to restore the reconstituted egg white to the pH desired

3. Storing the solids at freezer temperatures

4. Storing the aqueous egg white in frozen form

5. Using the egg white solids before they have deteriorated to the point where they lose a significant portion of the desired property of the egg white. This is generally dependent on protein solubility and foaming properties.

In this example then, one must consider the usual factors relative to the enzyme cost, the cost of enzyme processing, and what this superimposes on the drying costs (including processing time and tanks). One

Continued

Action	*Utility*
Starch liquification	Produces low D.E. corn syrups. These carbohydrates further degrade to glucose by subsequent action of acid or saccharifying enzymes.
Cleaves short chain glucose oligomers	Cleaves maltose and higher polymers to produce glucose from corn syrup
Converts a percentage of glucose to fructose (equilibrium limiting)	Converts corn syrup to a sweeter fructose-containing product
Converts H_2O_2 to $H_2O + O_2$	Removes residual peroxide from egg albumen following peroxide pasturization
Glucose converted to gluconic acid	Permits spray drying of eggs without browning on storage
Hydrolyzes pectin, cellulose, hemicellulose	Clarifies fruit juices and improves juice extraction yield. Prevents gelling of concentrates.
Cleaves lactose	Prevents sandiness (lactose crystallization) in milk based frozen desserts
Protein breakdown	Tenderizes meat
Protein breakdown	Prepares protein hydrolyzates for flavoring from gluten and other sources
Oxygen removal	Stabilizes citrus terpenes from light-catalyzed oxidation

must also consider the spoilage rate if the solids are not stabilized, the lowered flavor ratings for products stabilized by yeast fermentation, the loss in dry substance by fermenting the glucose to alcohol and CO_2 (by *S. cerevisiae*), the heterofermentative fermentation of *Aerobacter aerogenes* in which half the glucose is converted to volatiles, and the cost of discarding spoiled solids of that egg white produced without conversion or removal of the glucose.

Of course, even in the enzyme stabilization process, the time for glucose removal to a level that will inhibit significant Maillard browning is inversely related to the enzyme concentration, so that even there, one must balance enzyme cost against the cost of additional tank capacity. Often the amount of enzyme used is a function of both the rate at which the egg whites are received from the breaking room and the predetermined feed rate to the spray dryer as well as the available tankage in which the reaction is carried out. The same plant may use more than one desugarization schedule depending on whether it is the first vat in the morning or whether it is the last one prior to the dryer clean-up period.

One must also consider the financial advantage in having a process that can be easily controlled as to desugarization rate when compared with the poor control of a natural fermentation process that may have widely variable times, depending on factors beyond the control of the operator, such as ambient temperature, level of contamination, and nature of contaminant. The number of variables used in this example explains why economic justification is often not calculated on an industry-wide basis but on a plant-by-plant basis, depending on the particular local conditions that influence the components of the economic calculations.

Safety Factors. In addition to clearly economic considerations, several factors obliquely influence economics or otherwise determine commercial success or failure. One of the major inhibiting factors for the introduction of new enzymes in the U.S. is the acceptance of only three organisms as being *ipso facto* suitable as source organisms for enzymes to be used in food. These organisms are *Aspergillus niger, Aspergillus oryzae* and *Bacillus subtilis*. Enzymes from other organisms generally have to be cleared through the food additive petition route (*e.g.*, carbohydrases from *Rhizopus oryzae* and milk clotting enzymes from *Mucor miehei, Mucor pusillus, Bacillus cereus*, and *Endothia parasitica*. The cost of clearing a food additive through the food additive petition route has been estimated at $7–9 million, about the same as the cost of clearing a drug, including detailing efficacy. Depending on the organism from which the enzyme is derived and the scope of intended use, the cost could be a small fraction of the figure stated. However, the cost still could easily run into several hundred thousand dollars if the organism is new or one other than those listed above and one which has not been used before and has no basis for safety demonstration.

Safety precautions encompass selecting organisms that are known or shown to be nonpathogenic and which do not produce toxins or compounds such as beta-nitropropionic acid that would require testing for carcinogenicity according to the Delaney Amendment in the United States. In fact, the historical limitation in the U.S. to the three organisms mentioned has resulted in commercial enzymes that have a totally clean record of nonpathogenicity, freedom from mycotoxins, and freedom from carcinogens. New organisms must be screened not only for those hazards, but also for strain variability—as a living organism, the culture might undergo changes over time. One must be certain that the organism that will be used years afterward is no different from the organism that underwent testing and approval. Again, the organisms used by the United States enzyme industry have been commercially applied long enough to be recognized as intrinsically stable.

Enzymes must be manufactured under strictly controlled conditions, and in general the final products must be free of the organism that produced them. This is not the case where the entire culture is used directly as the enzyme preparation or where the intact cell containing the enzyme is used as a naturally immobilized enzyme. Safety is also dependent on the materials that are used in the concentration and purification of the enzyme. Where carryovers of materials or residues might persist, these must be of food grade. Safety often relates to the nature of activators. Some enzymes, such as glucose isomerase, require heavy metal activation, and these are then only usable in an intermediate processing step that completely removes a soluble metal activator prior to the enzyme reaction product becoming an article of commerce.

Stability in Storage. One must be able to manufacture the commercial enzyme preparation in one place, to put it into a form in which it can be readily shipped to the user, and then to store it for a reasonable period of time without significant change in activity.

Purity. The enzyme must be free of objectionable quantities of undesirable enzyme activities. The preparation must contain the necessary amounts of cooperative enzyme contaminants.

Availability. The source of enzyme—microbial, plant, or animal—is important. With animal sources, if the demand increases, growing the animals just for this product should be justifiable. Plant sources should not be subject to the vagaries of nature or to unusual labor requirements for collection.

Assayability. Enzyme activity must be determined in a manner that takes into account the changes brought about when the product is acted on. Proteases, for example, are often assayed in terms of milk clotting units for chill-proofing purposes, hemoglobin digestion units for use as digestive aids, and even actual use in standard hamburger compositions as tenderizers. Often enzymes that are even considered as single entities contain many isozymes that tend to have somewhat different properties. The balance between these isozymes, *e.g.*, papain, would effect the activity of the overall preparation in tenderizing meat, chill-proofing beer, or digesting protein that might respond to only one specific assay method. In other cases, the commerce enzymes represent mixtures in which only the principal enzyme activities are standardized or even thoroughly characterized. The effects of other enzymes normally present or present in variable amounts may also influence the application and the assay. The situation becomes even more complex when one considers enzyme systems (*see* Table I), where the mixture of enzymes in the preparation would vary according to the application.

Controllable Activity. This is another important attribute, as many reactions such as milk clotting, being only the first action of a proteolytic enzyme, must be stopped when they have accomplished the desired effect. This prevents them from carrying out undesirable transformations. To be commercially successful, an enzyme's action must be controllable both from the start and the stop. In some cases the stop is automatic because the substrate is exhausted. The enzyme is made inactive just as if it had been totally inactivated.

The above discussion is an oversimplification of the commercialization, but it presents basic factors common to virtually all commercialization problems. In addition, each particular situation has additional criteria and looping of problems (*e.g.*, the need to remove one trace enzyme may unstabilize the desired activity or may remove another trace enzyme important to the action of the primary one). Commercialization, while appearing to be a straightforward simple extension and scale up from fundamental research data, seldom is either simple or straightforward.

Considerations for Commercialization

The preceeding section dealt with characteristics necessary to bring enzymes into commercial use. There are less than 20 enzymes now commercially used on a scale that has significant impact on either the enzyme industry or the food industry. Figure 1 presents the number of enzymes known, *via* published lists, as a function of time in terms of the year of publication. It also indicates the number of enzymes in commercial use in the food industry at present. The difference speaks for itself. Using

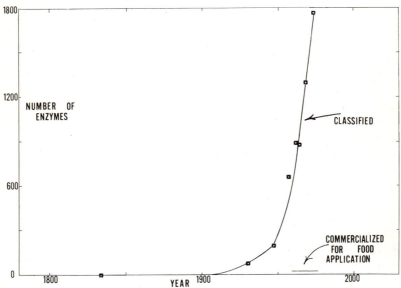

Figure 1. *The increase in the number of known and classified enzymes with time, compared to their commercialization in food systems. References:* (18, 19).

history as a basis, what are the prospects now for increased applications of enzymes to food problems? Further, what are the market considerations that will influence the commercialization of new opportunities for enzymes, now or in the future?

Identification of Applications. Commercial success presupposes the existence of a need that could be filled by a particular enzyme. In recent years, the development of soy protein and soy products has identified the need to eliminate the flatulence factor from these products. This led to the development of the means for producing the enzyme, stacchyase, by a commercial firm. (Interestingly, this enzyme preparation is now being promoted as an α-galactosidase to break down raffinose in beet sugar syrup.) Identification of the need therefore, preceeds commercialization, even where the specific enzyme is known to exist and where its scale-up is possible.

The present social, legislative, and technological environment has created a combination of circumstances likely to increase identification of needs, and, hence, the opportunities for the application of enzymes not now prepared on a commercial scale. Paramount in this is the trend toward greater consumption, in quantity and variety, of preprocessed foods. The preprocessing and subsequent need for storage stability under either ambient or frozen conditions increases requirements for the stability of the desirable attributes—odor, taste, texture, and appearance. The environmental factors necessitating disposal of waste streams in a non-polluting fashion has raised the cost of disposal to the point where it may be more economical, in many instances, to use presently discarded materials. These raw material streams would be subjected to further processing *via* enzymes, to either facilitate disposal or render them more valuable and usable. An example of this is the treatment of whey in the preparation of whey protein, a much sought after product. The use of lactase to hydrolyze lactose facilitates the separation of protein from nonprotein constituents, particularly lactose, through ultrafiltration. This results in a whey protein product with substantially higher protein content.

Availability of Application Data. While needs are being identified at a greater rate, the expansion of the number of enzymes known and the specific knowledge of each of these enzymes as to source, conditions of operation, details of the reactions catalyzed etc., increases the likelihood that the enzyme that is needed for a particular problem is already available. This is not to imply that one has the final answer immediately available, but rather that if one is looking to do something to a particular substrate, much information is currently available on so many substrates that there is a great likelihood of having information on that particular substrate.

For example, in the course of preparing a new food item it is found that ingredient A is unstable, and the product cannot be stored for a reasonable time after processing to even warrant distribution and subsequent consumption. The food technologists postulate that if A could be altered it might not exhibit the same properties. If an enzymatic approach to the solution to this problem were to be investigated, the first step would be to consider A as a substrate and determine what enzymes act on A. The literature would then show that, when acting on A, each of the effective enzymes would result in particular end products. The literature might disclose sources for these particular enzymes and conditions under which each enzyme could function (temperature, pH, etc.), as well as indicate additional substrates, cofactors, stabilizers, and inhibitors. The need for preparative action by using another enzyme to form an enzyme system might also be indicated. Impurities that accompany the particular enzymes might have a deleterious effect by acting on other substrates found within the food. The specificity of the enzyme preparation for the single substrate desired to be removed and the implications of nonspecificity by identification of other substrates for this enzyme that might also be present in the food products would perhaps be found in the literature as well. By looking at the literature details of the overall reaction catalyzed, the conditions, the enzyme source, the ease of control, and the factors mentioned, the food technologist may be able to select an enzyme to accomplish a particular goal. None of the work that may be reviewed in the literature may have had anything at all to do with foods except that a component now in a food has been identified as a substrate for the enzyme.

Enzyme Research Applied in Commerce. Eight years ago there might have been half as many enzymes known and perhaps a lot less was known about them compared with today. The information explosion, at least in this case, is desirable. Technological advances in enzyme manufacture involving ultrafiltration for purification and concentration and involving immobilization to permit reuse, as well as better control of the process, may now make feasible a process that was only a laboratory demonstration.

Some of the basic information on stabilizing sulfhydryl enzymes, has been responsible for their commercialization. Without the judicious use of reducing compounds throughout the processing of the papaya latex, it would not have been possible to maximize the proteolytic activity of commercial papain preparations. Examples of other studies of enzymes which have contributed to commercialization are: the determination of calcium ion as a requirement for amylase stability at high temperature, the difference in properties of catalases derived from bacterial, fungal, or liver sources (here cost, initial activity, heat stability, stability to peroxide, and pH characteristics all play a role), and the difference in properties of the plant proteases especially in relation to heat stability.

Safety Standards. The ever changing environment not only creates opportunities but often creates barriers as well. The development of a new analytical technique to detect infinitesimal amounts of a trace contaminant might result in the need for further purification of a particular preparation or the need to identify other substances that should be looked for. The justifiable focus on safety is such that at the start of a research project one does not know what the regulatory and legislative constraints will be when the project is completed. This uncertainty can result in enough variation in expenditure for clearance to make or break a project. Recently a strain of *Aerobacter aerogenes*, considered a source of pullulanase, was reclassified as a variety of *Klebsiella pneumoniae*. The implications of pathogenicity implicit in this new classification so raised the problems and cost of irrefutably demonstrating safety, that this organism is not now being even considered as a commercial source.

U.S. and Foreign Research Attitudes. Management styles as they differ from country to country also bear on the kinds of work that are done. For example, while in the United States promotional dollars are considered in terms of dollar return, in some countries, such as England, increased market share is often regarded as worthwhile enough that it need not have economic justification for the expenditure. The United States practice compares investment opportunities (*e.g.*, money in the bank *vs.* research investment) by using discounted cash flow and net present value. Any project that takes more than 5 or 6 years to complete will, on this basis, tend to be less attractive than earning interest in the bank. In Central Europe the attitudes towards beer are such that only those ingredients that have historically been used can continue to be used. Under these circumstances, it is unlikely that the enzymes that might be useful in beer processing would be investigated.

The system in Japan, where government, industry, and university team up to intensively explore an area or technology, has now shown what can be done if efforts are directed at a particular area such as the production and application of enzymes. Following the concentration on electronics, Japanese emphasis shifted to biochemical processes. As a result, there are more companies producing enzymes in Japan than in any other country. The recent advances in applied enzymology in the United States—microbial rennin, glucose isomerase, and α-galactosidase—were all initially Japanese exports. The work going on there now, relative to the mixed enzymes derived from *Trichoderma* and other cellulolytic-pectolytic-proteolytic etc. systems, is expected to yield processes of commercial interest in this country, such as the enzymatic peeling of fruits,

Table IV. Proposed or Emerging

Food	Enzyme(s)
Amino acids (L-methionine)	acylase (immobilized, microbial)
Bean products	stacchyase (fungal)
Beer	glucose oxidase/catalase (fungal)
	diacetyl reductase (yeast)
	tannase
Bread	lactase (fungal)
Chocolate-cocoa	amylase
Confections	amylase
	pepsin and trypsin
Corn syrup (high-maltose)	pullulanase (bacterial)
Fruit juices	amylase
	naringinase (fungal)
Fruits (candied)	mixed fungal
Gelatin	lipase (fungal, mainly)
Glucose	cellulase

Applications of Enzymes to Foods

Action	*Potential Utility*
Deacetylates L-methionine	Permits resolution of acylated racemic mixtures of methionine
Hydrolyzes stachyose to assimilable carbohydrates	Reduces carbohydrates passing to human cecum, reducing flatulance, especially with soy products
Removes oxygen	Prevents growth of aerobic organisms which can generate volatile acids and poor flavor
Converts diacetyl to acetoin (requires NADH)	Removes diacetyl which contributes undesirable flavor to beer
Tannin breakdown	Suggested for use to remove tannin and thereby reduce haze formation. Not currently practiced.
Cleaves lactose	Provides more fermentable sugar for yeast-raised bread dough from the milk solid in the mix
Liquifies starches	Could reduce viscosity of chocolate and permit reduction of cocoa butter content.
Hydrolyzes starches	Salvages sugar from candy scraps, especially with fondant creams from starch-molded candies and starch gum scrap. This facilitates solubility and filtration for reuse in sugar syrup for gum drop and other products.
Degrades protein, especially egg albumen	Permits filtration of syrup in scrap reclaimation. Protein otherwise foams and obstructs filtration.
Cleaves α-1,6 glucose linkages in pullulan, starch glycogen, etc.	With β-amylase, completely degrades starch to maltose. (Use of β-amylase alone will result in the accumulation of β-limit dextrins.)
Hydrolyzes starches	With pectinase to clarify fruit juices
Cleaves naringin	Removes bitterness from citrus juice and peels
Hydrolytic	Increases porosity of fruit to permit more rapid penetration of sucrose without osmotic shrinkage, speeding candying process
Cleaves fat	Increases yield of high quality gelatin from from bones by degreasing them
Cleaves cellulose	Hydrolyzes wood pulp or paper to make glucose or fermentable sugars. Bagasse from spent sugar cane has significant potential. Several processes potentially can solve or alleviate waste problems.

Table IV.

Food	Enzyme(s)
Honey (artificial) or syrups	invertase (yeast)
Milk	lactase (fungal)
Protein (fish)	protease, alkaline (bacterial)
Protein (rape-seed)	thioglucosidase (myrosinase)
Shellfish	amylase, cellulase, and protease systems
Starch	amylase from *Bacillus lichenifor-mis*
Sucrose (beet sugar)	α-galactosidase
Sucrose (cane sugar)	dextranase
Tea	tannase (fungal)
Vegetables	Macerating or cell separating enzyme systems (fungal)

enzymatic removal of the seed coat of soy beans, and cell separating enzymes (*see* Table I).

Work on tannase and other enzymes, further supports the premise that many previously unknown commercially feasible applications for enzymes exist if the effort is made to identify the need and seek a solution. While, at this time, there is insufficient data to determine whether the proportion of projects which were or will become commercially profitable, justifies the research investment, it seems likely with the continued inflation rate in Japan that this type of research, if not already uneconomical, will in the future be no more economically done in Japan than in the United States.

New Technology for Processed Foods. The trend toward convenience and processed foods has brought with it specific new technologies which have a bearing on the opportunities for enzyme use through processing.

Continued

Action	*Potential Utility*
Cleaves sucrose to invert sugar (equilibrium limiting)	With some carmelization and/or flavorings artificial honey and honey-like syrups can be made to meet growing shortages of honey.
Cleaves lactose	Permits commercialization of concentrated fresh milk, manufacture of milk for those with lactose intolerance, and conversion of a lactose fraction from whey to prepare a syrup.
Hydrolyzes proteins	Solubilizes fish and other proteins, improving yield of useable protein from food processing residues or fish meal
Cleaves thioglucosides releasing allyl isothiocyanates	Preparation of nontoxic protein from rapeseed and other *Brassica* sources
Catalyzes various hydrolytic reactions	Loosens shells, facilitates removal, dissolves visceral mass in clams
Liquifies starch at 100°C.	Facilitates jet cooker processing of starch
Hydrolyzes raffinose	Improves yield of sucrose from beet molasses by eliminating raffinose interference in crystallization of sucrose
Cleaves dextrans	Reduces viscosity of cane juices due to dextrans present as result of *Leuconostoc* action.
Cleaves tannin	Reduces precipitation of tannin (cloudiness) in tea
Hydrolytic	*See* Table I

One of the more dramatic growth areas within the food industry has been food fabrication, and this approach to food technology has placed a great demand not only for diverse food ingredients but more specifically for a broader and more complete line of flavored products. Dwivedi (24), has reviewed some of the results of applying enzymes to food flavors. Though a number of articles have been published in this area, there is a limited volume of commercial business.

Sometimes technological advances create new problems. As we find more ways to extend the shelf of life of processed foods, we then confront new problems causing product acceptability loss. These new problems may provide opportunities for the application of enzymes, new and old, to food products. Indirect technological advances such as those in the packaging industry can often have an effect on storage life of products, such as improved oxygen or moisture barriers or substantially less expensive barriers which may or may not meet the standards of the older more expensive ones. Another aspect of the processed food market is shown in food product standardization. When one buys a brand name jar of food in San Francisco, it should taste and perform the same as a jar of the same product purchased in New York, even though these products were manufactured at separate locations. Thus, there is a need for not only the standardization of food ingredients but for the standardization of processing conditions by which these finished products are made. The advancing technology in the food industry provides new opportunities in the application of enzymes. In all cases the requirements for performance and for economic justification must hold.

Proposed or Emerging Enzyme Applications. Several new enzymes and/or new applications are known today but are not commercialized or widely used. Induction time, changes in economics, food regulations, or processing practices could result in overnight acceptance of what now seem to be unfamiliar, marginal, or uneconomic applications. Textured soy was not accepted until the price of meat dramatically increased. Table IV lists examples of proposed enzymes and new or limited applications. However, it contains only a fraction of the possibilities.

Although many factors have been mentioned which could impede the growth of the food enzyme industry, there is a significant basis for optimism. It includes the number of enzymes from which to draw, the growing volume of basic knowledge about them, the increased opportunity brought about by technological changes in food processing, the new technologies being applied to enzyme manufacture and utilization, the dissipation of public ignorance of what an enzyme is, the number of new enzyme systems still emerging from worldwide (but especially Japanese) research, and the number of foods and enzymes for which applications are already proposed.

Acknowledgment

We acknowledge the generous consultation of: Richard Metzger, who searched out specific information from the literature; Ted Craig, who advised us on commercial practices; Ted Cayle, who advised us on commercial practices, especially in brewing; and John Kitt, who advised us on confectionery practices.

Literature Cited

1. Gould, R. F., Ed., *Advan. Chem. Ser.* (1969) **95.**
2. Dingle, J. T., Fell, H. B., Eds., "Frontiers of Biology: Lysosomes in Biology & Pathology," 14, Elsevier, New York, 1969.
3. Hori, S., Fugono, T., U.S. Patent **3,482,995** (1969).

4. Scott, D., *J. Agr. Food Chem.* (1953) **1**, 727–730.
5. Blakemore, S. M., U.S. Patent **3,031,307** (1962).
6. Toyoma, N., personal communication, 1973.
7. Lehninger, A. L., "Mitochondrion," W. A. Benjamin, Reading, Mass., 1964.
8. Rubin, M. M., Piez, K. A., Katchalsky, A., *Biochemistry* (1969) **8**, (9), 3628–37.
9. Lehner, A. G., Swiss Patent **540,013** (1973).
10. Reed, G., "Enzymes in Food Processing," Academic, New York, 1966.
11. Hoover, M. W., *Food Technol.* (1967) **21**, (3A Suppl.), 4–7.
12. Jacobs, M. B., "Chemistry & Technology of Food and Food Products," Interscience, New York, 1951.
13. Eskin, N. A. M., Henderson, H. M., Townsend, R. J., "Biochemistry of Foods," Chapter 1, Academic, New York, 1971.
14. *Ibid.*, 139.
15. *Ibid.*, 13.
16. *Ibid.*, 201.
17. Drawert, F., Heimann, W., Emberger, R., Tressel, R., *Naturwissenschaften* (1965) **52**, (11), 304–305.
18. Tauber, H., "Enzyme Technology," p. 420, John Wiley and Sons, New York, 1942.
19. Eapen, K., Tape, N. W., Sims, R. P. A., U.S. Patent **3,732,108** (1973).
20. Minifie, B. W., "Chocolate, Cocoa, and Confectionery: Science & Technology," p. 412, Avi, Westport, Conn., 1970.
21. *Ibid.*, 387.
22. Commission on Biochemical Nomenclature, "Enzyme Nomenclature," Elsevier, New York, 1972.
23. Barman, T. E., "Enzyme Handbook," **1**, **2**, Springer-Verlag, New York, 1969.
24. Dwivedi, B. K., *CRC Crit. Rev. Food Technol.* (1973) **3**, 457–478.

Annotated Bibliography

The following chapters in this book represent, in the main, the proceedings of a 1973 American Chemical Society symposium which focused on the fundamental protein chemistry and enzymology of many enzymes used in food processing. In addition to the primary literature, there are a large number of good books and review articles available, since the broad subject of food-related enzymes has been of both academic and commercial interest for many years.

In order to provide the interested reader with rapid access to the many other facets of this field, an annotated bibliography is provided. No specific attempt was made to include references to enzymes which occur naturally in food. The emphasis here is on commercial importance because that is where review literature is available.

General Enzymology.

1. "Enzymes," Dixon, M., Webb, E. C., Academic, New York, 1958, 1964, 1974.
   This book is one of the best known overall testaments of enzyme chemistry. It covers measurements, preparation, kinetics, nomenclature, specificity, mechanisms, inhibitors, cofactors, structures, biosynthesis, enzyme systems, and enzyme biology. In addition, a fairly broad list of classified enzymes and crystallized enzymes is appendicized. The book

is dedicated to principles rather than application and comparative enzymology rather than having individual enzyme chapters. Tabulated data throughout the text is excellent.

2. "Principles of Enzymology for the Food Sciences," Whitaker, J. R., Marcel Dekker, New York, 1972.
   This is a good teaching text which fulfills its stated purpose of dealing with the academic enzymology principles. Methodology, theory, and kinetics are developed in the first portion of the book and are then applied to elucidate the properties and chemistry of specific enzymes of importance. The book does not deal with specifics of using enzymes in food processing.

Enzyme Encyclopedias.

3. "Enzyme Handbook," Barman, T. E., 1, 2, Springer–Verlag, New York, 1969.
   Concise, basic enzyme data is recorded for some 800 enzymes including the Enzyme Commission number, trivial name, systematic name, reaction catalyzed, equilibrium constant, molecular weight, specific activity, specificity, Michaelis constants, inhibitors, light absorption data, and references. Where required, conditions, source, etc. are specified.

4. "Enzyme Nomenclature," (Commission on Biochemical Nomenclature), Elsevier, Amsterdam, 1973.
   This volume provides recommendations which serve to classify known enzymes by listing the Enzyme Commission number, the various names for each enzyme, the reaction catalyzed, brief comments, and selected references. There is no attempt to do more than classify.

Enzymes Indigenous to Food.

5. "Biochemistry of Foods," Eskin, Henderson, Townsend, Academic, 1971.
   This excellent book deals with biochemical changes in meat, fish, and plants; enzymes in the food industry; browning; and biodeterioration. It is unique in approaching this subject from the *in vivo* viewpoint and contains information not found in other texts. The coverage is narrow but the content of useful information is high (217 pages).

Enzyme Applications to Food.

6. "Enzymes in Food Processing," Reed, Academic, New York, 1966, 1974.
   This book provides an introduction to general enzymology for the industrial enzymologist and then deals, chapter by chapter, with enzymes or application areas specific to food technology. For each enzyme, basic data are provided on chemistry, kinetics, sources, pH optima, and the like. In addition, specific food application data are given. Production and legal aspects are also mentioned. This is a "How to Use Enzymes in Foods" book.

7. "Enzymes, Industrial," DeBecze, G. I., in Kirk-Othmer, "Encyclopedia of Chemical Technology," 2nd ed., 8, p. 173–230, John Wiley & Sons, Inc., New York, 1965.
   A brief and general discussion of the industrial use of enzymes, including general enzymology, specific applications, (not limited to foods) suppliers, properties, sources, production methods, and definitions of units.

8. "Food Enzymes," DeBecze, G. I., *Critical Reviews in Food Technology* (1970), 1, p. 479, CRC Press, Cleveland, Ohio.
   This is similar to the article above, but is more current. The specific manufacturer data may need verification.

9. "Production and Application of Enzyme Preparations in Food Manufac-
    turing," Society of Chemical Industry, London, 1961.
    Topics here are approached almost entirely from the application point
    of view and much good application data are presented. It must be
    remembered that this is the proceedings of a 1959 symposium. Infor-
    mation is good, old, and interestingly handled.

Patent Literature.

Patent literature often contains many details not found elsewhere and
represents a resource that should not be overlooked. However, not all issued
patents represent commercially used or usable processes. Patent style often
states as fact that which is only anticipated at the time of filing, and the reader
must refer to the details of examples in the patents for specific and evaluate
them for himself.

10. "Encyclopedia of Enzyme Technology," Meltzer, Y. L., Future Stochastic
    Dynamics, Inc., Flushing, New York, 1973.
    This book is based predominantly on the patent literature and pro-
    vides "how to" data regarding the production, purification, and applica-
    tion of commercial enzymes. Coverage is not limited to food appli-
    cations, and 70 subjects are grouped as: Enzymes, Enzymatic Processing,
    Enzyme Stabilization, Polymer–Enzyme Products, Cell Culture, Protein
    Analysis, Nucleic Acids etc., Amino Acids, Peptide Synthesis, and
    Applications. Indexing includes U.S. patent number, company and
    patent assignee, inventor, and subject.

11. "Enzymes in Food Processing and Products," Wieland, H., Noyes Data
    Corp., Park Ridge, New Jersey, 1972.
    This book, using as a basis U.S. patents since 1960, describes food
    processing practices which are dependent upon enzymes. The chapters
    are Fruit and Vegetable, Starch and Sugar Conversion, Baked Goods,
    Cheese-making, Meat Tenderization, Special Meat, Fish and Cereal
    Applications, Flavor Applications, Deoxygenating and Desugaring, and
    Enzyme Stabilization. A patent index is provided by company, inventor,
    and patent number, but not by subject or by class and sub-class.

12. "Commercial Food Patents, U.S.," 1970, '71, '72, '73, North, H. B. and
    North, O. S., Arlington, Virginia.
    These books contain selected abstract reprints from the U.S. Patent
    Gazette. They are organized into about 20 food categories and indexes
    are provided to assignee, inventor, and U.S. patent number. Those
    dealing with enzymes are not specifically identified nor is there a sub-
    ject index.

13. "Enzyme Technology Digest," Roberts, I. M., Neus, Inc., P.O. Box 1365,
    Santa Monica, Cal. 90406.
    This is an information exchange in enzyme technology issued three
    times a year since May 1, 1972, under the sponsorship of the National
    Science Foundation. It contains abstracts from meetings, agenda for
    meetings, news and reviews of books, patents, publications, and theses.
    Available by subscription only, beginning 1975.

14. International Patent Research Office, P.O. Box 1260, The Hague (2077)
    Holland.
    This organization provides a periodic list of patent titles from 14 indus-
    trial countries. The particular search bulletin related to this book is
    Enzymes, and although patents are subdivided by country, they are
    neither limited to foods nor subdivided by subject. Titles are not trans-
    lated into English. Available on a subscription basis.

Insolubilized Enzymes And Other Enzyme Engineering Sources.

This area is rapidly becoming an entire field unto itself, but to date it is limited to exploration in the form of academic and commercial feasibility studies. There is no question that this technology will become commercial —the only questions are: when and for what selected applications can insolubilization be justified?

"Biochemical Aspects of Reactions on Solid Supports," Stark, Academic, New York, 1971.

"Enzyme Engineering," Wingard, L. B. Jr., Interscience, New York, 1972.

"Immobilized Enzymes," Zoborsky, O. R., CRC Press, Cleveland, Ohio, 1973.

"Immobilized Enzymes—A Compendium of References from the Recent Literature," Two Volumes, Corning Glass Works, Corning, New York, 1972.

"Membrane Bound Enzymes," Porcellati, G., di Jeso, F., Plenum, New York, 1971.

RECEIVED March 6, 1974.

Analytical Applications of Enzymes

JOHN R. WHITAKER

Department of Food Science and Technology, University of California, Davis, Calif. 95616

Enzymes are useful and convenient in analytical chemistry because of their high specificity and sensitivity, which permits quantitative assays on crude materials under controlled, mild reaction conditions. Applications involve determining enzyme amounts in analyses for inborn errors of metabolism, nongenetically associated diseases, adequate heat treatment of foods, in quality control of foods and other agricultural products, and in measuring soil quality. Enzymes are used to determine quantitatively specific compounds which serve as substrates, activators, or inhibitors of enzymes. They are also used to determine isomeric configuration of compounds, the primary structures of complex molecules such as proteins, nucleic acids, carbohydrates, and triglycerides, the conformation of complex molecules, and the structures of cells and subcellular organelles.

Enzymes have been used in analysis for more than 125 years. Because of their specificity, it was recognized early that enzymes could be used to detect and determine the concentration of minute amounts of compounds in complex biological systems. For example, peroxidase was used to detect hydrogen peroxide at concentrations of 2 ppm by 1845 (*1, 2*), and the amylases were used to determine carbohydrates in foods by 1880 (*3*). However, the availability of well-characterized enzymes of high specificity in pure form during the last two decades has made enzymatic analyses more attractive. Major advances in instrumentation for detecting changes in compounds and structures induced by enzymatic action as well as better understanding and control of the factors which influence enzyme activity are associated with the availability of enzymes.

General Principles of Analysis with Enzymes

Enzymes are specific in their action. This specificity results from the necessity for correct binding of a compound in the active site of an enzyme before catalysis can take place. The binding locus of the enzyme active site has a definitive three-dimensional structure caused by the surrounding amino acid residues from various regions of the polypeptide chain. Although there may be some flexibility in the binding locus, it must fit a potential substrate closely so that there can be multiple interaction points *via* hydrogen bonds, hydrophobic bonds, van der Waals forces, or salt linkages. No covalent bonds are formed in binding. Proper alignment of the compound in the active site is essential to bring the susceptible bond into proper orientation with two or more groups involved in the catalysis step. Only a few compounds are likely to meet these requirements for any particular enzyme. For example, D and L isomers may meet the requirements for binding, but only one of the isomers will be properly positioned for catalysis to occur.

Specificity may be restricted to a single compound, such as the action of urease on urea, or may be broader as in the ability of α-chymotrypsin to hydrolyze esters, amides, hydroxamates, thiolesters, and peptides of L-tyrosine, L-tryptophan, and L-phenylalanine. Because of this high specificity crude extracts can usually be analyzed directly for a constituent by enzymes. Glucose can be determined quantitatively by using glucose oxidase even in the presence of large amounts of other carbohydrates. Of some 60 carbohydrates tested, only β-D-glucose, 2-deoxy-D-glucose, and 6-deoxy-D-glucose are oxidized at appreciable rates (relative rates are 100:3.3:10, respectively, Ref. *4, 5*). The anomer α-D-glucose is oxidized less than 0.06 times as fast as the β-anomer (*4*).

Enzymatic analyses are carried out near room temperature and neutral pH to permit detection and determination of labile substances and intermediates which often can be estimated only crudely by other methods. Enzymatic methods are quite sensitive and often detect as little as 1×10^{-12} to $10^{-15}M$ concentration of a substrate. For example, picogram amounts of ATP can be determined quite readily with luciferase.

Analytical Determinations Involving Rate Assays. Rate assays must be used to determine the concentration of enzyme, of an activator, or of an inhibitor, and they may be used to determine the concentration of a compound which serves as a substrate for an enzyme. Substrate concentration may also be determined by a total change method. The advantages of a rate assay over a total change method are its speed and the requirement for less enzyme.

The rate of an enzyme-catalyzed reaction is affected by the concentrations of enzyme, substrate, activators, and inhibitors and by pH and temperature. To a lesser extent, the rate may also be influenced by the ionic strength and dielectric constant of the medium as well as by the specific buffer used. For assays based on rate determinations these experimental parameters must be controlled precisely.

The first step in an enzyme-catalyzed reaction is the combination of enzyme with substrate to form an adsorptive complex often referred to as the Michaelis complex (Equation 1):

$$E + S \underset{k_{-1}}{\overset{k_1}{\rightleftharpoons}} ES \overset{k_{cat}}{\longrightarrow} E + P \tag{1}$$

where E is enzyme, S is substrate, ES is the enzyme–substrate complex, and P is the product of the reaction. The observed reaction rate, v, is given by the expression:

$$v = dP/dt = k_{cat}[ES] \tag{2}$$

The maximum velocity, V_{max}, will be attained when all the enzyme is forced into the ES form by an excess of substrate. The effect of substrate and enzyme concentrations on the observed velocity is given by the Michaelis-Menten expression (6, 7):

$$v = \frac{V_{max}[S]}{K_m+[S]} = \frac{k_{cat}[E_o][S]}{K_m+[S]} \tag{3}$$

where $[E_o]$ is the total enzyme concentration, and $K_m = (k_{-1} + k_{cat})/k_1$ is the substrate concentration at which $v = 0.5\ V_{max}$.

Under most conditions the initial rate, v_o, of the reaction is directly proportional to enzyme concentration (8). In assays to determine the amount of enzyme in a sample the initial substrate concentration should be at least 10 times K_m so that the reaction is zero order with respect to substrate concentration (Equation 3). At substrate concentrations less than 0.1 K_m the reaction follows a first-order rate process with the rate directly proportional to substrate concentration. Enzyme rate assays to determine the amount of a compound as substrate in a sample should be run under these conditions. At substrate concentrations greater than 0.1 K_m and less than 10 K_m the reaction follows a mixed-order process intermediate between first and zero order.

The progress of an enzymatic reaction is followed by the concentration change of either one of the substrates or one of the reaction products as a function of time. Concentration change may be determined by either chemical or physical methods. In a chemical method aliquots are withdrawn from the reaction periodically. The enzymatic action is stopped by adding an inhibitory substance, by adding strong acid or base, or by cooling the solution, and the concentration of substrate left or product formed is measured after suitable derivatization. For example, in the reaction shown in Equation 4:

$$p\text{-Nitrophenyl phosphate} + \text{water} \xrightarrow[\text{phosphatase}]{\text{acid}} \qquad (4)$$

$$p\text{-nitrophenol} + \text{phosphate}$$

the substrate and products are colorless at pH values below 6. When an aliquot is removed into alkali, the nitrophenolate ion ($pK = 7.2$) absorbs maximally at 403 nm.

It is advantageous to follow the progress of an enzymatic reaction continuously. Since p-nitrophenol has more absorbance at 340 nm than p-nitrophenyl phosphate, the acid phosphatase-catalyzed reaction (Equation 4) can be monitored continuously in a spectrophotometer at 340 nm. The alkaline phosphatase-catalyzed hydrolysis of p-nitrophenyl phosphate can be followed at 403 nm. Many analytical determinations using enzymes can be monitored continuously by the change in absorbance at 340 nm when NADH (or NADPH) is oxidized to NAD$^+$ (NADP$^+$) or vice-versa. NADH absorbs strongly (ϵ 6.2 $\times$ 10^3M^{-1}cm^{-1}) at 340 nm while NAD$^+$ has no absorbance at this wavelength. The primary enzyme involved may be a dehydrogenase requiring NADH (or NAD$^+$), or the primary enzyme system may be coupled to a second enzyme system (indicator system) requiring NADH (or NAD$^+$). For example, glucose can be determined by the coupled reaction involving hexokinase and glucose 6-phosphate dehydrogenase (9).

$$\text{Glucose} + \text{ATP} \xrightarrow[\text{Mg}^{2+}]{\text{hexokinase}} \text{glucose 6-phosphate} + \text{ADP} \qquad (5)$$

$$\text{Glucose 6-phosphate} + \text{NADP}^+ \xrightarrow[\text{dehydrogenase}]{\text{glucose-6-phosphate}} \qquad (6)$$

$$\text{6-phosphoglucono-}\delta\text{-lactone} + \text{NADPH} + \text{H}^+$$

If the amount of substrate in a sample is determined by a rate process *via* a coupled reaction, the second enzyme (indicator system) must be in high enough concentration so that the rate of the second reaction is not rate determining. To accomplish this, the enzyme concentration in the second reaction must be high enough to make the rate of the second reaction approximately 1000 times faster than that of the first reaction if they were carried out separately. Stated another way, the product of enzyme concentration and specific activity of the second enzyme should be at least 100 times that of the product of these two parameters for the primary reaction (*10*). In addition, the pH must be compatible for both reactions. For this reason, rate assays involving the coupling of more than two enzyme reactions are rarely used for analytical purposes. However, when the amount of compound present in a sample is determined by a total change method, more than two enzymes may be coupled, as in the determination of glucose by the use of hexokinase, pyruvate kinase, and lactate dehydrogenase (*11*).

$$\text{D-Glucose} + \text{ATP} \xrightarrow[\text{Mg}^{2+}]{\text{hexokinase}} \text{D-glucose 6-phosphate} + \text{ADP} \quad (7)$$

$$\text{ADP} + \text{phosphoenolpyruvate} \xrightarrow[\text{Mg}^{2+},\ \text{K}^+]{\text{pyruvate kinase}} \text{ATP} + \text{pyruvate} \quad (8)$$

$$\text{Pyruvate} + \text{NADH} + \text{H}^+ \xrightarrow[\text{dehydrogenase}]{\text{lactate}} \text{lactate} + \text{NAD}^+ \quad (9)$$

The amount of glucose is determined from the absorbance decrease at 340 mm caused by formation of NAD^+.

Enzymatic reactions may also be followed manometrically (evolution or uptake of a gas), polarimetrically, potentiometrically, fluorimetrically, and by use of ion selective electrodes. The glucose oxidase-catalyzed oxidation of glucose (Equation 10):

$$\beta\text{-D-Glucose} + \text{O}_2 \xrightarrow[\text{oxidase}]{\text{glucose}} \delta\text{-D-gluconolactone} + \text{H}_2\text{O}_2 \quad (10)$$

can be followed manometrically by the uptake of O_2, by use of an oxygen electrode, amperometrically by incorporating ferrocyanide in the system (*12*), or by coupling with peroxidase and chromogen to determine the hydrogen peroxide formed. Any reaction in which protons are taken up or liberated can be followed continuously in a pH-stat. Hydrolysis of sucrose by invertase (β-fructofuranosidase) is best followed polarimetrically. Fluorimetric techniques are two to three orders of magnitude more sensitive than spectrophotometric techniques. Whenever there is a difference in fluorescence between the substrate and products, this method is potentially useful. Fluorimetry must be used cautiously, however, because many compounds, particularly proteins, quench fluorescence.

The use of immobilized enzymes as analytical reagents is receiving much attention. Immobilized enzymes, because of their greater stability and ease of removal from the reaction, may be used repeatedly, thus eliminating the major cost factor in these assays. Glucose can be determined continuously by using columns of immobilized glucose oxidase (*14*), as shown in Figure 1. The sample with glucose and incorporated O_2 is poured through a column containing immobilized glucose oxidase.

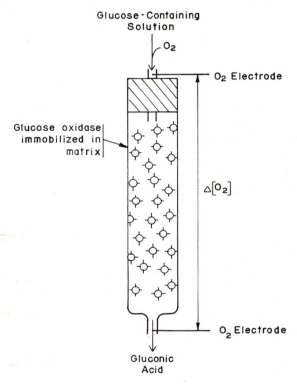

Figure 1.   Use of immobilized glucose oxidase
for determining glucose concentration

As the solution passes through the column, the glucose is enzymatically oxidized to δ-D-gluconolactone (which is hydrolyzed to gluconic acid nonenzymatically) with utilization of O_2. Change in O_2 concentration, related to the amount of glucose in solution, is determined with O_2-sensitive electrodes placed at the top and bottom of the column. The column may be used repeatedly.

In another advance in the use of immobilized enzymes in analysis, glucose oxidase was combined with an oxygen electrode to give an enzyme electrode for determining glucose concentration (*15, 16*). The enzyme electrode is an O_2-sensitive electrode surrounded by a semi-permeable gel membrane to hold glucose oxidase in place (Figure 2). When the electrode is immersed into a solution of glucose, the glucose diffuses into the gel membrane where it is converted to δ-gluconolactone

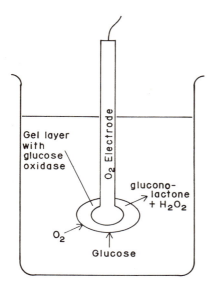

Gel layer
with
glucose
oxidase

O_2 Electrode

glucono-
lactone
+ H_2O_2

O_2

Glucose

Figure 2.   An enzyme electrode for determining glucose concentration

by glucose oxidase, with uptake of O_2. The rate of O_2 uptake gives a measure of glucose concentration. Incorporating catalase into the system to remove H_2O_2 might prolong the life of the electrode.

Enzyme electrodes for lactate determination using immobilized lactate dehydrogenase (16), for urea determination using immobilized urease (17), for L-amino acids using immobilized L-amino acid oxidase (18), and for various amines using the appropriate immobilized deaminase system (19) have also been prepared. A urease electrode is commercially available from Beckman.

Analytical Determinations Involving Total Change. The concentration of a compound which serves as a substrate for an enzyme may be determined by a rate assay or by a total change method. The advantages of the rate assay method are its speed, even at low enzyme concentrations, and its selectivity. With the usual spectrophotometric equipment a rate assay may be completed in one to five minutes. In the total change method the enzyme concentration is usually increased so that the reaction is complete in 10–30 minutes. Contamination by traces of other enzymes which convert the primary product to compounds not detectable by the assay procedure is more serious in the total change method than in the rate assay method. On the other hand, small variations in pH, temperature, ionic strength, dielectric constant, activators and inhibitors, and substrate concentration relative to K_m do not affect the results obtained by the total change method.

Enzyme-catalyzed reactions do not always proceed to completion because of an unfavorable equilibrium constant or insufficient reaction time. Nevertheless, a compound can be determined by a total change method. The reaction may be pulled to completion by trapping the products or by coupling one of the products to a second enzyme reaction. In determining lactate with lactate dehydrogenase the equilibrium lies far to the left in the direction of lactate at pH 9.5 ($K = 2.9 \times 10^{-12}M$ at 25°C).

$$\text{L-(+)-lactate} + \text{NAD}^+ \xrightarrow[\text{dehydrogenase}]{\text{lactate}} \text{pyruvate} + \text{NADH} + \text{H}^+ \quad (11)$$

By using $0.2M$ hydrazine and excess NAD^+ the reaction goes essentially to completion to the right ($K = 7 \times 10^2M$ at pH 9.5 and 25°C) and can be readily used for the specific determination of L-(+)-lactate (13). In the usual method for determining glucose by using glucose oxidase, peroxidase, and the leuco compound o-dianisidine:

$$\beta\text{-D-glucose} + \text{O}_2 \xrightarrow[\text{oxidase}]{\text{glucose}} \delta\text{-D-gluconolactone} + \text{H}_2\text{O}_2 \quad (12)$$

$$\text{H}_2\text{O}_2 + o\text{-dianisidine} \xrightarrow{\text{peroxidase}} \text{oxidized } o\text{-dianisidine} \quad (13)$$
$$\text{(yellow)}$$

the reaction is stopped after 10 minutes reaction at 25°C. Not all the glucose is oxidized because of the slow mutarotation rate of α-D-glucose to β-D-glucose. Nevertheless, the total concentration of glucose is determined by using a standard curve prepared with glucose under the same conditions.

Analytical techniques involving enzymes are important in food science and nutrition, in clinical medicine, in toxicology and pesticide analysis, in soil science, in microbiology as well as in biochemistry, chemistry, and physiology. Examples of applications in these areas are given throughout this discussion.

Enzyme assays may be designed:

(a) To determine the amount of enzyme in a sample

(b) To determine the amount of a specific compound which serves as substrate, activator, or inhibitor of an enzyme

(c) To determine the isomeric configuration of molecules

(d) To determine the structure of complex molecules such as proteins, carbohydrates, lipids, and nucleic acids

(e) To probe the conformation of complex molecules

(f) To probe the structure of cells and sub-cellular organelles.

Enzyme Analysis

Determination of enzyme activity level is important in many applications in clinical medicine, in food science and nutrition, in agricultural chemistry, in biochemistry, and in physiology. Because of the large number of these applications we are highly selective in the following discussion.

Inborn Errors of Metabolism. Through his extensive investigations of patients with alcaptonuria who excreted large amounts of homogentisic acid in the urine, Garrod realized in 1899 that this and other inborn errors of metabolism resulted from genetic defects involving specific enzymes (20, 21). However, the relationship between a defect in a gene and its expression as a defect in a specific enzyme was not made explicit until 1941 (22). Since then, knowledge in this field has increased rapidly. Wacker and Coombs listed 48 genetic diseases, each of which can be traced specifically to the absence of, or an abnormality in, one or more enzymes (23). In 1972, Raivio and Seegmiller included 81 genetic diseases in their review (24). A selected group of these genetic diseases is listed in Table I. Several excellent books have been published recently on this topic (25, 26).

Table I.   Some Genetic Diseases Caused by Enzyme Deficiencies

Disease	Defective enzyme
Phenylketonuria	phenylalanine hydroxylase
Alcaptonuria	homogentisic acid oxidase
Hyperammonemia I	ornithine transcarbamylase
Hyperammonemia II	carbamoylphosphate synthetase
McArdle's syndrome	muscle phosphorylase
Hypophosphatasia	alkaline phosphatase
Congenital lactase deficiency	β-galactosidase
Hereditary fructose intolerance	fructose 1-phosphate aldolase
Gout	hypoxanthine phosphoribosyl-transferase
Refsum's disease	phytanic acid α-oxidase

The greatest advances at the enzyme level have been in understanding the inborn errors associated with lipid metabolism (27). At least nine of these diseases have been described at the enzyme level (Table II). The first five diseases listed in Table II involve enzymes which successively degrade Cer-Glc-Gal-(NeuNAc)-GalNAc-Gal to ceramide, glucose, galactose, and neuraminic acid. No genetic diseases associated

Table II.   Genetic Lipid Storage Diseases Involving Sphingolipid Hydrolysis[a]

Disease	Enzyme	Defective reaction[b]
Generalized gangliosidoses	G_{M1} galactosidase	Cer—Glc—Gal—(NeuNAc)—GalNAc—Gal + H_2O → Cer—Glc—Gal—(NeuNAc)—GalNAc + Gal
Tay-Sachs disease	G_{M2} hexosaminidase	Cer—Glc—Gal—(NeuNAc)—GalNAc + H_2O → Cer—Glc—Gal—NeuNAc + GalNAc
Fabry's disease	ceramide trihexosidase	Cer—Glc—Gal—Gal + H_2O → Cer—Glc—Gal + galactose
Ceramide lactoside lipidoses	lactosylceramidase	Cer—Glc—Gal + H_2O → Cer—Glc + galactose
Gaucher's disease	glucocerebrosidase	Cer—Glc + H_2O → Cer + glucose
Metachromatic leukodystrophy	sulfatidase	Cer—Gal—3 SO_4 + H_2O → Cer—Gal + H_2SO_4
Krabbe's disease	galactocerebrosidase	Cer—Gal + H_2O → Cer + galactose
Niemann-Pick disease	sphingomyelinase	Sphingomyelin + H_2O → Cer + phosphorylcholine
Fucosidosis	α-L-fucosidase	?

[a] Adapted from Ref. 27.
[b] Cer = ceramide (N-acylsphingosine); NeuNAc = neuraminic acid; GalNAc = N-acetylgalactosamine.

with the removal of the N-acetyl group from N-acetylgalactosamine and removal of neuraminic acid from galactose have yet been described.

Diseases caused by inborn errors have distinctive clinical symptoms and are frequently diagnosed on the basis of accumulation of metabolic products. However, enzyme assays for the disease evaluation are many times more sensitive than clinical symptoms or measurement of metabolic products and permit differentiation among defects which have common clinical symptoms. For example, hemolytic anemia can be caused by defects in at least 10 different enzymes (Table III), and there are at least eight glycogen storage diseases, each caused by a defect in a different enzyme (24). The management of these diseases through the combined efforts of the medical profession, nutritionists, and food scientists can only be accomplished through monitoring of the enzymes involved. Enzyme analytical methods make feasible large-scale screening programs for detection of these diseases long before clinical symptoms appear.

Nongenetically Associated Diseases. Different organs differ quantitatively, and often qualitatively, in their enzyme makeup. When the cells of a specific organ are damaged through microbial infection or mechanical or chemical injury, the enzymes of those cells are liberated into the blood at an increased rate. The enzyme patterns found in blood serum following cellular damage are clearly characteristic of a given organ (28) and can be used for diagnostic purposes. Determination of certain enzyme activities is now a routine procedure in most clinical laboratories. Isoenzyme patterns also indicate the origin of the enzymes in the blood serum. For example, lactate dehydrogenase occurs in five forms which are readily separated by simple electrophoretic techniques. In heart, brain, and kidney the H_4 (heart-type) form is found in largest amount while in other tissues such as smooth muscle, testicle, ovary, lung, liver, and spleen, the M_4 (muscle-type) form dominates. In hepatitis the serum contains primarily M_4 isoenzyme while in myocardial infarction the H_4 isoenzyme dominates. Differences in heat stability (29), inhibition by sulfite ions (30), and differential behavior with coenzyme analogs (31) can also be used to assess the relative amounts of each isozyme.

While increases in the activity level of a single enzyme in the serum may be sufficient to pinpoint the distressed organ, it is usually insufficient to distinguish among different diseases of that organ. Determination of glutamate-oxaloacetate transaminase (GOT), glutamate-pyruvate transaminase (GPT), and alkaline phosphatase have proved particularly successful in the diagnosis and control of liver diseases. Some changes in enzyme levels in various liver diseases are shown in Figure 3. The relative levels of the five enzyme activities in a particular disease are more meaningful than the absolute levels, as the absolute levels vary greatly among individuals and with the time following initiation of disease (see acute viral hepatitis, in Figure 3).

Adequate Heat Treatment. In the food industry vegetables and fruits are blanched before freezing, and milk and milk products are pasteurized to destroy microorganisms and endogenous enzymes to prolong the storage life. Determining peroxidase activity in fruits and vegetables and alkaline phosphatase activity in milk and milk products is most commonly used for measuring the adequacy of heat treatment

Table III. Enzymes Involved in Different Types of Hemolytic Anemia [a]

Adenylate kinase
Diphosphoglycerate mutase
Glutathione peroxidase
Glutathione reductase
Glutathione synthetase
Hexokinase
Hexosephosphate isomerase
Phosphoglycerate kinase
Pyruvate kinase
Triosephosphate isomerase

[a] (24).

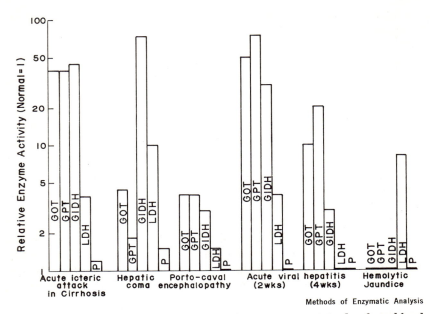

Methods of Enzymatic Analysis

Figure 3.   Schematic representation of five enzyme activity levels in blood serum as a result of various liver diseases. These data (28) do not show the large range in values among individuals. GOT, glutamate-oxaloacetate transaminase; GPT, glutamate-pyruvate transaminase; GlDH, glutamate dehydrogenase; LDH, lactate dehydrogenase; P, alkaline phosphatase.

Both of these enzymes are relatively heat-stable compared with microorganisms and other enzymes. By the time all peroxidase or alkaline phosphatase activity is destroyed, adequate heat treatment has been given. Too little heat treatment results in immediate detection of these enzymatic activities in the products. Marginal heat treatment results in no detectable activity immediately following heat treatment but detectable amounts one to two days later. It is important to use the exact amount of heat treatment because heat is detrimental to texture, flavor, and odor of a food.

Other Uses. Malt is sold and used on the basis of its diastatic power as determined by estimating the maltose formed through the action of endogenous β- and α-amylases during a standard incubation period.

Catalase activity is important in milk and milk products analyses. Since normal milk has no significant amount of catalase, catalase activity indicates the presence of leucocytes caused by disorders or diseases of the udder, of colostrum in the milk, or of bacterial contamination.

Determination of reductase activity in milk indicates whether there is bacterial contamination. The reductase activity is measured by the decolorization rate of methylene blue. Acceptable milk decolorizes methylene blue in about three hours at room temperature while a decolorization time of less than one hour indicates bacterial contamination.

Determining the behavior of various enzymes during the brewing process helps to maintain uniform beer quality (*32*). There are marked changes in the amounts of amylase, phosphofructokinase, malate dehydrogenase, and glucose-6-phosphate dehydrogenase, among other enzymes, during the ripening and malting of barley. During the fermentation process there are characteristic changes in the amounts of several enzymes produced by the yeasts, which depend in part on the yeast strain and the fermentation conditions.

The enzyme activity of a soil is a valuable criterion of its quality (*33*). The enzymes, derived from microorganisms living in the soil, can be assayed far more readily and more accurately than the microorganisms. The activities of invertase, β-glucosidase, amylase, and urease are most frequently measured in soils. In general, the higher the enzyme content, the better the soil quality. On the other hand, the presence of high levels of urease in soil is disadvantageous when urea is used as the nitrogen fertilizer. Too rapid hydrolysis of urea to ammonium carbonate results in a rise in pH of the soil. This may damage germinating seedlings and young plants and lead to nitrite toxicity and loss of nitrogen from the soil as gaseous ammonia (*34*). This has prompted a search for effective urease inhibitors that might be incorporated into the urea before application (*35*).

Substrate Analysis

Enzymes are useful in determining concentration, optical purity, structure, conformation, and cellular location of compounds.

Concentration. By far the greatest analytical use of enzymes has been in determining compound concentrations. Enzymes are useful for this purpose as their high specificity enables accurate determination of a compound concentration in crude mixtures. This avoids the need for elaborate fractionation techniques which frequently result in losses and changes in labile compounds. The concentration of thousands of compounds can be determined by using enzymes and many procedures for this purpose are available (*36–42*). We have selected only a few of these for illustration.

Assay methods are available for most of the carbohydrates ranging from glycogen, starch, and cellulose to the monosaccharides and their monophosphates. Concentration determination of complex carbohydrates usually depends on their enzymatic hydrolysis to monosaccharides and determination of monosaccharide concentration *via* an enzymatic reaction. For example, starch is hydrolyzed to glucose by glucoamylase, and the glucose concentration is determined by the combined use of glucose oxidase, peroxidase, and o-dianisidine (Ref. 43, Equations 12–14).

$$\text{Starch} + H_2O \xrightarrow{\text{glucoamylase}} \text{D-glucose} \tag{14}$$

The glucose concentration is determined from the absorbance change at 450 nm related to a standard curve prepared with D-glucose. A few of the many reactions available for determination of carbohydrates are given in Table IV.

D-Amino acids can be determined in the presence of L-amino acids with D-amino acid oxidase, which is found in the kidney and liver of all animals, particularly the sheep and pig. In this reaction, the D-amino acid is converted to an α-keto acid, ammonia, and hydrogen peroxide, with the uptake of O_2 (Equation 15).

$$R-CH(NH_2)-COOH + O_2 + H_2O \rightarrow \tag{15}$$
$$R-CO-COOH + NH_3 + H_2O_2$$

Table IV. Concentration Determination of Selected

Substrate	Enzymes involved
Glycogen	(a) acid-catalyzed hydrolysis to D-glucose; D-glucose determination with hexokinase-pyruvate kinase-lactate dehydrogenase
	(b) acid-catalyzed hydrolysis to D-glucose; D-glucose determination with glucose oxidase-peroxidase-o-tolidine
Starch	hydrolysis to D-glucose with glucoamylase, D-glucose determination with glucose oxidase-peroxidase-o-dianisidine
Cellulose	hydrolysis with cellulase, determination of solubilized material with dichromate
Raffinose	(a) directly with galactose oxidase-peroxidase-p-hydroxyphenylacetic acid
	(b) invertase-melibiase, then glucose oxidase-peroxidase-o-dianisidine
Sucrose	(a) invertase, then hexokinase-phosphoglucose isomerase-glucose-6-phosphate dehydrogenase
	(b) invertase, then glucose oxidase-peroxidase-p-hydroxyphenylacetic acid
Lactose	β-galactosidase (48), then hexokinase-glucose-6-phosphate dehydrogenase
Glucose	(a) glucose oxidase-peroxidase-o-dianisidine
	(b) D-glucose-6-phosphotransferase
	(c) glucose oxidase-peroxidase-homovanillic acid
Galactose	galactose oxidase-peroxidase-p-hydroxyphenylacetic acid
Frutose	hexokinase-phosphoglucose isomerase-glucose-6-phosphate dehydrogenase

The reaction can be followed by O_2 uptake (manometrically or oxygen electrode), by determination of ammonia, or by determination of hydrogen peroxide. A fluorimetric assay procedure for determining the hydrogen peroxide produced permits determinations of D-amino acids in the 1–100 μg/ml range (53).

L-Amino acid oxidase, from snake venom, performs the same function in Equation 15 with L-amino acids. As little as 0.01–5.0 μg/ml of several of the amino acids can be determined fluorimetrically by coupling the reaction with peroxidase-homovanillic acid (53).

Specific L-amino acid decarboxylases, produced by certain bacteria under selected growth conditions, catalyze the decarboxylation of certain amino acids (Ref. 54, Equation 16).

$$R—CH(NH_2)—COOH \rightarrow R—CH_2—NH_2 + CO_2 \qquad (16)$$

The carbon dioxide is measured manometrically, or the system is coupled to an amine oxidase-peroxidase-p-hydroxyphenylacetic acid system for determining the amine produced (37). Specific decarboxylases are avail-

Carbohydrates by Enzymatic Analysis

Sensitivity μg/ml	Reference
<1	(11)
	(44)
5–40	(43)
400–3000	(45)
0.1–50	(46)
5–40	
<10	(47)
2–100	(46)
~1	
5–40	(49) (50)
0.01–10	(51) (46)
1–5	(52)

able for L-tyrosine, L-histidine, L-ornithine, L-lysine, L-arginine, L-glutamic acid, and L-aspartic acid.

Amine concentrations can be determined by using either diamine oxidase or monoamine oxidase in which the amine is oxidized to an aldehyde, hydrogen peroxide, and ammonia with uptake of oxygen (Equation 17).

$$2R—CH_2—NH_3^+ + 2O_2 \rightarrow 2R—CHO + H_2O_2 + 2NH_3 \qquad (17)$$

The enzymes are rather nonspecific, so most amines in solution will be determined. The reaction can be monitored by oxygen uptake, aldehyde production, ammonia production, or hydrogen peroxide. By coupling the reaction to peroxidase-p-hydroxyphenylacetic acid the reaction can be followed fluorimetrically (37). The formation rate of ammonia can be determined conveniently by use of a Beckman 39137 cation selective electrode which permits the determination of 1–100 μg/ml of amine (19).

Urea can be readily determined by using urease. The ammonia produced is determined by Nessler's reagent (55), by coupling to the NADH dependent glutamate dehydrogenase reaction (56), or by the use of a cation sensitive electrode (57). The latter method permits urea determination over the range of 0.1–50 μg/ml in a few seconds (19).

Fluorimetric methods involving six dehydrogenases have been used to determine the concentrations of 21 of the more common organic acids (58). Sensitivity of the methods range from 0.02 μg/ml for determination of D-isocitric acid with isocitrate dehydrogenase to 200 μg/ml for determination of D-tartaric acid with malate dehydrogenase.

Determination of ethanol concentration is widely used in the wine, beer, and distilled beverage industries as well as in tissues. Ethanol is readily determined by the use of alcohol dehydrogenase by measuring the rate or extent of NAD$^+$ reduction spectrophotometrically at 340 nm.

$$CH_3—CH_2—OH + NAD^+ \rightarrow CH_3—CHO + NADH + H^+ \qquad (18)$$

By using fluorescence techniques as little as 0.1 μg/ml of ethanol can be determined (59). A number of other lower aliphatic alcohols may be determined with alcohol dehydrogenase or with alcohol oxidases (60).

Glycerol concentration can be determined with glycerol dehydrogenase (Ref. 60, Equation 19) or with glycerokinase coupled with glycerol phosphate dehydrogenase in the presence of NAD$^+$ (Ref. 61, Equations 20, 21).

$$Glycerol + NAD^+ \rightleftharpoons dihydroxyacetone + NADH + H^+ \qquad (19)$$

$$Glycerol + ATP \xrightarrow[Mg^{2+}]{} L(-)\text{-glycerol 1-phosphate} + ADP \qquad (20)$$

$$\text{L}(-)\text{-glycerol 1-phosphate} + NAD^+ \rightleftharpoons \text{dihydroxyacetone} \tag{21}$$
$$\text{phosphate} + NADH + H^+$$

In either method the equilibrium of the dehydrogenase reaction, which lies to the left, is displaced in the forward direction by working at pH 9.8 and by trapping the dihydroxyacetone (phosphate) with hydrazine. Triglyceride concentrations can also be determined by one of these methods. The triglycerides are hydrolyzed, either with lipase or non-enzymatically, and the glycerol concentration is determined.

A number of methods are available for determining inorganic compounds. Ammonia can be determined specifically with glutamate dehydrogenase by measuring the rate or extent of oxidation of NADH at 340 nm (Ref. 56, Equation 22).

$$NH_4^+ + H^+ + NADH + \alpha\text{-ketoglutarate} \rightleftharpoons \tag{22}$$
$$\text{glutamate} + NAD^+ + H_2O$$

Nitrate can be determined specifically by using nitrate reductase (*62*). Inorganic phosphate can be determined by using three enzymes (*63*). Phosphorylase a catalyzes the phosphorylation of glycogen to glucose-1-phosphate, which is then converted to glucose-6-phosphate by phosphoglucomutase. The glucose-6-phosphate is determined with glucose-6-phosphate dehydrogenase and $NADP^+$. The NADPH formed is fluorescent, and concentrations of inorganic phosphate at levels of 2×10^{-11}–$5 \times 10^{-10}M$ are easily determined.

The presence of enzyme activators such as metal ions and coenzymes can be determined by using enzymes. For the assay enzyme the metal ion or cofactor is essential for activity. The amount of activator in a sample is determined from the rate enhancement of the reaction. Guilbault (*37*) lists 12 metal ions and 10 cofactors which can be determined enzymatically. Three of these substances can be determined with luciferase (Equation 23).

$$\text{Luciferin} + O_2 + ATP \xrightarrow[\text{Mg}^{2+}]{\text{luciferase}} \text{oxyluciferin} + ADP + HPO_4^{2-} \tag{23}$$

As little as 10 ppb of Mg^{2+} (*37*), 10^{-3} mm of O_2 (*64*), and picomole quantities of ATP (*65*) can be determined by measuring chemiluminescence, according to this reaction.

Compounds which inhibit specific enzymes may also be determined by enzymatic methods. The inhibitor concentration is determined from the rate decrease when the sample is combined with the enzyme reaction. Assays of this nature are more complex than those described above since the type of inhibition—competitive, noncompetitive, or uncompetitive—must be established before the compound concentration can be satisfactorily measured. Nevertheless, 18 inorganic ions, as well as a number of organic compounds, can be determined by this method (37). One of the most useful applications of this technique involves determining pesticides in foods. Giang and Hall (66) first described the determination of organophosphorus inhibitors based on the *in vitro* inhibition of cholinesterase. The method has been used successfully for compounds like paraoxon, sarin, Systox, and TEPP and with the thiono- and dithio-phosphates following treatment of the inhibitor with dilute bromine water (67). Some cholinesterases are much more sensitive to inhibitors than others. For example, Sevin had an I_{50} of 0.04 and 5.0 for fly and human plasma cholinesterase, respectively (68).

DDT can be determined in microgram amounts by its inhibition of carbonic anhydrase (69). Several of the chlorinated insecticides as well as the carbamate insecticide, Sevin, are potent lipase inhibitors. As little as 0.1–1 μg/ml of Sevin, aldrin, heptachlor, lindane, 10 μg/ml of DDT, and 2,4-D can be determined fluorimetrically with 2–3% precision by the lipase inhibition when 4-methyl umbelliferone heptanoate is used as substrate (70). The lipase inhibition procedure for DDT determination is much easier to carry out than the carbonic anhydrase assay which involves monitoring gaseous CO_2 production.

Structural Analysis

Proteins. A protein is composed of approximately 20 α-amino and α-imino acids linked together by the peptide bond. Proteins are differentiated by their unique sequences of amino acids which in turn determine the secondary, tertiary, and even quaternary structures of the molecule. For many years determining the amino acid sequence of proteins was a difficult problem because 50 to 10,000 amino acid residues are involved, depending on the molecular weight of the protein. Fortunately, at least three factors are favorable for sequence work on proteins. First, proteins have unique sizes, and every molecule of a given protein is exactly identical in size and sequence. Second, proteins are hetero polymers composed of 20 different amino acids. Third, the use of well-characterized proteolytic enzymes with specific action toward one or only a few of the peptide bonds makes it possible to fragment a protein into segments that can be handled by traditional techniques. The procedure of initial fragmentation, developed during the classic work on insulin (71), is standard for protein sequence work, even with the availability of the automatic protein sequenator.

The rationale for the fragmentation approach is shown schematically in Figure 4. One aliquot of the protein is first treated with trypsin which hydrolyzes the peptide bonds in which lysine and arginine contribute the carbonyl moiety of the peptide bond. The tryptic peptides are separated by ion exchange chromatography and/or high voltage electrophoresis-chromatography, and each is subjected to chymotryptic digestion which hydrolyzes peptide bonds in which tyrosine, phenylalanine, and tryptophan contribute the carbonyl moiety of the peptide bond. The chymotryptic peptides are separated and the sequences of the small fragments

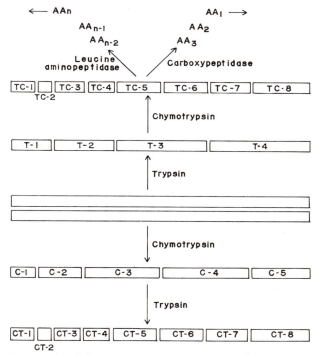

Figure 4. Schematic representation of selective, specific fragmentation of a protein by enzymatic hydrolysis

are determined by Edman degradation or by use of exo-splitting proteolytic enzymes. Carboxypeptidase A successively removes amino acids from the carboxyl (C-) terminal end of the peptide until the C-terminal residue is arginine, lysine, or proline. Carboxypeptidase B removes C-terminal arginine and lysine residues, and prolidase removes C-terminal proline residues. Aminopeptidase successively removes amino acid residues from the N-terminal end while cathepsin C removes successive dipeptidyl residues from the N-terminal end (72). Because of the availability of other methods and lack of sufficient quantities of highly purified exo-splitting proteolytic enzymes, the enzymatic methods involving end group hydrolysis now are of secondary importance in protein sequence work.

To establish the amino acid sequence unequivocally it is necessary to have peptides with overlapping sequences. This may be accomplished by determining the sequence of fragments obtained from treating a second aliquot of the protein with chymotrypsin. If these fragments are then treated with trypsin as a check, peptides identical to those obtained previously by successive treatment with trypsin and chymotrypsin are obtained. Other proteolytic enzymes, such as pepsin, subtilisin, and papain, with wider specificity than trypsin and chymotrypsin have proved useful in sequencing of some proteins.

Peptide mapping procedures are useful in establishing homology among proteins as well as in determining subunit protein structures. In the peptide mapping procedure the protein is digested with a proteolytic enzyme, most frequently trypsin, and the peptides are separated on a cellulose paper sheet in one direction by high voltage electrophoresis and in the second direction by chromatography with suitable solvents. The peptides are located by a general reagent, such as ninhydrin, and by reagents specific for selected amino acids. Peptides which differ in amino acid composition will be separated on the basis of their charge and/or solubility differences.

This generally useful technique was used first to detect the difference between normal and sickle cell hemoglobin (73). A typical result is shown in Figure 5. The single peptide difference between the two hemoglobins is indicated by the shaded and stippled spots. The difference

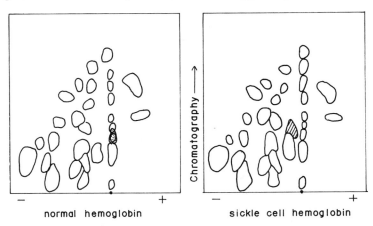

Figure 5. Peptide maps of tryptic digests of normal and sickle cell hemoglobins. Adapted from Ref. 73. The single peptide difference between the two hemoglobins is indicated by shading and by stippling.

between the two is caused by replacing a glutamyl residue with a valyl residue in each of the two β-chains. Thus a difference in two amino acid residues in 574 is readily detectable.

Many proteins are composed of more than one polypeptide chain (subunit). Peptide mapping procedures are often useful in determining how many subunits are present and whether the subunits are identical. Consider, for example, a hypothetical protein with 20 residues of arginine plus lysine per molecule as determined by amino acid analysis. (The molecular weight of the protein must be known.) If the molecule is composed of one subunit, treatment with trypsin would give 21 peptides observable by peptide mapping procedures. On the other hand, if the molecule is composed of four identical subunits, only six peptides would be obtained by tryptic digestion.

Enzymatic fragmentation methods are also useful in isolating peptides containing parts of the active site amino acid sequence and in determining regions of the molecule modified by chemical treatment. Following specific covalent labelling of the protein at the active site or chemical modification of selected groups on the molecule, the molecule is fragmented with trypsin, the peptides separated, and the amino acid sequence and location of the label is determined by the methods described above.

Glycoproteins. Structural analyses of several glycoproteins have been accomplished in the last 10 years due in part to the use of specific enzymes (74, 75, 76, 77). The glycoproteins are proteins containing various amounts of carbohydrate covalently linked to the protein through the amide group of asparagine or the hydroxyl groups of serine, threonine, hydroxylysine, and hydroxyproline. The amount of carbohydrate in glycoproteins ranges from less than 1% to more than 80%, and the carbohydrate may be attached as single, di-, or polysaccharide units. The amino acid composition of glycoproteins is not different from that of other proteins, and the techniques for determining amino acid sequences are the same as for nonconjugated proteins. The carbohydrate part is made up of a characteristic group of monosaccharides which include D-galactose, D-mannose, D-glucose, L-fucose, the amino sugars D-glucosamine and D-galactosamine, usually in the N-acetyl form, and the amino sugar derivatives, the sialic acids. Any given glycoprotein may contain from two to seven of these residues.

Chemical structure elucidation of the carbohydrate moiety of a glycoprotein is a difficult problem because there is often chain branching and because the individual carbohydrate chains are not always identical. Specific exoglycosidases which release monosaccharide units only from the terminal nonreducing end of the carbohydrate chain are most useful in structural analysis. Among the enzymes available for sequence analysis are neuraminidase, β-galactosidase, α-galactosidase, β-N-acetylglucosa-minidase, α-mannosidase, β-mannosidase, α-N-acetylgalactosaminidase, α-fucosidase, α-glucuronidase, and β-xylosidase. Still needed are enzymes which will cleave the glycopeptide bond in the intact glycoprotein. Enzymes capable of cleaving the glycopeptide bond in fragments of the glycoprotein include β-aspartyl-N-acetylglucosamine amidohydrolase (78) and α-N-acetylgalactosaminidase (79).

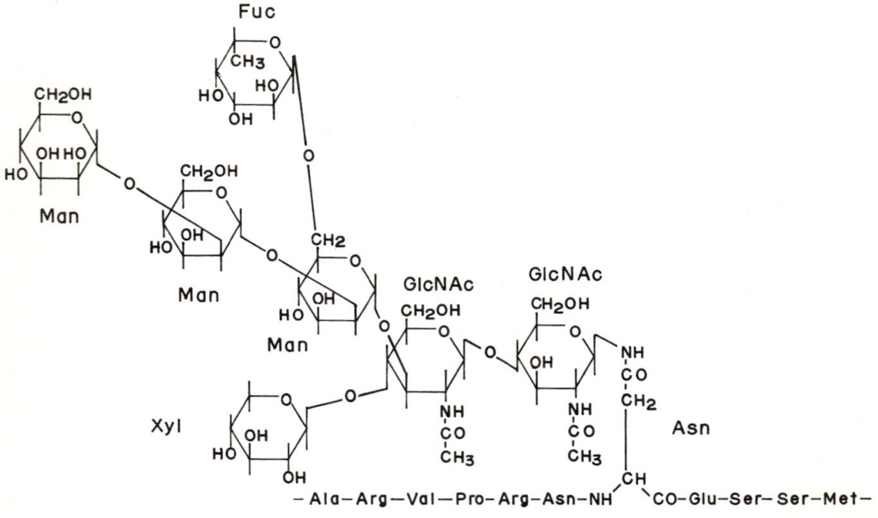

Biochemistry

Figure 6. Proposed structure of the carbohydrate moiety of stem bromelain (80)

To determine the structure of the carbohydrate moiety of a glyco-protein, the protein is exhaustively digested with a proteolytic enzyme, frequently pronase, to remove as much of the protein part as possible. The carbohydrate chains, still covalently attached to one or more amino acid residues, are separated, often by gel filtration, and the composition and sequence of monosaccharides in the chain are determined. The glycopeptide is treated exhaustively with one exoglycosidase to give maximal splitting of the susceptible terminal bond. The products are separated and the glycopeptide treated with another enzyme, etc. The procedure requires highly purified enzyme preparations and has the disadvantage that a trial and error combination method must be used with the enzymes leading to numerous permutations. For this reason, it is often useful to obtain preliminary data *via* treatment with dilute acid (0.05–0.1M H_2SO_4, 100°C) or with combined multiple enzyme systems. In these methods the relative release rates of the monosaccharides are determined.

An example of using enzymes to determine the carbohydrate structure of stem bromelain is shown in Figure 6 (*80*). The glycopeptide was
isolated following exhaustive digestion with pronase. Separate treatment
of aliquots of the glycopeptide with β-D-xylosidase, α-D-mannosidase, and
α-L-fucosidase liberated one mole of D-xylose, two moles of D-mannose,
and one mole of L-fucose, respectively. This indicates that a β-D-xylosyl
residue must be at one nonreducing end of the molecule, an α-D-mannosyl
residue joined to another α-D-mannosyl residue at another nonreducing
end, and an α-L-fucosyl residue at still another nonreducing end. Thus,
the structure must have at least two branch points. Treating the original
glycopeptide with α-L-fucosidase followed by α-D-mannosidase gave one
mole of L-fucose and three moles of D-mannose. Thus, the α-fucosyl resi-
duce must be attached to the third α-D-mannoyl residue from the non-
reducing end. Treating the product obtained from digestion of the original glycopeptide with β-D-xylosidase, α-D-mannosidase, and α-L-fucosidase
(or following a Smith degradation with periodate) with β-D-*N*-acetyl-
glucosaminidase released one mole of β-D-*N*-acetylglucosamine while
treatment of the product obtained from digestion of the original glycopeptide with α-D-mannosidase and α-L-fucosidase gave no release of β-D-
N-acetylglucosamine. Thus, the β-D-xylosyl residue must be attached to
the β-D-*N*-acetylglucosamine. One residue of β-D-*N*-acetylglucosamine
remained attached to asparagine following all the above enzymatic
treatments.

Nucleic Acids. Specific enzymes are most important in determining
the structure of nucleic acids. This area of research has grown rapidly
following the elucidation of the alanine transfer RNA structure in 1965
(*81*).

The chain length of a polynucleotide can be determined by using
alkaline phosphatase (*82, 83*) or nucleotide phosphorylase (*9*). In the
alkaline phosphatase method the terminal phosphate is removed by the
enzyme and the concentration of released phosphate is determined. The
total phosphate is determined following ashing of the sample with concentrated sulfuric acid. The polymer chain length is determined from the
ratio of total to end terminal phosphorus. The method may be used on
polymers up to several hundred monomers; however, it is more satisfactory for short chains. Method sensitivity could be increased by successive
treatment with alkaline phosphatase and polynucleotide kinase. The
terminal phosphates of the polynucleotide are first removed with alkaline
phosphatase, then the dephosphorylated polynucleotide is treated with
polynucleotide kinase and AT^{32}P. The γ-phosphate of ATP is transferred
to the 5′-hydroxyl terminal of the polynucleotide. Following treatment
again with alkaline phosphatase, the chain length is determined from the
ratio of total phosphorus to ^{32}P. Recently this technique of molecular

weight determination has been brought to a high degree of sensitivity by using neutron activation for analysis of the terminal phosphate released by treatment with calf thymus alkaline phosphatase and the total phosphate (85). Chain lengths up to 3.6×10^4 mononucleotides (the size of a single strand of DNA, MW = 2×10^7) were readily estimated. Nucleotide phosphorylase is an exo-splitting nuclease which degrades, *via* phosphorolysis, a polynucleotide chain from the 3'-OH end sequentially to mononucleotides and the terminal dinucleotide. Chain length is determined from the concentration of dinucleotide formed and total weight.

The 5'-terminal residue of a polynucleotide can be determined readily by using enzymes. In one method (86), the 5'-hydroxyl-RNA (or 5'-hydroxyl-DNA) is treated with polynucleotide kinase and AT^{32}P to give 5'-phosphate (^{32}P)-RNA (or 5'phosphate DNA). The polynucleotide is then completely digested to mononucleotides with snake venom phosphodiesterase, and the mononucleotides are separated. The radioactive 5'-terminal mononucleotide is readily detected. In a second method, the polynucleotide is degraded with nucleotide phosphorylase (84) to give the 5'-terminal dinucleotide, whose sequence is readily determined following separation from the mononucleotides. The 3'-terminal residue can be determined following successive treatment of the polynucleotide with alkaline phosphatase and snake venom phosphodiesterase (87). Snake venom phosphodiesterase degrades polynucleotides from the 3'-terminal end to form 5'-mononucleotides. The digest consists of 5'-mononucleotides and the 3'-terminal nucleoside (from prior alkaline phosphatase action) which are readily separable.

The procedure for determining the base sequence of nucleic acids is similar to that for proteins (*see* above). The large molecule is first fragmented with a highly specific enzyme, the fragments are separated and fragmented further with an enzyme of different specificity, etc. What at first appeared to be almost insurmountable obstacles—large size and a polymer composed primarily of four different bases—yielded to analytical study because of the availability of well-characterized, highly specific enzymes. The complete base sequence analysis of alanine transfer RNA, with 77 nucleotides (26,000 molecular weight), involved the use of ribonuclease T_1 and pancreatic ribonuclease A (81). Ribonuclease T_1 is specific for bonds involving guanine so that a guanylate residue (-Gp) is the 3'-terminal end of nucleotide digests with this enzyme. Ribonuclease A is specific for bonds involving pyrimidines so that uridylate (-Up) and cytidylate (-Cp) are the 3'-terminal residues following digestion. Overlapping sequences were obtained by partial digestion with the enzymes.

Advances in enzymatic digestion methodology and nucleotide separation have made it possible to study the primary structures of the
messenger RNAs ($\sim$ 3000–4000 residues) as well as of DNAs (10^4–10^6
residues).

Workers in Sanger's laboratory (*88, 89, 90*) have elucidated the primary sequence of the genome of *Escherichia coli* bacteriophage R17. The
RNA molecule contains approximately 3500 nucleotides and acts as a messenger for biosynthesis of three proteins: the coat protein (which is the
main protein coat of the virus), the A or "maturation" protein, and the
RNA replicase. The generalized approach to this problem is schematically
illustrated in Figure 7. The uniformly ^{32}P-labelled RNA was treated
with ribonuclease T_1 under conditions leading to hydrolysis of only part
of the guanylate bonds so as to give large fragments. Following separation of the fragments by polyacrylamide gel electrophoresis, separate
aliquots of a fragment were treated with ribonuclease A, with ribonuclease A following treatment of nucleotide with a water soluble carbodimide,
and with ribonuclease U_2. Ribonuclease A hydrolyzes bonds involving
uridylate and cytidylate residues to give nucleotides in which these residues are at the 3'-terminal end. Treatment with a water soluble carbodiimide modifies uridylate and guanylate residues (*91*). The phosphate
bond involving the modified uridylate residue is no longer hydrolyzed
by ribonuclease A so only bonds involving cytidyate residues are hydrolyzed in the modified nucleotide. Ribonuclease U_2 cleaves specifically
adjacent to the purines guanine and adenine to give nucleotides in which
guanylate and adenylate are the 3'-terminal residues (*92*). Following
ribonuclease T_1 digestion only the bonds involving adenylate residues
remain to be hydrolyzed by ribonuclease U_2.

The oligonucleotides are separated by electrophoresis on cellulose
acetate or by homochromatography, and the nucleotide sequence of each
is determined following partial digestion with spleen phosphodiesterase
and definitive separation of the digestion products by a two-dimensional
ionophoretic procedure at pH 1.9 (*93*). In this separation procedure any
nucletotide will move faster than a corresponding nucleotide having the
same structure but with one extra residue added to its 5'-terminal end,
i.e., X-Y-Z will move faster than W-X-Y-Z, etc.. Therefore, the various
degradation products from spleen phosphodiesterase will be arranged in
order of their size on fractionation on DEAE-cellulose paper. The distance between any two products which differ by only one residue depends
on the nature of that residue. From the distance one may deduce the
nature of the various products. Thus, it is possible to determine the
complete sequence of an oligonucleotide up to 8–10 residues from a
single partial degradation and DEAE-cellulose paper fractionation. The

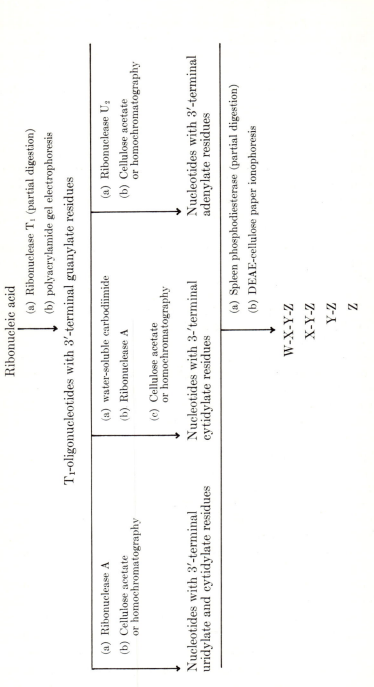

Figure 7.   Sequence determination of a ribonucleic acid using specific enzymes.   A tetranucleotide illustrates the procedure of partial digestion with spleen phosphodiesterase.

complete nucleotide sequence of bacteriophage MS2 RNA has been determined using techniques similar to these (*94*).

The same general procedure is available for DNA sequence determination. Initial fragmentation is obtained with a deoxyribonuclease, preferably with specificity for the deoxythymidylate residues. The partial nucleotide sequences of DNA from bacteriophages λ and 424 have been reported (*95*). As shown schematically in Figure 8, DNA is partially degraded with pancreatic deoxyribonuclease. The terminal phosphates of the fragments are removed with *Escherichia coli* alkaline phosphatase. The fragments are then treated with polynucleotide kinase and AT^{32}P to produce the labelled oligonucleotides which are more readily detected during the separation procedures. The complete sequence of each separated oligonucleotide is then determined with the aid of spleen and snake venom phosphodiesterases.

The preferred sites of hydrolysis by a ribonuclease from *Enterobacter sp.* are between the adenylate and cytidylate residues (-A-C-, *96*). No

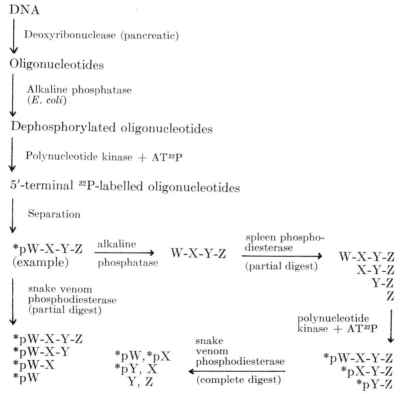

Figure 8.   An enzymatic method for determining the nucleotide sequence of a deoxyribonucleic acid

hydrolysis occurred between two purine residues, two pyrimidine residues, or between uridylate and guanylate residues. Undoubtedly, other nucleases with quite limited specificities will be discovered which will aid nucleotide sequence determinations.

Methylases specific for each of the four bases of sRNA (97) and for adenine and cytosine of DNA (98) are available. Selective modification of a nucleic acid or its oligonucleotide fragments before digestion with a specific nuclease gives additional dimensions for selective degradation. Use of ^{14}C-methyl labelled S-adenosylmethionine permits selective isotopic labelling by these enzymes.

Carbohydrates. Considerable information about the fine structure of complex carbohydrates has been obtained through the specific action of well-characterized pure glycosidases.

Evidence of heterogeneity of amylose has been obtained by using β-amylase. β-Amylase is an exoglucosidase with specificity for the α-D-$(1 \rightarrow 4)$ glucosidic linkage involving the pentultimate glucose from the nonreducing end of the molecule. The enzyme successively removes maltose units from the nonreducing end of α-D-$(1 \rightarrow 4)$ glucosidic polymers until it comes to a branch point in the molecule where its action stops. Theoretically, amylose, a linear α-D-$(1 \rightarrow 4)$ glucose polymer, should be converted 100% to maltose by β-amylase. In practice, amyloses from different sources are converted to maltose in 80–100% yield (99, 100), which indicates limited branching. The extent of amylose branching increases as the plant matures, as determined by the β-amylosis values (100, 101). Branching in some amyloses is caused by α-D-$(1 \rightarrow 6)$ glucosidic linkages, since treatment with pullulanase leads to 100% conversion to maltose (102).

The average molecular weight of amylose and other linear maltodextrins can be determined by using β-amylase (103). Amylose is treated with β-amylase under conditions in which enzymatic degradation is complete. Molecules with an even number of glucose units give maltose; those with an odd number of glucose units give maltose plus one glucose unit per chain. The total amylose concentration is determined using glucoamylase, which converts all the amylose to glucose. In each case the amount of glucose formed is estimated using glucose oxidase. Assuming that the sample contains equal number of molecules having odd and even numbers of glucose units, the average degree of polymerization, and thus molecular weight, can be calculated. The degree of polymerization can be estimated with 5% accuracy even for molecules with more than 1000 glucose units.

Data on the fine structure of amylopectin from several sources have been obtained through the action of β-amylase and pullulanase (104, 105, 106). Pullulanase removes specifically the α-D-$(1 \rightarrow 6)$ linkages in amylopectin (*107*) and certain glycogens (*108*). One of the main characteristics of an amylopectin is its average unit chain length.. This value can be estimated by periodate oxidation (*109*). However, it is difficult to make adequate corrections for overoxidation and for amylose-like contaminants. Moreover, the enzymatic method gives data on the extent of branching and the average length of the inner and outer parts of the chains. The

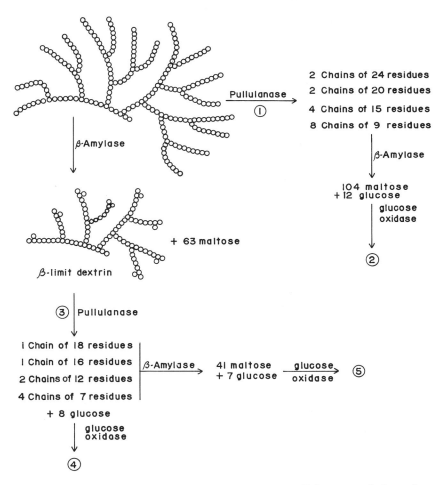

Figure 9.   Schematic representation of action of pullulanase and β-amylase on amylopectin. Data obtained at numbered points: (1) average size of unit chains from increase in reducing groups and average molecular weight; number of branched points; (2) average size of unit chains from amount of glucose and maltose formed; (3) average size of inner chain portion from increase in reducing groups and molecular weight of β-limit dextrin; number of branched points; (4) number of unit chains in amylopectin; (5) number of chains and average size of inner chain portion.

extent of branching in amylopectin is determined from the increase in reducing groups following debranching with pullulanase. The debranched chains are subjected to complete hydrolysis with β-amylase to determine the average chain length as described above for amylose. The results obtained by the enzymatic method on average length of the unit chain are more reproducible and the values comparable to those obtained by periodate oxidation (*105, 109*). The average length of the inner chain portion is determined by subjecting amylopectin directly to exhaustive hydrolysis with β-amylase to give maltose and β-limit dextrin. The average length of the inner chain portion is determined on the β-limit dextrin following pullulanase and β-amylase digestion as described above (Figure 9). The average length of the outer chain portion is calculated from the difference between the average length of unit chains and the inner chain portion. Alternatively, the average length of the unit chains and the inner and outer chain portions can be estimated from the size of the amylopectin and β-limit dextrin (determined by gel filtration) and the number of reducing groups produced by the pullulanase action.

The best evidence for the fine structure of glycogen (Figure 10) came from selective enzymatic degradation (*110*). Glycogen was first exhaustively treated with muscle phosphorylase, which removes D-glucose from the external chains to within a few units of a branch point. The limit dextrin was separated and treated with amylo-($1 \rightarrow 6$) glucosidase [$1(\rightarrow 6)$-glucan 6-glucanohydrolase] which removes exposed α-D-($1 \rightarrow 6$)-linked glucose units. The product formed was separated and the new limit dextrin subjected again to phosphorylase. By this procedure limit dextrins representing 64, 38, 23, and 12% of the total liver glycogen were obtained by successive treatments with the two enzymes, thus supporting the proposed tree like structure of glycogen. The combined action of α-amylase and R-enzyme has also been used to determine the structure of glycogen (*111*). α-Amylase hydrolyzes α-D-($1 \rightarrow 4$) linkages randomly while R-enzyme (amylopectin 6-glucanohydrolase) hydrolyzes the α-D-($1 \rightarrow 6$) linkages in the smaller fragments produced by α-amylase. More recently, pullulanase (*112*) and isoamylase from yeast (*113, 114*), which hydrolyzes α-D-($1 \rightarrow 6$) linkages, have been used for debranching glycogen. The average length of the unit chain and the outer and inner portions can be determined for glycogen using the procedures outlined above for amylopectin. This method is useful in determining the structure of glycogen formed in Type IV glycogen storage disease (*108*).

Specific enzymatic degradation has been useful in determining the structures of pectins, gums and other plant polysaccharides (*115*), the mucopolysaccharides of higher animals (*116*), and other complex carbohydrates (*117*) including those from microorganisms (*118, 119, 120*). In these heterogeneous polymers enzyme treatment is most useful for deter-

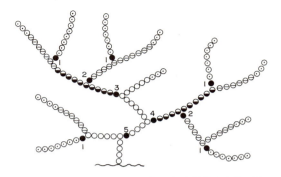

Journal of Biological Chemistry

Figure 10.   Fine structure of liver glycogen as revealed by selective enzymatic degradation (110). ⊙, ⊖, and ◑ represent glucose residues removed by first, second, and third treatment with muscle phosphorylase; ● represents glucose units removed by amylo-(1 → 6) glucosidase (the numbers beside these units indicate the number of treatment following phosphorylase action).

mining the monosaccharide sequences (*see* glycoprotein discussion) as well as the type of linkages involved.

Enzymatic analysis may prove to be useful in determining the extent of starch modification during processing and storage of foods. Gluco-amylase hydrolyzes the glucosidic linkage only when there has been no modification of the glucose moiety. Difference in values between the anthrone method and enzymatic method would reflect the degree of modification. Starch in several foods undergoes retrogradation during storage. Retrograded starch, in which the polymer chains are aligned and hydrogen-bonded to each other, is degraded much more slowly by amylase than is starch, in which the chains are randomly oriented.

Triglycerides. Structural analysis of the natural fat triglycerides is particularly difficult because of the many possible molecular species which have very similar chemical and physical properties. Major advances in separation and analytical techniques since 1955 have revolutionized this field. The chemically different triglycerides are now separable by gas–liquid chromatography, by thin-layer chromatography, and/or by permeation chromatography. By use of selective enzymatic deacylation and phosphorylation techniques, the positional isomers can be separated and characterized (*121*).

Pancreatic lipase was first shown to be a selective deacylation reagent for triglycerides in 1955 (*122, 123, 124, 125*). Mammalian pancreatic lipase has nearly absolute (> 97%) specificity for removal of fatty acids from the *sn*-1,3-positions, thus making it ideal for analytical application.

Any release of sn-2-position fatty acids is generally attributed to acyl migration or to contamination with a nonspecific lipase. Limited treatment of a triglyceride with pancreatic lipase produces a mixture of sn-1,2- and sn-2,3-diglycerides and sn-2-monoglycerides which can be separated by TLC. The sn-2-monoglycerides have been widely used in studying the positional distributions of fatty acids in natural fats. Milk lipase (126, 127) and Rhizopus arrhizus lipase (128), which have the same specificity as pancreatic lipase, have also been used in the structural analysis of triglycerides. The mold, Geotrichum candidum, produces an exocellular lipase which is highly specific for fatty acids with cis-9-unsaturation regardless of their position on the triglyceride (129, 130) thus making it particularly applicable in triglyceride analysis. It remains to be seen whether Ricus lipase (131) is the nonspecific lipase needed to produce representative sn-1,3-diglycerides and sn-1- and sn-3-monoglycerides.

Treatment with pancreatic lipase can be used to determine the fatty acids at the sn-2- and the combined sn-1,3-positions of a triglyceride. A procedure capable of distinguishing between the sn-1- and sn-3-positions, which are stereochemically distinct, has met with remarkable success (132). In this procedure, as depicted in Figure 11, the triglycerides are degraded to representative diglycerides with pancreatic lipase. The sn-1,2- and sn-2,3-diglyceride fraction is separated from the sn-2-monoglyceride fraction by preparative TLC and reacts with phenyl dichlorophosphate to produce a mixture of sn-1,2-diacyl-3-phosphatidylphenol and sn-2,3-diacyl-1-phosphatidylphenol. Treatment with phospholipase A liberates the fatty acid from the 2-position of the sn-3-phosphatide but not from the sn-1-phosphatide. Separation and fatty acid analysis of all the products, including the sn-2-monoglyceride from pancreatic lipase action, permits unequivocal assignment of fatty acids to all three positions of the original triglyceride.

Lands et al. (133) have suggested an alternative method for the stereospecific analysis of triglycerides. Following pancreatic lipase treatment and separation of the sn-1,2- and sn-2,3-diglyceride fraction, the diglyceride fraction is treated with diglyceride kinase from Escherichia coli. The diglyceride kinase stereospecifically phosphorylates the sn-1,2-diglyceride to sn-1,2-diacyl-3-phosphatidate but has no effect on the sn-2,3-diglyceride. Separation and fatty acid analysis of all the products permits specific assignment of fatty acids on the original triglycerides.

When the isomeric triglycerides contain only one residue of cis-9-unsaturated fatty acid, there is an advantage in using Geotrichum candidum lipase rather than pancreatic lipase to produce the representative diglycerides (134, 135). For example, with the triglyceride myristo-palmito-olein, which has six possible isomers, treatment with Geotrichum candidum lipase gives maximally only six diglycerides for analysis while

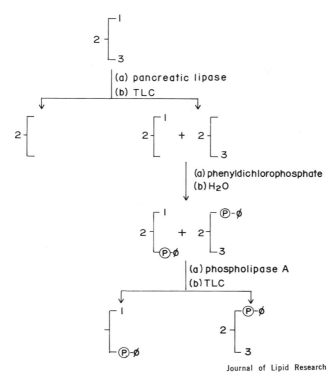

Journal of Lipid Research

Figure 11.   Stereospecific analysis of triglycerides by the Brockerhoff method (132). The numbers 1, 2, and 3 on the glycerol symbol refer to the fatty acids at the sn-1-, sn-2-, *and* sn-3-positions, *respectively, of the original triglyceride; P-φ, phenylphosphate group.*

pancreatic lipase gives rise maximally to 12 diglycerides and three *sn*-2-monoglycerides.

Dissection of Multifunctional Molecules. Protein molecules generally have more than one antigenic site and occasionally more than one distinctly different biologically active site. Whenever these sites are discretely separated spatially and are joined by only one or a few interconnecting peptide chains, proteolytic enzymes may separate the sites while preserving biological activities. Proteolytic scission is most successful when the molecule is elongated or composed of more than one polypeptide chain.

Proteolytic dissection has contributed substantially to our knowledge of the structure and function of the γ-globulins. The γG-immunoglobulins constitute the largest portion, about 80% or more, of the γ-globulin system and have MW of about 160,000. The molecules are elongated, about 235 × 44 A. A major advance in understanding the structure and

mechanism of γG-immunoglobulins came when Porter (136) found that
partial digestion with cysteine-activated papain gave three fragments
separable by CM-cellulose chromatography. Fragments I and II each
contained one antigen binding site while fragment III had no binding
activity. Later work showed that the allotypic factors of rabbit γG-
immunoglobulins are also located on fragments I and II (137) while
other biological properties such as the capacity to fix complement (138),
to bind to skin (139), to transfer across the placental barrier (140), and
to react with the rheumatoid factor (141) are all associated with fragment
III. Fragments I, II, and III are derived from γG-immunoglobulins
(Figure 12) by splitting one or two especially labile peptide bonds in
the hinge region of the molecule while reducing the interchain disulfide
bond in this region by cysteine (included as activator of papain).

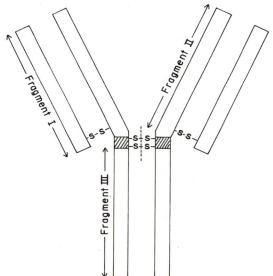

Figure 12. Schematic structure of human γG-immuno-
globulin molecule. The
hinged region attacked by
papain is indicated by the
diagonal lines.

The γM-immunoglobulins constitute a second group of γ-globulins
(142), with MW = 850,000–900,000. Proteolytic digestion of the γM-
immunoglobulins with cysteine-activated papain gives fragments similar
to those obtained from γG-immunoglobulin while mild reduction alone
leads to subunits resembling the γG-immunoglobulins. The γM-immunoglobulins appear to be composed of five subunits resembling the γG-im-

munoglobulins arranged in a circular array (Figure 13). The subunits
are held together by disulfide bonds.

Myosin is one of the contractile proteins of the muscle and has MW
= 485,000 (*143*). It is an elongated molecule 1600 A long and 15–20 A

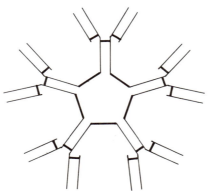

Figure 13. Schematic structure of hu-
man γM-immunoglobulin molecule. The
heavy lines between each γG-immuno-
globulin like subunit, as well as within
each subunit, represent disulfide bonds.

wide at the tail and 40–50 A wide at the head (Figure 14). It is composed of two polypeptide chains of approximately 200,000 MW and two
to three relatively low molecular weight chains noncovalently bound to
the molecule (*144*). The molecule has two biological activities associated
with the contractile process: the adenosinetriphosphatase activity associated with energy provision and the interaction with a second muscle
protein, actin, which provides the mechanical basis for contraction. Brief
treatment of myosin with trypsin (*145*), or chymotrypsin (*146*), as well
as other proteolytic enzymes, specifically cleaves myosin into two parts,
producing light (LMM) and heavy (HMM) meromyosins (Figure 14).
HMM contains both the adenosinetriphosphatase and actin-binding properties of the molecule. HMM can be specifically split by proteolytic
enzymes, particularly insolubilized papain, into subfractions HMM S-1
and HMM S-2. Thus, since myosin has three regions particularly sensitive
to proteolysis, selective splitting can lead to seven fragments of myosin.
All these fragments have been isolated and their physical, chemical, and
biological properties studied (*147*). LMM can be further fragmented by
trypsin, which suggests that there are five repeating units along the LMM
segment (*148*).

Rabbit muscle glyceraldehyde-3-phosphate dehydrogenase (MW =
140,000) catalyzes the oxidation and phosphorylation of glyceraldehyde-

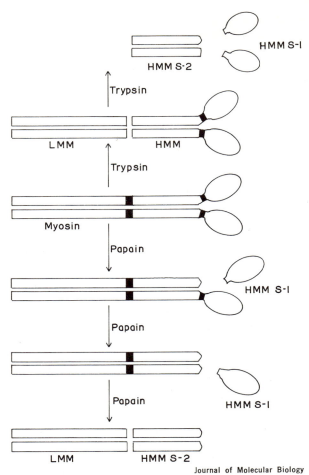

Figure 14. Myosin fragmentation by papain and trypsin (147)

3-phosphate to 1,3-diphosphoglycerate in two steps. The substrate is first oxidized to 3-phosphoglycerate *via* participation of the bound cofactor, NAD$^+$, to form the phosphoglyceryl enzyme intermediate. In a second step, the enzyme is deacylated with inorganic phosphate serving as the physiologically important acceptor. Limited chymotrypsin digestion of the enzyme gives a fragment of MW = 60,000–70,000 which retains its ability to oxidize the substrate but has no phosphorylating activity (*149*). On treatment with substrate the amount of acylenzyme formed was considerably larger with digested enzyme than with native enzyme. The native enzyme also has phosphatase activity which is retained in the digested enzyme.

The two enzymatic activities, histidinol dehydrogenase and imidazolylacetolphosphate-L-glutamate aminotransferase, associated with separate proteins, can be incorporated into a single polypeptide chain by genetic manipulation of the two adjacent genes of *Salmonella typhimurium*. When the double headed enzyme was digested with trypsin, the aminotransferase activity was destroyed rapidly while at least 70% of the histidinol dehydrogenase activity survived (*150*). Gel electrophoresis in the presence of sodium dodecylsulfate gave MW = 85,000 and 45,000 for the double headed enzyme and the histidinol dehydrogenase active fragment, respectively. The latter molecular weight is identical to that of normal histidinol dehydrogenase.

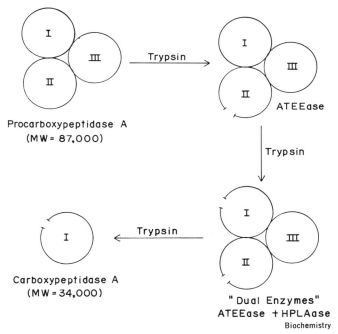

Figure 15. *Conversion of procarboxypeptidase A to carboxypeptidase A by trypsin* (151). *The subunits are not cyclic polypeptides. ATEEase and HPLAase represents activities of activated subunits I and II on acetyl-1-tyrosine ethyl ester and hippuryl phenyllactic acid, respectively.*

Bovine carboxypeptidase A is produced in the pancreas as a zymogen, procarboxypeptidase A, MW = 87,000. The proenzyme is composed of three polypeptide chains (*151*, schematically shown in Figure 15). On limited digestion with trypsin one or more peptide bonds in subunit II is split resulting in its conversion to an enzyme (ATEEase) having activity on acetyl-L-tyrosine ethyl ester similar to that of chymotrypsin. Continued treatment with trypsin hydrolyzes a peptide bond in subunit I causing carboxypeptidase A activity. The unit, still close to MW = 87,000, has two enzymatic activities. More extensive trypsin treatment destroys both subunits II and III retaining only the activated subunit I which is carboxypeptidase A, MW = 34,500.

Monitors of Conformational Changes

Proteins. The splitting rate of a particular peptide bond in a protein by a proteolytic enzyme is determined by the nature of its chemical environment and, to a larger extent, by the nature of its three-dimensional surrounding. The nature of the chemical environment is determined by the linear sequence of amino acids surrounding the scissible bond while the three-dimensional surroundings of a bond result from spatial arrangement of all the groups. It is generally agreed that proteolytic enzymes hydrolyze specific peptide bonds in a protein at an appreciable rate only when it is denatured or there is a segment of the protein chain exposed to the surface which approaches the denatured state. Therefore, proteolytic enzymes can be used as molecular probes to determine the surface topography of a protein and how this topography changes as a function of pH, temperature, dielectric constant, denaturing agents, or binding of ligands. The extreme sensitivity to steric factors, as a result of critical orientations necessary for binding of enzyme and substrate, makes the proteolytic method comparable to methods based on chromophore perturbation in the protein for detecting conformational changes in protein structure. The proteolytic method often has a wider distribution of susceptible peptide bonds than there are chromophores.

The use of proteolytic enzymes to show conformational changes caused by thermally induced unfolding of proteins is well documented. Ribonuclease A undergoes a thermal transition that is pH dependent and which can be followed by optical rotation changes or by spectral changes caused by changes in the environment of two of the buried tyrosines (152). At room temperature ribonuclease A is not attacked by chymotrypsin, trypsin, carboxypeptidase A, or leucine aminopeptidase (Figure 16). But, as shown first by Scheraga and co-workers (153, 154), in the transition temperature zone it becomes increasingly more susceptible to attack by these enzymes. Native lysozyme is not attacked by trypsin at room temperature but is hydrolyzed at an increasing rate as the temperature is raised to that critical for thermal unfolding of the molecule (156). The activation energy for the hydrolysis process is 57 kcal/mole which indicates that the rate-limiting step is the molecule unfolding. Thermal unfolding of collagen (157) and fetuin (158) have also been measured by using proteolytic enzymes.

RELATIVE RATE OF HYDROLYSIS

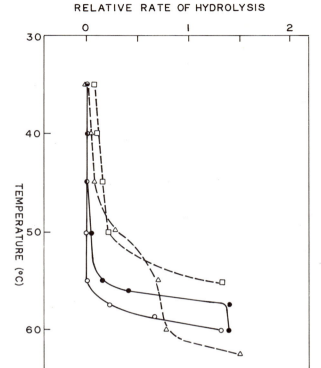

Biochemistry

Figure 16. Hydrolysis rate of bovine ribonuclease relative to that of performic acid-oxidized ribonuclease as a function of temperature (155). The following enzymes were used: ○, *aminopeptidase;* ●, *carboxypeptidase A;* △, *trypsin; and* □, *chymotrypsin.*

Proteolytic enzymes are useful in studying changes in protein conformation following removal of a portion of the peptide chain. Subtilisin, under certain conditions, specifically cleaves the Ala_{20}-Ser_{21} peptide bond in ribonuclease A, transforming it into ribonuclease-S. Ribonuclease A and ribonuclease-S are indistinguishable by a variety of physicochemical, enzymatic, and immunological properties. However, they are distinctly different with respect to digestibility by trypsin (159) indicating that the conformations must be different. More likely, the conformation elucidated by x-ray diffraction or by immunology is that caused by the major population of molecules. Digestibility, on the other hand, may detect loosened, susceptible conformations involving a small number of the molecules at any one time. There are many well-documented cases involving zymogen activation where limited proteolysis causes the remainder of the molecule to resist proteolysis (160).

Binding of ligand to protein can bring about marked changes in conformation as probed by proteolytic enzymes. In the presence of excess NAD^+ glyceraldehyde-3-phosphate dehydrogenase from muscle completely resisted attack by trypsin (161). On removal of NAD^+ the digestion rate was increased about threefold. Denaturation in urea caused a 14-fold increase in digestion rate. Hexokinase (162) and creatine phosphokinase (163) are also stabilized against proteolysis by the presence of substrate or substrate analogs. Chicken egg white conalbumin, free of Fe^{3+}, is readily digested by a number of proteolytic enzymes; when two Fe^{3+} are bound, it becomes completely resistant to attack by all proteolytic enzymes (164). On the other hand, the trypsin digestion rate of NADP-dependent succinic semialdehyde dehydrogenase from *Pseudomonas sp.* is increased more than 20-fold in the presence of 2mM $NADP^+$ (165). Carboxypeptidase A removes histidine and tyrosine from the carboxyl terminal end of the β-chain of oxygenated hemoglobin four times faster than from deoxygenated hemoglobin (166). These observations agree with x-ray crystallographic data which show the C-terminal ends

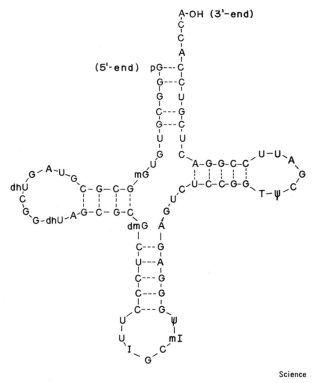

Science

Figure 17. The primary structure of yeast alanine transfer RNA showing paired and unpaired bases (81)

on the surface of oxyhemoglobin but tucked into the interior of the molecule in deoxygenated hemoglobin.

Nucleic Acids. Enzymes are useful in conformational analysis of *t*RNA (*162*). The *t*RNA molecule, consisting of some 80 nucleotides, contains regions of paired (stacked) bases as well as regions of unpaired bases (Figure 17), giving the molecule tertiary structure. Some *t*RNA molecules exist in at least two conformations, one of which is biologically active (form I) and the other inactive (form II) (*168, 169*). Form I is more compact and contains more base pairing than form II and is specifically acylated and deacylated by a specific aminoacyl-*t*RNA synthetase. On the other hand, form II is readily digested by ribonuclease T_1 while form I is not. Polynucleotide phosphorylase is also useful for distinguishing different conformations of *t*RNA (*170*). Polynucleotide phosphorylase is an exo-splitting enzyme which degrades a polynucleotide chain from the 3'-OH end sequentially to the terminal dinucleotide. When a solution of *t*RNA is treated with nucleotide phosphorylase in the presence of excess phosphate, some of the *t*RNA molecules are completely degraded while others resist degradation. Whether the two forms represent native and denatured *t*RNA or some other conformational forms is not clear.

Probes of Cellular Structure. The membrane matrix of animal cells contains, among other compounds, enzymes, proteins, phospholipids, and triglycerides. The rate and extent of release of these compounds from the membrane during selective enzymatic digestion can be used to probe the degree of exposure of these compounds to the surface environment.

As shown by electron microscopy, selective digestion by papain releases the small protruding knobs from the membrane surface of the brush border of intestinal epithelial cells while leaving the unit membrane intact. After separation of the knobs and membranes by differential centrifugation, all the invertase and maltase activities of the brush border were found in the knobs (*171*).

The membrane structure of red blood cells has been extensively investigated, often with the aid of proteolytic enzymes. Treatment of intact red blood cells with pronase releases two well-defined glycoproteins which retain the full biological activity of the M and N blood group antigens (*172, 173*). Erythrocyte membranes can, under appropriate conditions, form right side out or inside out vesicles. The type of vesicles formed can be determined by studying the types of proteins released from the membrane surface by proteolytic enzymes (*174*). Whole ghosts and right side out vesicles were much more susceptible to proteolysis than were inside out vesicles. Use of proteolytic enzymes as probes permitted partial answers to whether certain proteins exist on one side of the membrane only, whether any major protein molecules penetrate from one surface to the other, and whether there is uniform dispersion of a given protein over the membrane surface.

Specific glycosidases are useful in determining the chemistry and fine structure of yeast cell walls (118, 119), in comparative studies of the cell wall composition of different organisms (175), and in the preparation of spheroplasts and protoplasts (118, 176). Use of selected specific enzymes for wall analysis avoids many of the difficulties encountered on treatment with strong acids and alkalies. These include selective breakage of only a few bonds so that major fragments of the wall are isolated, retention of activity by the enzymes released, and evidence for the nature of the chemical bonds holding the different components together. Some 10 enzymes are in the cell wall of various yeasts (118). Their relative release rates from the cell wall indicate their location and attachment in the cell wall.

Since the cell walls of many yeasts are composed of a rigid glucan layer surrounded by a thick mannan-protein coat, the β-glucanases [β-$(1 \to 3)$ and β-$(1 \to 6)$] and mannosidases are most useful in structural analyses (118, 119, 177). In a comparative study on yeasts (175), the different genera were divided into four groups based on the action of β-$(1 \to 3)$ and β-$(1 \to 6)$ glucanases on the cell wall. Selective removal of the cell wall by enzymes to form protoplasts permits analytical studies on the structure of the membrane (178).

Histochemical techniques for detecting substrate before and after enzyme treatment are extremely useful in studies on cellular structure. One of the oldest histochemical tests utilized saliva to identify suspected glycogen or starch. More definitive results are obtained when thin sections of a tissue are incubated in a buffered solution of purified amylase and stained for poly-*vic*-glycols. Material stained by periodic acid-Schiff reagent in the control, but not in the section exposed to amylase, is assumed to be glycogen or starch. Two more of the numerous histochemical techniques associated with localization of substrate are—using hyaluronidase to locate hyaluronic acid and chondroitin 4- and 6-sulfates (179) and using neuraminidases to locate sialomucins (180). By use of electron microscopy in combination with the histochemical technique subcellular localization can be obtained.

There is considerable interest in reagents which covalently attach labels to cell surfaces. To be successful, intact cells and reagents too large to penetrate the cell surface must be used. To date most attention has been focused on the lactoperoxidase-catalyzed iodination of exposed tyrosine residues. The first work, done with human erythrocytes (181), has been extended to splenic lymphocytes (182).

Cytochemical determination of the subcellular localization of enzymes has been very successful in studies on cellular structure (183) and organized enzyme systems (*184*). Description of the cell structure based on activity of localized enzymes in the presence of selected fluorescence probes (*185*) will undoubtedly be possible. Cytochemical techniques for localizing enzymes in cells have the advantage that the normal morphological cell organization is preserved. However, there is the disadvantage that it is often qualitative in nature. This is because the method for tissue fixing prior to enzyme detection often destroys much of the enzymatic activity. Nevertheless, the technique is extremely useful and complimentary to tissue fractionation studies (*see* below). Cytochemical techniques involving light microscopy have been worked out for more than 80 enzymes, and subcellular localization of some 20 enzymes is possible by cytochemical techniques using electron microscopy (*186*). To be useful in either cytochemical or fractional detection of specific organelles, the marker enzyme must be found in substantial amounts only in one organelle. This is the case for a number of enzymes as shown in Table V (*187*).

Much of the information on subcellular organization comes from tissue fractionation techniques. The cell membranes are ruptured selectively to leave the organelle membranes intact. The organelles are then separated by differential centrifugation. Enzyme analyses hold the key to determining which organelles have been separated (Table V) since organelles from different cells have different sedimentation properties. Enzyme analyses are also important in determining the purity of an organelle preparation as well as assigning location to enzymes. For

Table V.   Marker Enzymes Useful in Cytochemical Detection and Fractional Separation of Specific Organelles[a]

Organelle	Marker enzyme
Mitochondria	Succinic dehydrogenase
	Cytochrome oxidase
Lysosomes	Acid phosphatase
	Aryl sulfatase
Peroxisomes	Catalase
	α-Hydroxy acid oxidase
Golgi apparatus	Thiamine pyrophosphatase
	Nucleoside diphosphatase
Endoplasmic reticulum	Glucose 6-phosphatase
Plasma membranes	Nucleoside phosphatase
	Alkaline phosphatase
Melanosomes	Tyrosinase
Motor endplate	Acetylcholinesterase
Nucleus	DNA-dependent RNA polymerase
Soluble fraction	Lactate dehydrogenase
	Glucose-6-phosphate dehydrogenase

[a] Adapted in part from Ref. *187*.

example, the best preparations of lysosomes still contain more than 50% light mitochondria. Assignment of an enzyme location is made on the basis of its relative concentration in the various organelles in relation to the concentration of a marker enzyme known to be located in a specific organelle.

Literature Cited

1. Osann, G., *Poggendorfs Ann.* (1845) **67**, 372.
2. Schönbein, C. F., *J. Prakt Chem.* (1851) **53**, 69.
3. *see* H. Stetter, "Enzymatic Analysis," Verlag Chemie, Weinheim/Bergstr., 1951.
4. Keilin, D., Hartree, E. F., *Biochem. J.* (1948) **42**, 221.
5. Pazur, J. H., Kleppe, K., *Biochemistry* (1964) **3**, 578.
6. Michaelis, L., Menten, M. L., *Biochem. Z.* (1913) **49**, 333.
7. Briggs, G. E., Haldane, J. B. S., *Biochem. J.* (1925) **19**, 338.
8. Whitaker, J. R., "Principles of Enzymology for the Food Sciences," p. 211, Marcel Dekker, New York, 1972.
9. Slein, M. W., in "Methods of Enzymatic Analysis," H. U. Bergmeyer, Ed., p. 117, Academic, New York, 1965.
10. Bergmeyer, H. U., in "Methods of Enzymatic Analysis," H. U. Bergmeyer, Ed., p. 3, Academic, New York, 1965.
11. Pfleiderer, G., in "Methods of Enzymatic Analysis," H. U. Bergmeyer, Ed., p. 59, Academic, New York, 1965.
12. Blaedel, W. J., Olson, C., *Anal. Chem.* (1964) **36**, 343.
13. Hohorst, H. J., in "Methods of Enzymatic Analysis," H. U. Bergmeyer, Ed., p. 266, Academic, New York, 1965.
14. Updike, S. J., Hicks, G. P., *Science (Washington)* (1967) **158**, 270.
15. Updike, S. J., Hicks, G. P., *Nature (London)* (1967) **214**, 986.
16. Williams, D. L., Doig, A. R., Jr., Korosi, A., *Anal. Chem.* (1970) **42**, 118.
17. Guilbault, G. G., Montalvo, J. G., Jr., *J. Amer. Chem. Soc.* (1969) **91**, 2164.
18. Guilbault, G. G., Hrabankova, E., *Anal. Chem.* (1970) **42**, 1779.
19. Guilbault, G. G., Smith, R. K., Montalvo, J. G., Jr., *Anal. Chem.* (1969) **41**, 600.
20. Garrod, A. E., *Proc. Roy. Soc. Med. Chir.* (1899) **2**, 130.
21. Garrod, A. E., Lancet (1908) **2**, 1.
22. Beadle, G. W., Tatum, E. L., *Proc. Nat. Acad. Sci. U.S.* (1941) **27**, 499.
23. Wacker, W. E. C., Coombs, T. L., *Annu. Rev. Biochem.* (1969) **38**, 539.
24. Raivio, K. O., Seegmiller, J. E., *Annu. Rev. Biochem.* (1972) **41**, 543.
25. "Lipid Storage Diseases: Enzymatic Defects and Clinical Implications," J. Bernsohn, H. J. Grossman, Eds., Academic, New York, 1971.
26. "Lysosomes and Storage Diseases," H. G. Hers, F. van Hoof, Eds., Academic, New York, 1973.
27. Brady, R. O., *Fed. Proc.* (1973) **32**, 1660.
28. Schmidt, E., Schmidt, F. W., Horn, H. D., Gerlach, U., in "Methods of Enzymatic Analysis," H. U. Bergmeyer, Ed., p. 651, Academic, New York, 1965.
29. Plagemann, P. G. W., Gregory, K. F., Wróblewski, F., *Biochem. Z.* (1961) **334**, 37.
30. Wieland, Th., Pfleiderer, G., Ortanderl, F., *Biochem. Z.* (1959) **331**, 103.
31. Kaplan, N. O., Ciotti, M. M., Hamolsky, M., Bieber, R. E., *Science (Washington)* (1960) **131**, 392.
32. Piendl, A., *Amer. Soc. Brew. Chem., Proc.* (1973), 60.

33. Hofmann, E., in "Methods of Enzymatic Analysis," H. U. Bergmeyer, Ed., p. 720, Academic, New York, 1965.
34. Gasser, J. K. R., *Soil Fert.* (1964) **27**, 175.
35. Bremner, J. M., Douglas, L. A., *Soil Biol. Biochem.* (1971) **3**, 297.
36. Bergmeyer, H. U., "Methods of Enzymatic Analysis," Academic, New York, 1965.
37. Guilbault, G. G., "Enzymatic Methods of Analysis," Pergamon, New York, 1970.
38. Guilbault, G. G., *Anal. Chem.* (1966) **38**, 527R.
39. *Ibid.*, (1968) **40**, 459R.
40. *Ibid.*, (1970) **42**, 334R.
41. Fishman, M. M., Schiff, H. F., *Anal. Chem.* (1972) **44**, 543R.
42. Colowick, S. P., Kaplan, N. O., *Methods Enzymol.* (1955-1973), **1-28**.
43. Thivend, P., Mercier, C., Guilbot, A., *Methods Carbohyd. Chem.* (1972) **6**, 100.
44. Rerup, C., Lundquist, I., *Acta Pharmacol. Toxicol.* (1967) **25**, 41.
45. Halliwell, G., in "Methods of Enzymatic Analysis," H. U. Bergmeyer, Ed., p. 64, Academic, New York, 1965.
46. Guilbault, G. G., Brignac, P. J., Jr., Juneau, M., *Anal. Chem.* (1968) **40**, 1256.
47. Bergmeyer, H. U., Klotzsch, H., in "Methods of Enzymatic Analysis," H. U. Bergmeyer, Ed., p. 99, Academic, New York, 1965.
48. Hu, A. S. L., Wolfe, R. G., Reithel, F. J., *Arch. Biochem. Biophys.* (1959) **81**, 500.
49. Martinsson, A., *J. Chromatog.* (1966) **24**, 487.
50. Bergmeyer, H. U., Moellering, H., *Clin. Chim. Acta* (1966) **14**, 74.
51. Guilbault, G. G., Brignac, P., Jr., Zimmer, M., *Anal. Chem.* (1968) **40**, 190.
52. Schmidt, F. H., *Klin. Wochschr.* (1961) **39**, 1244.
53. Guilbault, G. G., Hieserman, J. E., *Anal. Biochem.* (1968) **26**, 1.
54. Gale, E. F., *Methods Biochem. Anal.* (1957) **4**, 285.
55. Johnson, M. J., *J. Biol. Chem.* (1941) **137**, 575.
56. Kaltwasser, H., Schlegel, H. G., *Anal. Biochem.* (1966) **16**, 132.
57. Katz, S. A., *Anal. Chem.* (1964) **36**, 2500.
58. Guilbault, G. G., Sadar, M. H., McQueen, R., *Anal. Chim. Acta* (1969) **45**, 1.
59. Guilbault, G. G., Kramer, D. N., *Anal. Chem.* (1965) **37**, 1219.
60. Guilbault, G. G., Sadar, S. H., *Anal. Lett.* (1969) **2**, 41.
61. Wieland, O., *Biochem. Z.* (1957) **329**, 313.
62. Egami, F., Iida, K., Doke, T., Taniguchi, S., *Bull. Chem. Soc. Japan* (1954) **27**, 619.
63. Schulz, D. W., Passonneau, J. V., Lowry, O. H., *Anal. Biochem.* (1967) **19**, 300.
64. Chase, A. M., *Methods Biochem. Anal.* (1960) **8**, 61.
65. Wahl, R., Kozloff, L. M., *J. Biol. Chem.* (1962) **237**, 1953.
66. Giang, P. A., Hall, S. A., *Anal. Chem.* (1951) **23**, 1830.
67. Fallscheer, H. O., Cook, J. W., *J. Ass. Offic. Agr. Chem.* (1956) **39**, 691.
68. Zweig, G., Archer, T. E., *J. Agr. Food Chem.* (1958) **6**, 910.
69. Keller, H., *Naturwissenschaften* (1952) **39**, 109.
70. Guilbault, G. G., Sadar, M. H., *Anal. Chem.* (1969) **41**, 366.
71. Sanger, F., in "Currents in Biochemical Research," D. E. Green, Ed., p. 434, Wiley (Interscience), New York, 1956.
72. McDonald, J. K., Zeitman, B. B., Reilly, T. J., Ellis, S., *J. Biol. Chem.* (1969) **244**, 2693.
73. Ingram, V. M., *Nature (London)* (1956) **178**, 792.
74. Spiro, R. G., *Methods Enzymol.* (1966) **8**, 3.
75. *Ibid.*, (1972) **28**, 3.

76. Spiro, R. G., *Annu. Rev. Biochem.* (1970) **39**, 599.
77. "Glycoproteins. Their Composition, Structure, and Function," A Gott-schalk, Ed., 2nd ed., Elsevier, New York, 1972.
78. Mahadevan, S., Tappel, A. L., *J. Biol. Chem.* (1967) **242**, 4568.
79. Weissmann, B., Hinrichsen, D. F., *Biochemistry* (1969) **8**, 2034.
80. Yasuda, Y., Takahashi, N., Murachi, T., *Biochemistry* (1970) **9**, 25.
81. Holley, R. W., Apgar, J., Everett, G. A., Madison, J. T., Marquisee, M., Merrill, S. H., Penswick, J. R., Zamir, A., *Science (Washington)* (1965) **147**, 1462.
82. Torriani, A., *Biochim. Biophys. Acta* (1960) **38**, 460.
83. Garen, A., Levinthal, C., *Biochim. Biophys. Acta* (1960) **38**, 470.
84. Tener, G. M., *Methods Enzymol.* (1968) **12**, 220.
85. Clerici, L., Sabbioni, E., Campagnari, F., Spadari, S., Girardi, F., *Biochemistry* (1973) **12**, 2887.
86. Hurwitz, J., Novogrodsky, A., *Methods Enzymol.* (1968) **12**, 207.
87. Fraenkel-Conrat, H., *Methods Enzymol.* (1968) **12**, 224.
88. Jeppesen, P. G. N., *Biochem. J.* (1971) **124**, 357.
89. Jeppesen, P. G. N., Barrell, B. G., Sanger, F., Coulson, A. R., *Biochem. J.* (1972) **128**, 993.
90. Rensing, U. F. E., *Biochem. J.* (1973) **131**, 593.
91. Gilham, P. T., *J. Amer. Chem. Soc.* (1962) **84**, 687.
92. Arima, T., Uchida, T., Egami, F., *Biochem. J.* (1968) **106**, 609.
93. Sanger, F., Brownlee, G. G., Barrell, B. G., *J. Mol. Biol.* (1965) **13**, 373.
94. Min Jou, W., Haegeman, G., Ysebaert, M., Fiers, W., *Nature (London)* (1972) **237**, 82.
95. Murray, K., *Biochem. J.* (1973) **131**, 569.
96. Marotta, C. A., Levy, C. C., Weissman, S. M., Varricchio, F., *Biochemistry* (1973) **12**, 2901.
97. Hurwitz, J., Gold, M., *Methods Enzymol.* (1968) **12**, 480.
98. Gold, M., Hurwitz, J., *Methods Enzymol.* (1968) **12**, 491.
99. Banks, W., Greenwood, C. T., *Staerke* (1967) **19**, 197.
100. Geddes, R., Greenwood, C. T., MacKenzie, S., *Carbohyd. Res.* (1965) **1**, 71.
101. Greenwood, C. T., Thomson, J., *Biochem. J.* (1962) **82**, 156.
102. Banks, W., Greenwood, C. T., *Arch. Biochem. Biophys.* (1966) **117**, 674.
103. Banks, W., Greenwood, C. T., *Carbohyd. Res.* (1968) **6**, 177.
104. Abdullah, M., Lee, E. Y. C., Whelan, W. J., *Biochem. J.* (1965) **97**, 10P.
105. Lee, E. Y. C., Whelan, W. J., *Arch. Biochem. Biophys.* (1966) **116**, 162.
106. Mercier, C., *Staerke* (1973) **25**, 78.
107. Abdullah, M., Catley, B. J., Lee, E. Y. C., Robyt, J., Wallenfels, K., Whelan, W. J., *Cereal Chem.* (1966) **43**, 111.
108. Mercier, C., Whelan, W. J., *Eur. J. Biochem.* (1970) **16**, 579.
109. Adkins, G. K., Banks, W., Greenwood, C. T., *Carbohyd. Res.* (1966) **2**, 502.
110. Larner, J. B., Illingworth, B., Cori, G. T., Cori, C. F., *J. Biol. Chem.* (1952) **199**, 641.
111. Whelan, W. J., Roberts, P. J. P., *Nature (London)* (1952) **170**, 748.
112. Brown, D. H., Illingworth, B., Kornfeld, R., *Biochemistry* (1965) **4**, 486.
113. Gunja, Z. H., Manners, D. J., Maung, K., *Biochem. J.* (1961) **81**, 392.
114. Gunja-Smith, Z., Marshall, J. J., Mercier, C., Smith, E. E., Whelan, W. J., *FEBS Lett.* (1970) **12**, 101.
115. Aspinall, G. O., in "The Carbohydrates, Chemistry and Biochemistry," W. Pigman, D. Horton, A. Herp., Eds., 2nd ed., Vol. IIB, p. 515, Academic, New York, 1970.
116. Jeanloz, R. W., in "The Carbohydrates, Chemistry and Biochemistry," W. Pigman, D. Horton, A. Herp, Eds., 2nd ed., Vol. IIB, p. 590, Academic, New York, 1970.

117. "The Carbohydrates, Chemistry and Biochemistry," W. Pigman, D. Horton, A. Herp, Eds., 2nd ed., Vol. IIB, Academic, New York, 1970.
118. Phaff, H. J., in "The Yeasts," H. H. Rose, J. S. Harrison, Eds., Vol. 2, p. 135, Academic, New York, 1971.
119. MacWilliam, I. C., *J. Inst. Brew.* (1970) **76**, 524.
120. Troy, F. A., Koffler, H., *J. Biol. Chem.* (1969) **244**, 5563.
121. Litchfield, C., "Analysis of Triglycerides," Academic, New York, 1972.
122. Mattson, F. H., Beck, L. W., *J. Biol. Chem.* (1955) **214**, 115.
123. *Ibid.*, (1956) **219**, 735.
124. Savary, P., Desnuelle, P., *Compt. Rend.* (1955) **240**, 2571.
125. Savary, P., Desnuelle, P., *Biochim. Biophys. Acta* (1956) **21**, 349.
126. Gander, G. W., Jensen, R. G., *J. Dairy Sci.* (1960) **43**, 1762.
127. Jensen, R. G., Sampugna, J., Pereira, R. L., Chandan, R. C., Shahani, K. M., *J. Dairy Sci.* (1964) **47**, 1012.
128. Laboureur, P., Labrousse, M., *Bull. Soc. Chim. Biol.* (1966) **48**, 747.
129. Alford, J. A., Pierce, D. A., Suggs, F. G., *J. Lipid Res.* (1964) **5**, 390.
130. Marks, T. A., Quinn, J. G., Sampugna, J., Jensen, R. G., *Lipids* (1968) **3**, 143.
131. Noma, A., Borgström, B., *Biochim. Biophys. Acta* (1971) **227**, 106.
132. Brockerhoff, H., *J. Lipid Res.* (1965) **6**, 10.
133. Lands, W. E. M., Pieringer, R. A., Slakey, P. M., Zschocke, A., *Lipids* (1966) **1**, 444.
134. Jensen, R. G., Sampugna, J., Quinn, J. G., *Lipids* (1966) **1**, 294.
135. Sampugna, J., Jensen, R. G., *Lipids* (1968) **3**, 519.
136. Porter, R. R., *Biochem. J.* (1959) **73**, 119.
137. Kelus, A., Marrack, J. R., Richards, C. B., *Protides Biol. Fluids, Proc. Colloq.* (1961) **8**, 176.
138. Amiraian, K., Leikhim, E. J., *Proc. Soc. Exp. Biol. Med.* (1961) **108**, 454.
139. Ovary, Z., Karush, F., *J. Immunol.* (1961) **86**, 146.
140. Brambell, F. W. R., Hemmings, W. A., Oakley, C. L., Porter, R. R., *Proc. Roy. Soc.* (1960) **B151**, 478.
141. Goodman, J. W., *Proc. Soc. Exp. Biol. Med.* (1961) **106**, 822.
142. Metzger, H., *Advan. Immunol.* (1970) **12**, 57.
143. Godfrey, J. E., Harrington, W. F., *Biochemistry* (1970) **9**, 894.
144. Dreizen, P., Gershman, L. C., *Trans. N.Y. Acad. Sci.* (1970) **32**, 170.
145. Mihályi, E., Szent-Györgyi, A. G., *J. Biol. Chem.* (1953) **201**, 189.
146. Gergely, J., Gouvea, M. A., Karibian, D., *J. Biol. Chem.* (1955) **212**, 165.
147. Lowey, S., Slayter, H. S., Weeds, A. G., Baker, H., *J. Mol. Biol.* (1969) **42**, 1.
148. Bálint, M., Szilágyi, L., Fekete, G., Blazsó, M., Biró, N. A., *J. Mol. Biol.* (1968) **37**, 317.
149. Krimsky, I., Racker, E., *Biochemistry* (1963) **2**, 512.
150. Kohno, T., Yourno, J., *J. Biol. Chem.* (1971) **246**, 2203.
151. Brown, J. R., Yamasaki, M., Neurath, H., *Biochemistry* (1963) **2**, 877.
152. Hermans, J., Jr., Scheraga, H. A., *J. Amer. Chem. Soc.* (1961) **83**, 3293.
153. Rupley, J. A., Scheraga, H. A., *Biochemistry* (1963) **2**, 421.
154. Ooi, T., Rupley, J. A., Scheraga, H. A., *Biochemistry* (1963) **2**, 432.
155. Klee, W. A., *Biochemistry* (1967) **6**, 3736.
156. Gorini, L., Felix, F., Fromageot, C., *Biochim. Biophys. Acta* (1953) **12**, 283.
157. von Hippel, P. H., Gallop, P. M., Seifter, S., Cunningham, R. S., *J. Amer. Chem. Soc.* (1960) **82**, 2774.
158. Verpoorte, J. A., Green, W. A., Kay, C. M., *J. Biol. Chem.* (1965) **240**, 1156.
159. Allende, J. E., Richards, F. M., *Biochemistry* (1962) **1**, 295.
160. Mihályi, E., "Application of Proteolytic Enzymes to Protein Structure Studies," p. 179, CRC, Cleveland, 1972.

161. Elödi, P., Szabolcsi, G., *Nature (London)* (1959) **184**, 56.
162. Trayser, K. A., Colowick, S. P., *Arch. Biochem. Biophys.* (1961) **94**, 169.
163. Lui, N. S. T., Cunningham, L., *Biochemistry* (1966) **5**, 144.
164. Azari, P. R., Feeney, R. E., *J. Biol. Chem.* (1958) **232**, 293.
165. Nirenberg, M. W., Jakoby, W. B., *Proc. Nat. Acad. Sci. U.S.* (1960) **46**, 206.
166. Zito, R., Antonini, E., Wyman, J., *J. Biol. Chem.* (1964) **239**, 1804.
167. Ishida, T., Arceneaux, J. L., Sueoka, N., *Methods Enzymol.* (1971) **20**, 98.
168. Gartland, W. J., Sueoka, N., *Proc. Nat. Acad. Sci. U.S.* (1966) **55**, 948.
169. Lindahl, T., Adams, A., Geroch, M., Fresco, J. R., *Proc. Nat. Acad. Sci. U.S.* (1967) **57**, 178.
170. Thang, M. N., Beltchev, B., Grunberg-Manago, M., *Methods Enzymol.* (1971) **20**, 106.
171. Johnson, C. F., *Science (Washington)* (1967) **155**, 1670.
172. Bender, W. W., Garan, H., Berg, H. C., *J. Mol. Biol.* (1971) **58**, 783.
173. Cook, G. M. W., Eylar, E. H., *Biochim. Biophys. Acta* (1965) **101**, 57.
174. Steck, T. L., Fairbanks, G., Wallach, D. F. H., *Biochemistry* (1971) **10**, 2617.
175. Tanaka, H., Phaff, H. J., Higgins, L. W., *Abh. Deut. Akad. Wiss. Berlin Kl. Med.* (1966) **6**, 113.
176. Sommer, A., Lewis, M. J., *J. Gen. Microbiol.* (1971) **68**, 327.
177. Lampen, J. O., *Antonie van Leeuwenhoek* (1968) **34**, 1.
178. Kaback, H. R., *Methods Enzymol.* (1971) **22**, 99.
179. Leppi, T. J., Stoward, P. J., *J. Histochem. Cytochem.* (1965) **13**, 406.
180. Quintarelli, G., Tsuiki, S., Hashimoto, Y., Pigman, W., *J. Histochem. Cytochem.* (1961) **9**, 176.
181. Phillips, D. R., Morrison, M., *Biochemistry* (1971) **10**, 1766.
182. Vitetta, E. S., Baur, S., Uhr, J. W., *J. Exp. Med.* (1971) **134**, 242.
183. Shnitka, T. K., Seligman, A. M., *Annu. Rev. Biochem.* (1971) **40**, 375.
184. Seligman, A. M., Karnovsky, M. J., Wasserkrug, H. L., Hanker, J. S., *J. Cell Biol.* (1968) **38**, 1.
185. Brand, L., Gohlke, J. R., *Annu. Rev. Biochem.* (1972) **41**, 843.
186. Pearse, A. G. E., "Histochemistry, Theoretical and Applied," 3rd ed., p. 475, Little, Brown, Boston, 1968.
187. Sobel, H. J., in "Pathology Annual," S. C. Sommers, Ed., p. 57, Appleton, New York, 1968.

RECEIVED September 17, 1973.

Cellulases

G. H. EMERT, E. K. GUM, JR., J. A. LANG, T. H. LIU,
and R. D. BROWN, JR.

Department of Biochemistry and Nutrition, Virginia Polytechnic Institute and
State University, Blacksburg, Va. 24061

The active cellulase of Trichoderma viride *has been fractionated into several component enzymes. Three forms of cellobiohydrolase and a cellobiase have been purified to electrophoretic homogeneity. These enzymes, together with β-(1 → 4)-endo-glucanases, comprise the components of the cellulase complex necessary for the conversion of cellulose to glucose. We propose that the function of the cellobiohydrolases is cleavage of cellooligosaccharides at the site of their production (i.e., the cellulose surface). The cellobiase then catalyzes glucose formation and controls the level of cellobiose, which inhibits cellobiohydrolase action. A clear understanding of the functional components of cellulases should allow replacement of C_1–C_x terminology by more precise nomenclature and permit the intelligent use of cellulases, component enzymes, or cellulolytic microorganisms in producing or modifying foodstuffs.*

The study of enzymatic cellulose hydrolysis *in vitro* dates from the investigations of snail digestive juice by Seillière (*1*) in 1906, but only within the past decade have the activities of cellulases been observed in chemical detail using highly purified preparations. For forty years after their discovery cellulase enzymes were studied primarily in regard to their biological sources and range of substrates, especially commercial cellulosic materials. In the early 1950's Reese and co-workers proposed a model of cellulase action (*2, 3*) which focused attention on the relationship between the multi-enzymic nature of cellulases produced by some organisms and the capacity of such systems or complexes to degrade the most resistant or crystalline forms of cellulose. Useful information is obtained continually on biological sources, enzyme production and control, microbial deterioration of cellulosic substances, and the ecological roles of the cellulases. However it has become evident that only by isolation, purification, and chemical and physical characterization of the individual enzymes will the mode of action of cellulases be understood clearly. Since the ACS symposium on "Cellulases and Their Applications" in 1968 (4), excellent reviews of recent research on cellulases have appeared, covering substrate diversity (5), modes of action (6), and comparative properties (7, 8).

Cellulases are formed by many bacteria, fungi, higher plants, and some invertebrate animals (7, 9). The physiological roles of these enzymes include the extracellular formation of soluble oligosaccharides and glucose for use as a carbon source by microorganisms and for modifying cell walls during growth and morphogenesis in higher plants. The economic effects of cellulolysis range from beneficial (e.g., rumen microorganisms) to destructive (e.g., plant pathogens and wood decay organisms). The significance of the process to the biosphere is suggested by the estimate that one-third of the photosynthetically fixed carbon dioxide, some 10 billion tons of carbon annually, is converted to cellulose, which in turn must be degraded to maintain the carbon balance. From economic considerations alone it would be important to control cellulase activity to minimize the destruction of cellulosic commodities and foodstuffs and enhance degradation of cellulosic wastes. However there has been increasing interest in the application of cellulase enzymes for the controlled modification of foodstuffs, the saccharification of waste cellulose, and the production of protoplasts from plant cells (7).

Cellulase Substrates

Before considering the cellulase enzymes and their properties, it is necessary to survey the types of substrates which are commonly used in assaying cellulase activity. Table I lists several classes of cellulose and cellulosic derivatives with degrees of polymerization ranging from tens of thousands for native cellulose to six or less for the soluble oligosaccharides. Within each type many variant forms could be listed, such as those with different degrees of crystallinity and crystal lattice type or those with different degrees of substitution. The first three types are essentially insoluble in water and are relatively resistant to enzymatic degradation. Amorphous swollen cellulose, although insoluble, is much more susceptible and represents an intermediate state between insoluble and soluble forms. The soluble derivatized cellulose most often utilized as a substrate is carboxymethylcellulose (CM-cellulose). Although the negative charge and unspecified distribution of carboxymethyl substituents make precise interpretation of hydrolysis rates difficult (7, 10), under carefully defined conditions the results of the viscosimetric assay have been expressed in standard units (*11*). The cellooligosaccharides and their derivatives, such as borohydride reduction products and methyl β-(1 → 4)oligoglucosides, are each specific, soluble compounds in contrast to the inexactly defined polymers whose molecular and supramolecular structures are only described statistically. As substrates the small molecules yield products which may be analyzed quantitatively to permit exact evaluation of kinetic parameters and bond cleavage specificity. In addition to the general classes of substrates shown, chromogenic substrates with a dye covalently bound to cellulose (*12*), have been used to screen for cellulase-producing organisms.

Table I. Types of Cellulose Used as Substrates for Various Cellulase Assays

Types	Examples	DP
Crystalline cellulose	Hydrocellulose, Avicel	100–200
Native cellulose	Cotton, dewaxed cotton	8,000–100,000
"Purified" wood cellulose	Filter paper, cellulose powder, Solka-floc, α-cellulose	200–700
Amorphous (Swollen) cellulose	Mercerized cellulose (25% NaOH)	4,000
	Walseth cellulose (85% H_3PO_4)	1,200–1,400
Derivatized (Soluble) cellulose	Carboxymethylcellulose (CM-cellulose)	370–2,140
	Hydroxymethylcellulose Methylcellulose	
Cellooligosaccharides	Cellobiose through cellohexaose	2–6

Cellulase Terminology

The term cellulase has been applied both to pure, well-characterized enzymes and to mixtures of enzymes produced by organisms which can degrade cellulose (Table II). These mixtures are sometimes called cellulase "complexes" or "systems." This implies a cooperative effect in the degradation of cellulose—not an integral, physical entity containing definite proportions of component enzymes. Instead of the trivial name "cellulase," the Commission on Enzymes of the International Union of Biochemistry has assigned the systematic name "β-(1 → 4)glucan 4-glucanohydrolase" (EC 3.2.1.4), which indicates a random or *endo*-β-(1 → 4)glucanase activity. The designation for β-glucosidases is EC 3.2.1.21, which has also been applied to "cellobiase" and *exo*-β-(1 → 4)-glucanase enzymes. Criteria have been suggested for distinguishing *exo*-from *endo*-glucanases (*13*) and *exo*-glucanases from β-glucosidases (*14*).

Both *exo*- and *endo*-glucanases catalyze hydrolysis of the polymeric substrates more rapidly than the corresponding oligomers. The *exo*-

hydrolases effect a configuration inversion at the anomeric glycosyl carbon atom, whereas the *endo*-hydrolases cause net retention. The *endo*-enzymes are often qualitatively distinguished from the *exo*-enzymes by the former's relatively high ratio of change in specific fluidity of a substrate to the release of reducing sugars. More direct evidence for the random hydrolysis of a substrate includes comparison of DP_w and DP_n during the initial phase of hydrolysis and the observation of the higher cellooligosaccharides formed during hydrolysis. *Endo-* and *exo*-glucanase activity have been distinguished by analyzing the initial hydrolysis products of terminally substituted cellooligosaccharides to determine frequency of bond cleavage. The β-glucosidases, on the other hand, are characterized by transferase activity, configuration retention at the anomeric glycosyl carbon atom, greater reactivity with dimers and trimers than with polymeric substrates, and high sensitivity to inhibition by D-gluconic acid δ-lactone. The trivial name cellobiase has been applied to widely differing enzymes. Some of these are also active toward aryl-β-glucosides but not polymeric substrates, whereas others cleave various oligosaccharides but not aryl-β-glucosides (*15, 16*). With respect to cellulase enzyme components the term is applied here to an enzyme whose function is to degrade cellobiose produced during polymeric substrate degradation. Such an enzyme has been termed cellobiase by Selby and Maitland (*17*), β-glucosidase by Wood (*18, 19*), and exoglucanase by Li, Flora, and King (*20*). Since the hydrolytic enzymes may be described as *exo-* or *endo*-glucanases and β-glucosidases, the former term, C_x may be abandoned (*8*).

The last two enzymes listed in Table II are termed C_1 factors by Reese since they are supposed, by some means of primary attack on the

Table II.   Cellulase Enzyme Components

Enzyme	*Substrate*	*Assay*
Endo-glucanase EC 3.2.1.4	$-G-G-G-G\overset{\downarrow}{-}G-G-G-$	Decrease in η of CM-cellulose
Exo-glucanase	$G\overset{\downarrow}{-}G-G-G-G-G$	Reducing sugars from CM-cellulose
β-Glucosidase EC 3.2.1.21	$G\overset{\downarrow}{-}O-R$	Release of *p*-nitrophenol
Cellobiase	$G\overset{\downarrow}{-}G$	1. Glucose oxidase 2. Polarimetry
C_1 (hydrocellulase or Avicelase)	Hydrocellulose or dewaxed cotton	1. Decrease in turbidity 2. Solubilization
C_1 (cellobiohydrolase)	G G G G G G	Solubilization of Walseth cellulose

substrate, to confer the capability of degrading crystalline cellulose on the cellulase complex in which they exist (Figure 1). When assayed in combination with the hydrolytic (or C_x) enzymes, C_1 activity can be followed by measuring the decrease in turbidity of a substrate suspension,

$$\text{Crystalline Cellulose} \xrightarrow{\quad C_1 \quad} \text{Hydrated Polyanhydro-} \xrightarrow{\quad C_x \quad}$$
$$\text{glucose chains}$$

$$\text{Cellobiose} \xrightarrow{\quad \beta\text{-Glucosidase} \quad} \text{Glucose}$$

Figure 1. C_1 and C_x cellulase hypothesis

the loss in dry weight of dewaxed cotton, or by the reducing sugars produced. The terms hydrocellulase or Avicelase refer to the type of substrate degraded by the enzyme, either alone or as a required component of a cellulase system. Halliwell *et al.* (21) and Wood and McCrae (22) have found that C_1 components from *Trichoderma viride* and *T. koningii*, respectively, have cellobiohydrolase activity which may be assayed by measuring the release of cellobiose from cellulosic substrates such as Walseth cellulose. It is not clear why the same enzyme from *T. koningii* was not found to degrade Walseth cellulose a few years earlier (23).

Biosynthesis and Control of Cellulase Enzymes

Both the quantity and properties of cellulases produced by microorganisms depend on the culture conditions. Commonly, cellulases are produced by culture of the organism either (a) in a liquid medium, which may be stationary, shaken, or submerged with aeration, or (b) by a Koji process on a solid substrate such as wheat bran (7). The complexity of the crude cellulosic carbon source usually leads to the production of a mixture of hydrolytic enzymes which may include amylases, proteases, chitinases, etc., in addition to the cellulases. Separation of proteins from culture filtrates by high resolution techniques such as chromatography, electrophoresis, or electrofocusing often reveals a number of enzyme species which may differ in specificity toward cellulosic substrates. These forms may represent:

(a) Complexes of enzyme with carbohydrate as seen in *Myrothecium verrucaria* (7), *Stereum sanguinolentum* (24), or *Penicillium notatum* (25).

(b) Degradation products caused by protease or carbohydrase action (26).

(c) Species which have undergone a change in isoelectric point during growth as demonstrated in *Chrysosporium lignorum* (27).

(d) The formation of essentially different enzymes with distinct structures and functions as typified by *Trichoderma viride* (20).

Although fungal cellulases are often more active, bacteria, particularly those of the rumen, have high cellulolytic activity (28). Hungate recently reported (29) that the cellulase of *Ruminococcus album* is partially inhibited by oxygen. This may account for the difficulty in isolating active cellulases from rumen organisms.

Earlier studies on production by fungi and control of cellulases pointed to the importance of cultural conditions (30) or the necessity of keeping soluble sugar levels low in the medium (31, 32) or of providing modified substrates which are utilized slowly. For the fungus *Trichoderma viride* QM6a mutants have been developed by Mandels and parent strain (33). Hulme has suggested that the control of cellulase production is primarily by glucose or catabolite repression. In continuous culture of *Myrothecium verrucaria* he has demonstrated that on limiting substrate, the production of cellulase is essentially independent of the carbon source (34). The most effective inducer postulated for cellulase biosynthesis is sophorose (β-D-glucopyranosyl-($1 \rightarrow 2$)-D-glucopyranose), which stimulates cellulase production in *Trichoderma viride* (35).

The relationship between catabolite repression by glucose and induction of cellulase by sophorose has been studied in *T. viride* by Nisizawa and co-workers (36, 37). The induction by sophorose ($10^{-3}M$) was competitively repressed by glucose and other metabolites such as pyruvate. Since glucose was an effective repressor when added one hour after the previous addition of actinomycin D, it was concluded that the repression takes place at the translational level. Previous work indicated (26) that the sophorose induction led to the formation of a cellulase component designated FII, which is the source of cellulase II discussed below. In higher plants indoleacetic acid (38) and abscisic acid (39) have been shown to stimulate cellulase production.

The activity of cellulases is inhibited by various natural products (40), the most important of which is the product cellobiose. By preventing overproduction of extracellular soluble sugars, this "feedback" inhibition is of competitive advantage to the organism.

Many microorganisms can grow only on degraded, swollen, or derivatized forms of cellulose whereas truly cellulolytic organisms can grow on native, highly crystalline cellulose such as cotton fibers. The former group of organisms has been termed noncellulolytic (2, 3) or pseudocellulolytic (41) and produces only β-glucanases and β-glucosidases capable of hydrolyzing soluble or hydrated poly-β-glucan chains or amorphous regions of partially crystalline cellulose. Such enzymes were called C_x cellulases by Reese (2, 3), and the existence of another factor or enzyme termed C_1 was postulated to carry out preliminary modification of crystalline cellulose. The precise function of this factor was unspecified, but it has evoked many speculations, including the activities of a "hydrogen bondase" which converts crystalline regions to amorphous ones (*42*), an "affinity factor" which binds a hydrolytic factor to the native crystalline cellulose (*43*), and a "cellobiohydrolase" which cleaves cellobiose units from the nonreducing end of β-$(1 \rightarrow 4)$glucans (*21, 22, 43*). The capacity to degrade crystalline cellulose actively, although evidenced during growth, was missing in the culture filtrates of many cellulolytic organisms as well as the pseudo-cellulolytic species. Some of the most prominent of the cellulolytic group are *Trichoderma viride, Trichoderma koningii, Fusarium solani, Fusarium javanicum* (*41*), and *Chrysosporium lignorum* (*44*).

Preparations said to contain C_1 from culture filtrates of either *T. koningii* or *F. solani*, when added to culture filtrates of various pseudo-cellulolytic organisms, were able to confer the capacity for extensive solubilization of dewaxed cotton fibers (*41*). Thus the purification of C_1 factors and elucidation of their function have concerned several laboratories during the past few years.

Purification and Characterization of Cellulase Enzymes

The information which is available now regarding the nature of highly purified enzymes capable of degrading native cellulose or its derivatives is reviewed in this section. In purifying enzymes from others which have similar or overlapping specificities, extreme care must be taken to ensure the chemical and physical identity of each enzyme component to assess its role accurately in the multistep process leading from crystalline cellulose to glucose. This entails verification of physical purity by different techniques and use of an appropriate variety of substrates to define enzymatic function.

Several cellulase enzymes have been purified from organisms whose culture filtrates have a limited capacity to solubilize crystalline cellulose. The three well-characterized cellulases, whose amino acid compositions are shown in Table III, are representative of this group of enzymes. Pettersson's results with the *Penicillium notatum* enzyme (*25, 44*), those of Eriksson's group with *Stereum sanguinolentum* (*16, 45*), and those of Whitaker and co-workers with the *Myrothecium verrucaria* cellulase (*46*) reveal many similarities in composition and function. Although the molecular weights vary from 21,200 to 49,000, each protein is rich in acidic and aromatic amino acids and contains low levels of basic amino acids. Each enzyme seems to be the principal or only cellulase elaborated by the organism from which it is obtained. The carbohydrate content

Table III. Amino Acid Compositions of Fungal Cellulases

Amino Acid	Stereum[a] sanguinolentum	Penicillium[b] notatum (moles/mole enzyme)	Myrothecium[c] verrucaria
Lysine	2	13	19
Histidine	1	3	7
Arginine	1	6	17
Aspartic acid and Asparagine	20	40	53
Threonine	20	29	41
Serine	23	27	39
Glutamic acid and Glutamine	15	29	32
Proline	6	11	25
Glycine	19	28	46
Alanine	15	33	38
Half-cystine	1	2	14–16
Valine	12	18	24
Methionine	1	8	4
Isoleucine	11	15	14
Leucine	10	21	26
Tyrosine	6	15	19
Phenylalanine	8	13	16
Tryptophan	14	13	16

[a] Molecular weight 21,200 (16,45).
[b] Molecular weight 35,500 (25,44).
[c] Molecular weight 49,000 (46).

of each enzyme is less than 1% although carbohydrate complexes have been reported for the enzymes from *Stereum* and *Myrothecium* (24, 47). These three enzymes have little or no activity toward cellobiose, and they solubilize native cellulose slowly. They do catalyze the hydrolysis of higher cellooligosaccharides, swollen cellulose, and carboxymethylcellulose (CM-cellulose) in a random manner. The identity of these cellulases with the C_x enzymes of cellulolytic organisms was reinforced by Wood's report (23) that the addition of C_1 enzyme from *T. koningii* or *F. solani* to a *M. verrucaria* culture filtrate caused a greater solubilization of dewaxed cotton fibers than the sum of that caused by the individual enzyme activities.

Purification and Characterization of Trichoderma Cellulase Components

With the example provided by the characterization of these chemically and enzymologically pure cellulases, we decided to purify and describe the enzyme components of a cellulolytic organism, *Trichoderma viride*. Brief descriptions of the properties of partially purified *exo-β-*$(1 \rightarrow 4)$glucanase (20), C_1 or hydrocellulase (48), and *endo-β-*$(1 \rightarrow 4)$-

glucanase (*20*) had been published by King's group. The reconstitution experiments using the above component enzymes are summarized in reference (*48*), and an uncharacterized enzyme Z is described as a thermostable glucanase associated with the hydrocellulase. *Exo*-glucanase aided in the formation of glucose from the cellooligosaccharides arising from the activity of the other enzymes. The hydrocellulase was less active in solubilizing cotton than was the crude enzyme, and the product obtained from hydrocellulose changed from 82% glucose using the crude enzyme to almost 97% cellobiose for the purified hydrocellulase or C_1. The *endo*-β-(1 → 4)glucanase did not seem to contribute to the degradation of the crystalline substrate.

Cellobiase. Figure 2 outlines the general purification steps used to isolate a pure cellobiase (named for its function in the cellulase complex) and three forms of the hydrocellulase. In purifying cellobiase it was expedient to replace the adsorption or affinity column with a batch separation on DEAE-Sephadex A-50 and to complete the purification with cation exchange chromatography on SP-Sephadex (*49, 50*). Table IV is a summary of the purification of the cellobiase and co-purification of

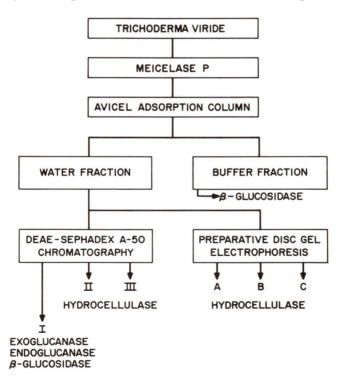

Figure 2. Purification scheme

Table IV.   Summary of Purification of Cellobiase

Procedure	Protein (mg)	Specific Activity (units/mg protein) Cellobiase	Glucosidase
Meicelase P	12,477	0.63	1.4
Ultrafiltration	6,969	0.87	2.2
DEAE-Sephadex batch	506	10.0	25
DEAE-Sephadex chromatography	59	34	70
SP-Sephadex chromatography	28	71	145

aryl-β-glucosidase activity. The difference between the 114-fold purification of the cellobiase and the 104-fold purification of the aryl-β-glucosidase is the loss of two minor components having aryl-β-glucosidase activity but little or no cellobiase activity during the third step. The protein purity was assessed both by sedimentation equilibrium studies in the ultracentrifuge and by disc gel electrophoresis at pH 4.5 and 9.5 (Figure 3). The protein was pure as judged by both criteria. The amino acid composition of the cellobiase was, like that of most cellulases, rich in acidic and hydroxylated amino acids. The amino acid composition is in reasonable agreement with that found (20) previously for the exo-glucanase, but a carbohydrate content of 10.3% and six glucosamine residues per enzyme molecule is reported, in contrast to the earlier report in which the carbohydrate content was less than 2%. The pH optima for both the aryl-β-glucosidase and cellobiase activity were 4.8–5.0.

Since the enzyme cleaved both cellobiose and p-nitrophenyl-β-glucoside, its preference for substrate was explored using cellotriose and p-nitrophenyl-β-cellobioside. In the latter case the susceptibility of the glycoside and holoside linkages was tested. Kinetic parameters are summarized in Table V. Michaelis constants for cellobiose and cellotriose are close to those reported earlier (20), and the maximum velocities indicate a preference for the larger substrates. Cellobioside was cleaved preferentially at the holoside linkage as described for a β-glucosidase from *Irpex lacteus* (6) and for phenyl-α-maltoside cleavage by the glucoamylase from *Rhizopus delemar* (51). D-Gluconic acid δ-lactone was strongly inhibitory ($K_i = 3.2 \times 10^{-5}M$) toward glucosidase and cellobiase activity, which is often taken as an indication of glucosidase action. King and co-workers (20), however, reported earlier that the cleavage of the glycosidic bond of cellobiose leads to inversion at the anomeric carbon atom which is characteristic of *exo*-glucanase action. The trivial name cellobiase will be used temporarily to describe its probable role in cellulolysis. A similar enzyme having both β-glucosidase and cellobiase activities has been reported from *Trichoderma koningii* by Wood (18). The relatively high K_m for cellobiose and the fact that substrate inhibition

and *p*-Nitrophenyl-β-D-Glucosidase

Total Activity (units)		Yield (%)	
Cellobiase	Glucosidase	Cellobiase	Glucosidase
7,120	17,440	100	100
6,052	15,330	85	88
5,055	12,650	71	72
1,994	4,130	28	24
1,850	4,060	26	23

Figure 3. Analytical disc gel electrophoresis of purified cellobiase in high and low pH gel systems. A total of 35 μgrams of purified cellobiase was used as the enzyme sample in each gel. The gels were stained for protein with Coomassie blue. The right-hand gel was obtained at a running pH of 4.5 toward the cathode and the left-hand gel at a running pH of 9.5 toward the anode (49).

was found for both cellobiose and cellotriose, but not for p-nitrophenyl-β-D-glucoside, may imply an important control mechanism for the cellulase system. By this mechanism the cellobiose level is kept high enough to inhibit cellulolysis should excessive amounts of cellobiose be produced. The product of p-nitrophenyl-β-D-glucoside cleavage, glucose had a K_i equal to 1.01mM. For technical reasons this could not be determined for the cellobiase reaction. Estimates of the enzyme activity on the cello-oligosaccharides, CM-cellulose, and Walseth cellulose indicate that the cellobiase is quite effective at converting the low molecular weight substrates to glucose.

Table V. Kinetic Constants of the Cellobiase

Substrate	K_m (mM)	V_{max} (μmoles substrate/ mg protein/ minute)	K_{is} (mM)
p-Nitrophenyl-β-D-glucoside	0.17 ± 0.008	118 ± 20	—
Cellobiose	1.80 ± 0.086	97 ± 20	31.5 ± 7.0
Cellotriose	0.22 ± 0.010	185 ± 30	26.6 ± 3.5
p-Nitrophenyl-β-D-cellobioside	0.20 ± 0.040	169 ± 30	—

Hydrocellulases. One of the most debated activities associated with cellulase action is that of the C_1 factor or enzyme which is supposed to carry out some initial function which renders crystalline cellulose susceptible to hydrolysis. After removal of unadsorbed β-glucosidase and *exo-* and *endo-*glucanase activity in the "buffer fraction," tightly bound enzymes of the "water fraction" are eluted from the affinity column with distilled water (Figure 2). These enzymes, which are active in the hydrolysis of crystalline cellulose or hydrocellulose, may be further purified by anion exchange chromatography on DEAE-Sephadex A-50 (52) or, with higher resolving power, by preparative disc gel electrophoresis (53). The chromatographic procedure also provides enzyme fractions which are needed for recombination with the relatively inactive hydrocellulase to restore activity toward crystalline substrates.

Such recombination studies in our laboratory and others (17, 48, 54) have shown clearly the multi-enzymic character of the complete cellulase system. When recombined with fraction I enzymes, fraction II or fraction III enzymes from chromatography of the water fraction proteins on DEAE-Sephadex caused the degradation of hydrocellulose.

Analytical disc gel electrophoresis showed that fraction III contained a mixture of proteins whereas fraction II contained one highly purified protein. We therefore decided to use preparative disc gel electrophoresis to resolve the active hydrocellulase proteins of the water fraction. This process resolved three purified proteins (Figure 4) with little activity toward CM-cellulose or crystalline cellulose. Each enzyme, when added to the fraction I enzymes, yielded a system which actively degraded hydrocellulose, suggesting a role like that of the hypothetical C_1 (53). These proteins were termed hydrocellulases A, B, and C in the order of their emergence from the electrophoretic system. Hydrocellulase C is identical to the protein in fraction II from chromatography of the water fraction on DEAE-Sephadex A-50. The carbohydrate content of the hydrocellulases A, B, and C was 10.4, 13.8, and 15.2%, respectively.

Although several properties of each of the three proteins have been investigated, most of our work has been done with the C form because: (a) it has the highest affinity for crystalline cellulose; (b) it is the predominant form in most, but not all, *T. viride* cellulase preparations; (c) it contains more carbohydrate than the A and B forms, suggesting that they may be degradation products from or intermediates in the biosynthesis of hydrocellulase C; and (d) it forms a slightly more active hydrocellulase system when recombined with the enzymes of fraction I.

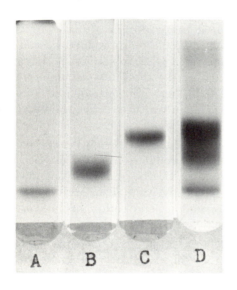

Figure 4. Analytical disc gel electrophoresis of the hydrocellulases A, B, and C. Hydrocellulases A, B, and C were run in gels A, B, and C respectively. A total of 30 μgrams of protein was used as the enzyme sample in each gel. Gel D is the pattern obtained when 100 μgrams of "water fraction protein" was used as the sample. Gels were stained with "Amido-Schwarz" (53).

Occurrence and Role of C_1 Factors (= Cellobiohydrolases) in Cellulose Degradation

A major question confronting cellulase researchers has been, "How many distinct enzyme components are necessary to permit reconstitution of a completely active system which will degrade native or crystalline cellulose?"

In 1964 King reported (48) that four enzymes from *Trichoderma viride*—C_1, *endo*-glucanase, *exo*-glucanase, and Z (a heat-stable glucanase)—were required for maximum glucose production from hydrocellulose. In 1965 the same group reported (20) that the first three components could account for the principal characteristics of the crude enzyme. However most of the observed activity could be accounted for by the C_1 enzyme alone; the *exo*-glucanase apparently served only to convert oligosaccharides to glucose.

Two years later Selby and Maitland (17) resolved the *T. viride* system into three components which they called C_1, CM-cellulase, and cellobiase. The principal advance was the purification of the C_1 to the point that it had little effect on cotton. This permitted a large increase in total solubilization as the result of recombination of individually ineffective components. In a similar fashion in 1969 Wood (54) reported resolution of the *Fusarium solani* complex into three components which he called C_1, C_x, and β-glucosidase.

Karl-Erik Eriksson (42) reported that he had prepared an enzyme from *Chrysosporium lignorum* which, when combined with each of three CM-cellulase enzymes, increased the hydrolysis rate of cotton tenfold. When he incubated this enzyme, which he believed to be of the C_1 type, with cellohexaose and observed cellobiose as the only product, he concluded that it was an *exo*-glucanase which split off cellobiose units. Later (27) Eriksson and Rzedowski reported that the three CM-cellulase enzymes from *C. lignorum* contained 13, 10, and 7% carbohydrate and catalyzed the hydrolysis of cotton and cellodextrins to produce cellobiose and glucose in the approximate ratio of 3:1.

In 1968 Nisizawa's laboratory (55) purified three cellulase components from Meicelase, which is a commercial cellulase from *Trichoderma viride*. These three components—Cellulases II, III, and IV—contained 16.8, 15.6, and 10.4% carbohydrate, respectively, and were active in hydrolyzing cellooligosaccharides, CM-cellulose, and cotton. They were inactive toward cellobiose and p-nitrophenyl-β-D-glucoside.

In 1972 Ogawa and Toyama (56) purified three components—A-I-a, A-I-b, and A-II-1—which were adsorbed on a gauze column during purification from Cellulase Onozuka P1500, a commercial preparation of *T. viride* cellulase. These three components had molecular weights of 32,000, 48,000, and 48,000 as determined by gel filtration and contained 7–16% carbohydrate. Each is reported to carry out the random hydrolysis of CM-cellulose and to degrade hydrocellulose (Avicel) and cellooligosaccharides except for cellobiose. The order of reactivity toward either cotton or Avicel was A-II-1 > A-I-b > A-I-a. The proteins adsorbed on cellulose comprised 38% of the total cellulase protein.

In 1971 and 1972 Halliwell and co-workers (*21, 57*) separated the *T. viride* cellulase into four fractions—C_1, C_2, CM-cellulase, and cellobiase. It was suggested that the C_1 enzyme was a cellobiohydrolase since the principal product of its action was cellobiose. Addition of cellobiase with the C_1 enzyme permitted 70% solubilization of cotton after 21 days. No evidence of the enzyme purity was presented.

Wood and McCrae (*22*), in 1972, reported the purification and some properties of the C_1 component of *T. koningii*. They obtained two protein peaks by chromatography and isoelectric focusing. The larger peak appeared asymmetric, suggesting the presence of two or more species. These protein components each had little activity toward CM-cellulose or crystalline forms of cellulose. However the major component catalyzed the hydrolysis of swollen cellulose, cellotetraose, and cellohexaose almost exclusively to cellobiose. As a result this enzyme was identified as a β-($1 \rightarrow 4$)glucan cellobiohydrolase. Wood (*58*) has also reported the resolution of the *Fusarium solani* C_1 component into three peaks by isoelectric focusing. The proteins in peaks derived by isoelectric focusing of C_1 components from either *F. solani* or *T. koningii* are associated with carbohydrate. Native cellulose is hydrolyzed rapidly when any of these proteins is present simultaneously with other enzymes of the cellulose complex.

Work in several laboratories (*22, 27, 55, 56*) has shown a pattern of cellulase action in cellulolytic organisms which requires at least one of a set of three closely related enzymes in order to hydrolyze crystalline cellulose effectively. These enzymes often possess little ability to degrade either CM-cellulose (as measured viscosimetrically) or crystalline cellulose. Nevertheless they are characterized by the capacity to cleave swollen cellulose or cellooligosaccharides almost entirely to cellobiose by virtue of their β-($1 \rightarrow 4$)glucan cellobiohydrolase activity. Recognition of this pattern has been difficult because prior to this report the three enzymes had not been purified and characterized apart from contaminating enzyme activity.

Enzymes from Trichoderma as Cellobiohydrolase Forms

For comparison some properties of hydrocellulases A, B, and C are presented in Table VI with those of the three cellulases, II, III, and IV, prepared in Nisizawa's laboratory (*55*). The carbohydrate contents appear similar with increasing levels in the order hydrocellulase A < B ⩽ C and Cellulase IV < III ⩽ II. The hydrocellulase carbohydrate content was determined as glucose and thus should be reduced slightly for comparison with the cellulase values determined as mannose. In each

Table VI.   Properties of Cellulases

Hydrocellulase [a]

	A	B	C
Molecular weight (sedimentation equilibrium)	46,100 ±1,580	47,500 ±1,580	49,700 ±1,730
Molecular weight (disc gel electrophoresis)	34,500	51,000	55,000
Carbohydrate (weight percent)	10.4	13.8	15.2
CM-cellulase activity (μmoles glucose equivalents/min/mg protein)	0.014	0.018	0.034
Activity toward cellobiose	nil	nil	nil
Product distribution from cellulodextrin (DP = 25)			
Product distribution from swollen Avicel (DP = 100–200)	—	—	G2>>G1
p-Nitrophenol release from p-Nitrophenyl-β-D-cellobioside (μmoles/min/mg protein)	—	—	8×10^{-4}

[a] From (53) E. K. Gum, Jr., unpublished results.

case mannose was the predominant constituent as was true for the *Chrysosporium* enzymes (43).

A low, but measurable, level of reducing sugar was released from CM-cellulose by the three hydrocellulases and by Cellulase IV. If the mode of action is that of β-(1 → 4)glucan cellobiohydrolase, as seems likely, then the higher levels of activity toward CM-cellulase exhibited by Cellulases II and III probably are the result of a contamination with *endo*-glucanase. This is also evidenced by the greater specific activity in the viscosimetric assay for Cellulases II and III, 27 and 42 times respectively, relative to that of Cellulase IV. No activity could be found in the viscosimetric CM-cellulose assay for the hydrocellulases (53) which agreed with Wood's results (22) for the *T. koningii* enzyme.

The molecular weights determined by ultracentrifugation were quite similar among either the hydrocellulases or Cellulases, but the values for the two sets do not agree. This probably reflects the different assumptions for partial specific volume and the fact that the Cellulase II, III, and IV molecular weights were estimated from sedimentation coefficients whereas the hydrocellulase molecular weights were obtained by sedimentation equilibrium studies using the meniscus depletion technique of Chervenka (59). Since results of this method suggested that some aggregation was occurring for hydrocellulases B and C, molecular weights

Isolated from *Trichoderma viride*

	Cellulase[b]	
IV	*III*	*II*
58,900	62,400	60,200
	(sedimentation velocity)	
10.4	15.6	16.8
0.007	0.185	0.326
nil	nil	nil
G2 > G1	G2 = G1	G2 = G1
12×10^{-4}	50×10^{-4}	12×10^{-4}

[b] From (6).

were also estimated on the basis of size (Stokes radius) using the disc gel electrophoretic method of Hedrick and Smith (60). The molecular weights calculated for hydrocellulases B and C were quite different from that obtained for hydrocellulase A. This suggests that the carbohydrate moieties may play an important role in determining protein conformation. Using gel filtration Toyama (56) has observed a similar distribution of apparent molecular weights.

None of the six enzymes cleaved cellobiose, but in comparing the products released from swollen Avicel by hydrocellulase C with those from cellulodextrin by the Cellulases, further evidence of contamination of cellulases II and III was found. For the latter cases both glucose and cellobiose are produced whereas both hydrocellulase C and Cellulase IV yield predominantly cellobiose. Cellulase IV produced much more cellobiose than glucose from cellotetraose whereas the other Cellulases yielded only slightly more cellobiose than glucose.

The amino acid contents of the proteins are generally similar. Each enzyme has high levels of aspartic acid and glycine but less of the basic amino acids. The compositions of hydrocellulase C and Cellulase IV closely resemble each other, with 15 of the 16 amino acids reported having the same relative abundance in the respective proteins.

Hydrocellulases A, B, and C correspond to Cellulases IV, III, and II, respectively, on the basis of carbohydrate content. Hydrocellulase C and Cellulase II have the highest affinity for the cellulosic affinity columns used in their purification (56, 61). The members of the two sets of enzymes, hydrocellulases and Cellulases, are probably β-$(1 \rightarrow 4)$glucan cellobiohydrolases, but the probable contamination of Cellulases III and II with *endo*-glucanase has obscured their cellobiohydrolase activity and led to the conclusion that "All three cellulase components showed both C_1 and C_x activities in a different ratio" (55).

The C_1 enzymes, hydrocellulases, and Avicelases may now be correctly and more specifically termed cellobiohydrolases because of their action on cellooligosaccharides and Walseth cellulose. The C_x enzymes may now be classified as *endo*- or *exo*-glucanases, or β-glucosidases. We therefore suggest abandoning the C_1–C_x terminology in favor of a more precise one based on the hydrolytic activities of the component enzymes.

Mode of Action of the Cellulase Complex

Interaction of the cellobiohydrolase with other enzymes of the cellulase complex may include binding to *endo*-glucanases since these activities are difficult to separate. The enzymes are adsorbed together tightly on the cellulosic substrate since this is the basis of the affinity methods of purification (55). We have determined (61) that the rate of cellobiose release from hydrocellulose is too low to account for the degradation rate catalyzed by the intact complex. However Cellulase IV catalyzes the hydrolysis of cellooligosaccharides 300 times faster than the hydrolysis of Avicel (55). Thus the cellobiohydrolase acting at the site of *endo*-glucanase activity (presumably the cellulose surface) could function at rates commensurate with that observed for crystalline cellulose degradation. There is no significant effect on the initial rate of hydrocellulose hydrolysis when the cellobiase is added with the cellobiohydrolase (49). The enhancement effect of such combinations reported by Halliwell (57) in 7- to 21-day assays may be caused by removal of the inhibitory effect of accumulated cellobiose.

Although the significance of the existence of three forms of cellobiohydrolase is not clear, there are several possible advantages of this mode of hydrolysis.

(a) Cellobiose may be transported or metabolized more slowly than glucose, thus minimizing catabolite repression.

(b) Cellobiose provides a specific feedback inhibitor characteristic of cellulose hydrolysis.

(c) The flow of a predominant product such as cellobiose may be modulated because of the kinetic properties of the cellobiase providing information to the cellulolytic system regarding demand for, and to the cell regarding supply of, glucosyl units.

(d) It would be a competitive advantage to a cellulolytic organism to produce a less universally utilized carbon source, such as cellobiose, rather than glucose.

Applications of Cellulases

Cellulases have been limited to a few specific applications, but economic and ecological factors have increased interest in their potential value. They have been used mainly as a component in digestive aids with other hydrolytic enzymes (Table VII). Toyama reported (62) in 1968 that commercial cellulase preparations were exported from Japan at a rate of 500 kg per month.

Many experiments have demonstrated the capacity of cellulases and of mixtures of cellulases, pectinases, and hemicellulases to break down or to soften plant cell walls. For human diets this would be beneficial in preparing infant or geriatric foods where reduced fiber content is desired. Recently treatment of wheat bran was found to increase the *in vitro* protein digestibility by 35% (63) and to increase weight gain of rats fed a bran-containing ration. The aleurone cell wall was the primary substrate for these enzymes.

Table VII.   Cellulase Applications

Digestive aids
Food processing:
    Nutrient availability
    Texture modification
    Pulp removal
Microbial growth for single cell protein
Saccharification of cellulose
Plant cell wall removal:
    Tissue culture
    Fusion of protoplasts

An indirect use of cellulase enzymes is the growth of cellulolytic microorganisms on waste cellulose to obtain "single cell protein" (64). The use of a combined culture of the bacteria *Cellulomonas*, which breaks down cellulose to cellobiose, and *Alcaligenes*, which can utilize this inhibitory product, resulted in rapid growth. The increasing need for protein and the availability of waste cellulose have focused much attention on this area.

The use of cellulases to saccharify waste cellulose has been studied for the past few years by M. Mandels and her colleagues at the U.S. Army Natick Laboratories (65). Rates of glucose production achieved are 2–4 grams/liter/hour, which is half the rate reported in commercial starch hydrolysis (4–8 grams/liter/hour). However starch is not a waste material as are the forms of cellulose which are potential sources of glucose. Table VIII indicates the effectiveness of *Trichoderma viride* cellulase in saccharifying various sources of waste cellulose (66). The importance of mechanical pretreatment of the substrate is obvious. Economic considerations are central to the commercial exploitation of the process.

Cellulases and other carbohydrases are being used to isolate higher plant cells and to produce protoplasts (67, 68). A new era in plant breeding is envisioned as cell lines are manipulated genetically by mutagens, protoplast fusion, or possibly transformed by DNA transfer.

Table VIII.   Hydrolysis of Cellulose Materials by *Trichoderma viride* Cellulase[a]

Cellulose, 10%	Saccharification at 48 hours (%)
Absorbent cotton	9.9
Cotton, ball milled	55.0
Rice hulls	2.0
Rice hulls, ball milled	24.0
Bagasse	6.0
Bagasse, ball milled	48.0
Newspaper, ball milled	41.0
Corrugated fiberboard, ball milled	78.0
Computer print-out, ball milled	77.0
Milk carton, ball milled	53.0

[a] Ten grams dry sample plus 90 ml (2 mg protein) of *T. Viride* QM9123 cellulase, incubated on shaker at 50°C, pH 4.8 (66).

Conclusion

As understanding of the components of the cellulase complex has increased, we can better define the mode of action by which native cellulose is depolymerized. The C_1–C_x model has directed attention to the specific steps in the crystalline cellulose breakdown. The use of nomenclature describing the catalytic function of each specific enzyme should encourage higher standards of purity. The ubiquity of the substrate invites imaginative applications of cellulases.

Acknowledgments

The authors gratefully acknowledge the encouragement and support of Kendall W. King, Elwyn T. Reese, and Mary Mandels. For skillful assistance in the characterization experiments we are indebted to L. B.

Barnett for molecular weight determination, Blanche Hall for amino acid analyses, and Betty W. Li for disulfide bond analyses.

Literature Cited

1. Seillière, G., *Compt. Rend. Soc. Biol.* (1906) **61**, 205.
2. Reese, E. T., Siu, R. G. H., Levinson, H. S., *J. Bacteriol.* (1950) **59**, 485.
3. Reese, E. T., Levinson, H. S., *Physiol. Plantarum* (1952) **5**, 345.
4. ADVAN. CHEM. SER. (1969) **95**.
5. Wood, T. M. in "World Review of Nutrition and Dietetics," Vol. 12, G. H. Bourne, Ed., pp. 227-265, Karger, Basel, 1970.
6. Nisizawa, K., *J. Ferment. Technol.* (1973) **51**, 267.
7. Whitaker, D. R. in "The Enzymes," Vol. V, P. D. Boyer, Ed., pp. 273-290, Academic, New York, 1971.
8. Reese, E. T., Mandels, M. in "High Polymers," Vol. V, Part V, N. M. Bikales, L. Segal, Eds., pp. 1079-1094, Wiley, New York, 1971.
9. Gascoigne, J. A., Gascoigne, M. M., "Biological Degradation of Cellulose," Butterworths, London, 1960.
10. Almin, K. E., Eriksson, K.-E., *Biochim. Biophys. Acta* (1967) **139**, 238.
11. Hulme, M. A., *Arch. Biochem. Biophys.* (1971) **147**, 49.
12. Poincelot, R. P., Day, P. R., *Appl. Microbiol.* (1972) **23**, 875.
13. Whitaker, D. R. in "Enzymatic Hydrolysis of Cellulose and Related Materials," E. T. Reese, Ed., p. 63, Pergamon, London, 1963.
14. Reese, E. T., Maguire, A. H., Parrish, F. W., *Can. J. Biochem.* (1968) **46**, 25.
15. Youatt, G., Jermyn, M. A. in "Marine Boring and Fouling Organisms," D. L. Ray, Ed., p. 397, University of Washington, Seattle, 1959.
16. Bucht, B., Eriksson, K.-E., *Arch. Biochem. Biophys.* (1969) **129**, 416.
17. Selby, K., Maitland, C. C., *Biochem. J.* (1967) **104**, 716.
18. Wood, T. M., *Biochem. J.* (1968) **109**, 217.
19. *Ibid.* (1971) **121**, 353.
20. Li, L. H., Flora, R. M., King, K. W., *Arch. Biochem. Biophys.* (1965) **111**, 439.
21. Halliwell, G., Griffin, M., Vincent, R., *Biochem. J.* (1972) **127**, 43P.
22. Wood, T. M., McCrae, S. I., *Biochem. J.* (1972) **128**, 1183.
23. Wood, T. M., *Biochim. Biophys. Acta* (1969) **192**, 531.
24. Eriksson, K.-E., Pettersson, B., *Arch. Biochem. Biophys.* (1968) **124**, 142.
25. Pettersson, G., Eaker, D. L., *Arch. Biochem. Biophys.* (1968) **124**, 154.
26. Tomita, Y., Suzuki, H., Nisizawa, K., *J. Ferment. Technol.* (1968) **46**, 701.
27. Eriksson, K.-E., Rzedowski, W., *Arch. Biochem. Biophys.* (1969) **129**, 683.
28. King, K. W., *Va. Agric. Exper. Sta. Tech. Bull.* **127** (1956).
29. Smith, W. R., Yu, I., Hungate, R. E., *J. Bacteriol.* (1973) **114**, 729.
30. Mandels, M., Reese, E. T., *J. Bacteriol.* (1957) **73**, 269.
31. *Ibid.* (1960) **79**, 816.
32. Reese, E. T., Lola, J. E., Parrish, F. W., *J. Bacteriol.* (1969) **100**, 1151.
33. Mandels, M., Weber, J., Parizek, R., *Applied Microbiol.* (1971) **21**, 152.
34. Hulme, M. A., Stranks, D. W., *Nature* (1970) **226**, 469.
35. Mandels, M., Parrish, F. W., Reese, E. T., *J. Bacteriol.* (1962) **83**, 400.
36. Nisizawa, T., Suzuki, H., Nakayama, M., Nisizawa, K., *J. Biochem. (Tokyo)* (1971) **70**, 375.
37. Nisizawa, T., Suzuki, H., Nisizawa, K., *J. Biochem. (Tokyo)* (1972) **71**, 999.
38. Fan, D. F., Maclachlan, G. A., *Plant Physiol.* (1967) **42**, 1114.
39. Cracker, L. E., Abeles, F. B., *Plant Physiol.* (1969) **44**, 1144.
40. Mandels, M., Reese, E. T. in "Advances in Enzymic Hydrolysis of Cellulose and Related Materials," E. T. Reese, Ed., p. 115, Pergamon, London, 1963.

41. Siu, R. G. H., "Microbial Decompcsition of Cellulose," Reinhold, New York, 1951.
42. Eriksson, K.-E., ADVAN. CHEM. SER. (1969) **95**, 58-59.
43. Eriksson, K.-E., *Arch. Biochem. Biophys.* (1969) **129**, 689.
44. Pettersson, G., *Arch. Biochem. Biophys.* (1969) **130**, 286.
45. Björndal, H., Eriksson, K.-E., *Arch. Biochem. Biophys.* (1968) **124**, 149.
46. Datta, P. K., Hanson, K. R., Whitaker, D. R., *Can. J. Biochem. Physiol.* (1963) **41**, 671.
48. King, K. W., *J. Ferment. Technol.* (1965) **43**, 79.
49. Emert, G. H., Ph.D. Dissertation, Virginia Polytechnic Institute and State University, 1973.
50. Emert, G. H., Brown, R. D., Jr., *Fed. Proc.* (1973) **32**, 528.
51. Suetsugu, N., Hirooka, E., Yasui, H., Hiromi, K., Ono, S., *J. Biochem. (Tokyo)* (1973) **73**, 1223.
52. Liu, T. H., Ph.D. Dissertation, Virginia Polytechnic Institute and State University, 1969.
53. Lang, J. A., Ph.D. Dissertation, Virginia Polytechnic Institute and State University, 1970.
54. Wood, T. M., *Biochem. J.* (1969) **115**, 457.
55. Okada, G., Nisizawa, K., Suzuki, H., *J. Biochem. (Tokyo)* (1968) **63**, 591.
56. Ogawa, K., Toyama, N., *J. Ferment. Technol.* (1972) **50**, 236.
57. Halliwell, G., Riaz, M., *Arch. Mikrobiol.* (1971) **78**, 295.
58. Wood, T. M. in "Fermentation Technology Today," G. Terui, Ed., p. 711, The Society of Fermentation Technology, Osaka, 1972.
59. Chervenka, C. H., *Anal. Biochem.* (1970) **34**, 24.
60. Hedrick, J. L., Smith, A. J., *Arch. Biochem. Biophys.* (1968) **126**, 155.
61. Gum, E. K. Jr., unpublished results.
62. Toyama, N., ADVAN. CHEM. SER. (1969) **95**, 359.
63. Saunders, R. M., Connor, M. A., Edwards, R. H., Kohler, G. O., *Cereal Chem.* (1972) **49**, 436.
64. Srinivasan, V. R., Han, Y. W., ADVAN. CHEM. SER. (1969) **95**, 447.
65. Mandels, M., Kostick, J., Parizek, R., *J. Polymer Sci.* (1971) Part C **36**, 445.
66. Mandels, M., personal communication.
67. Cocking, E. C., *Ann. Rev. Plant Physiol.* (1972) **23**, 29.
68. Carlson, P. S., *Proc. Nat. Acad. Sci. U.S.* (1973) **70**, 598.

RECEIVED September 17, 1963. Work supported in part by U.S. Department of Agriculture Cooperative State Research Service Special Grant 916-15-27.

Pectic Enzymes

JAMES D. MACMILLAN and MARK I. SHEIMAN

Department of Biochemistry and Microbiology, Rutgers University,
New Brunswick, N. J. 08903

Pectic substances, largely polygalacturonic acid or its methyl ester, are found in the primary cell walls of plants and in the middle lamella. Enzymes from plants, fungi, and bacteria degrade pectic substances and cause plant tissue maceration by esterase action on the methyl ester groups or by glycosidase or lyase action on the polygalacturonate chain. Numerous enzymes are distinguished from one another by mechanism (hydrolysis or trans elimination), specificity (pectin, polygalacturonate, oligogalacturonates, Δ4:5 unsaturated oligogalacturonates), action pattern (endo or exo), and stimulation by cations. Pectic enzymes are used commercially in producing fruit juices and wines. Fundamental properties of the enzymes are discussed with reference to their applications in food processes.

This review is primarily concerned with fundamental properties of the pectic enzymes with some reference to their role in nature and to their applications in food processing. Because of the vast amount of literature in this field the reader is directed to other articles and books on the chemistry of pectic substances and the enzymes which degrade them (1–21).

Pectic substances are found in the tissues of higher plants where they have an important structural role. They are deposited in the primary cell wall during the early stages of growth and in the middle lamella where they act as "intercellular cement." The enzymatic breakdown of pectic substances in the middle lamella leads to plant tissue maceration. Therefore the enzymes that degrade pectin are involved in natural processes such as the ripening of fruit (22, 23), abscission of leaves and plant organs (24), invasion of tissues by plant pathogens (6), and in the spoilage of fruits and vegetables (25, 26, 27). They are essential in the rapid decay of dead plant material and undoubtedly assist in recycling carbon compounds in soil, water, atmosphere, microorganisms, plants, and animals.

Pectic Substances

The pectic substances are polysaccharides composed principally of α-1,4 linked galacturonic acid or its methyl ester. The term pectin refers to material with a significant amount of esterification in contrast to unesterified polymers called pectic or polygalacturonic acids (Figure 1). The esterification degree of a particular pectin depends upon its source and extraction method. Some pectic substances are partially acetylated at the C_2 and C_3 positions of galacturonide residues.

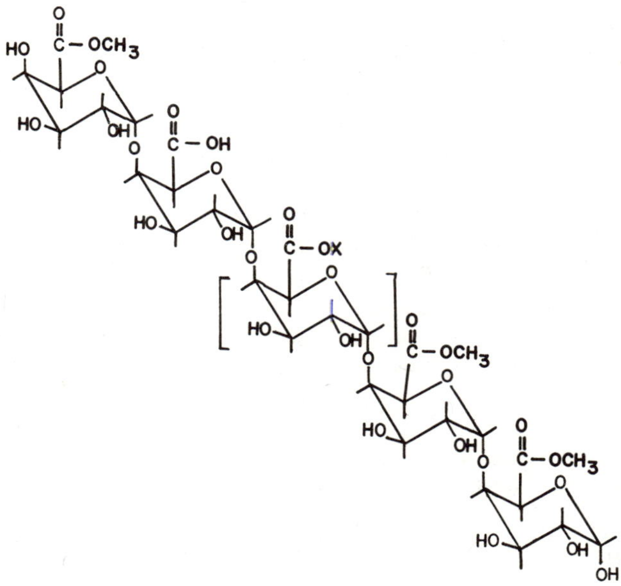

Figure 1. Structure of pectin. X is H or CH_3. The amount of esterification varies with the source and method of extraction.

Pectins contain varying amounts of neutral carbohydrates such as L-rhamnose, L-arabinose, and D-galactose, along with traces of other neutral sugars (28). Analysis of chemical and enzymatic degradation products of pectin indicate that it has a chain of α-D-(1,4)-galacturonic acid as a backbone structure. In some cases, as with lemon pectin, the C_1 of galacturonosylpyranose is linked to C_2 of L-rhamnosylpyranose, and blocks rich in L-rhamnose may exist. Other neutral sugars such as arabinose, galactose, and xylose are probably only found in side chains attached to C_2 or C_3 of galacturonides. The molecular weight of pectin varies greatly with source and extraction method; values from 30,000 to 120,000 have been reported (*29*). Polygalacturonic acid exists in solution in a symmetry which approximates a threefold screw axis (*30, 31*). Polygalacturonic acid esters, on the other hand, have twofold screw symmetry because of the ester groups on adjacent galacturonide units which project out on opposite sides of the main chain. The tendency for the hydrophilic ester groups on each side of the chain to come together forces the twofold screw symmetry (*32*). Upon heating there is a rapid, irreversible decrease in viscosity without any noticeable change in the methyl ester content or in reducing power. This implies that the secondary structure is responsible for the high viscosity of pectin solutions (*33*).

Pectic materials are most stable at pH 3–4, and outside this range chemical degradation occurs (*17*). Under alkaline conditions pectins (but not polygalacturonic acid) rapidly undergo a trans elimination reaction which results in the cleavage of the glycosidic bond with elimination of a proton at C_5 on the residue contributing the hydroxyl to the glycosidic bond. This causes a $\Delta4:5$ double bond to form in that residue (*34*). Deesterification will also occur under alkaline conditions (*35*).

Pectic Enzymes

Although pectins can contain other materials, the term pectic enzymes usually refers to those enzymes which catalyze the degradation of the uronide portions of the molecules. In the early literature only two major types of enzymes were thought to be involved in pectin degradation. The ester linkage was first hydrolyzed by pectinesterase (EC 3.1.1.11), and the glycosidic linkage was then hydrolyzed by polygalacturonase (EC 3.2.1.15) (Figure 2). Many early studies were conducted with impure commercial preparations of fungal polygalacturonases or substrates. As the emphasis shifted to studies of well-defined species of microorganisms and techniques were developed for enzyme purification, it became apparent that more than one type of depolymerizing enzyme existed. For example, Roboz, Barratt, and Tatum (*36*) reported an enzyme which attacked the glycosidic bonds of pectin in preference to those of polygalacturonic acid. Other workers showed differences in the action pattern, and exo- and endo-splitting enzymes were described.

Much of the confusion in pectic enzyme terminology was clarified by Demain and Phaff in 1957 (*12*) and Deuel and Stutz in 1958 (*11*). Their classification of depolymerases was based on the substrate attacked (pectin *vs.* non-esterified polygalacturonic acid), the mode of attack (random or endo attack *vs.* terminal or exo attack), and the optimal pH for enzyme activity. However it was assumed that hydrolysis was the

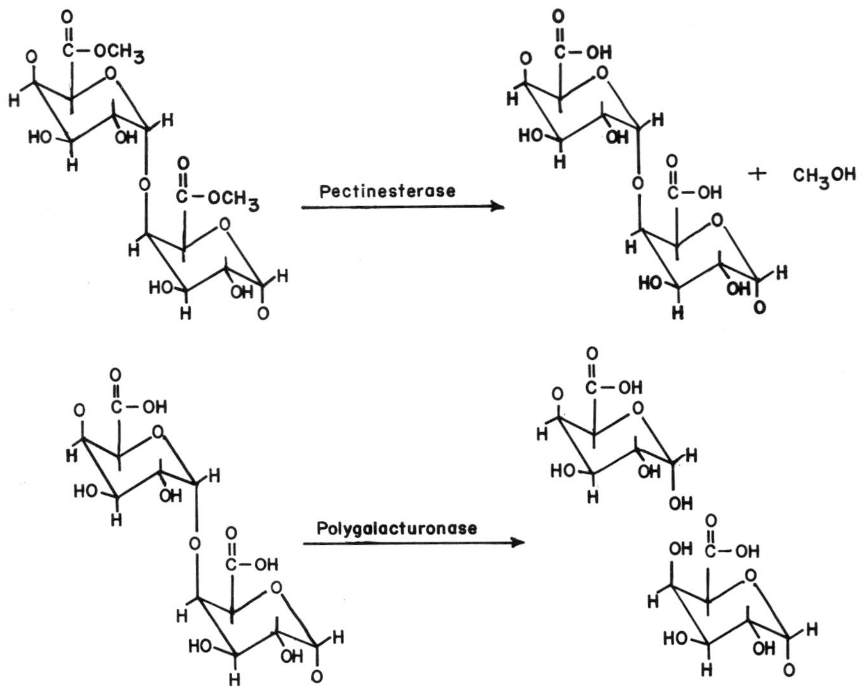

Figure 2. Reactions catalyzed by pectinesterase and polygalacturonase

only mechanism for cleavage of the glycosidic bond. In 1960 Alber-sheim, Neukom, and Deuel (*37*) found a pectic enzyme in a commercial preparation of fungal origin which split the α-1,4 bonds of pectin by a trans elimination mechanism rather than by hydrolysis (Figure 3). This degradation mechanism causes a Δ4:5 double bond to form at the non-reducing end of a chain, as in alkaline degradation of pectin. Edstrom and Phaff (*38, 39*) found a similar enzyme produced by *Aspergillus fonsecaeus*. The Δ4:5 unsaturated uronide products characteristically absorb ultraviolet light with a maximum at 235 nm. The enzyme was originally called pectin *trans*-eliminase, but the more acceptable trivial name is pectin lyase. Although the first lyases were specific for pectin, enzymes capable of degrading polygalacturonate by this mechanism were soon found, and both endo- (*40, 41, 42*) and exo- (*43–49*) action patterns were encountered. A classification scheme incorporating the trans elimination attack mechanism was proposed by Bateman and Millar (*6*).

More recently another group of pectic enzymes has been described which degrade oligogalacturonides in preference to the longer chain molecules of polygalacturonic acid. In one study the attack rate was inversely proportional to chain length (*50*). The mechanism, as with the other depolymeriases, can be either hydrolysis or trans elimination. This type of enzyme has been produced by *Bacillus* species (50, 51, 52), *Erwinia carotovora* (53), *Erwinia aroideae* (54), and a *Pseudomonas* species (55). The oligogalacturonate lyase from *Erwinia aroideae* showed a preference for substrates having an unsaturated galacturonide residue at the non-reducing end compared with those with normal termini (54). General properties of the various pectic enzymes are shown in Table I.

Pectinesterase

Pectinesterase action was first reported in 1840 by Frémy (56). He noted that the addition of carrot juice to a pectin solution caused formation of a gel. This gel was produced by the enzymatic deesterification of pectin followed by precipitation of the resulting polygalacturonic acid as calcium pectate (20). Various names such as "pectin methoxylase" and "pectin methyl esterase" have been applied to the enzyme, but pectinesterase is the preferred trivial name (57).

Pectinesterases are formed by numerous bacteria, fungi, and higher plants. They are usually found in conjunction with a depolymerizing enzyme such as polygalacturonase or polygalacturonate lyase. Fungal pectinesterases generally have a lower pH optimum and are less influenced by salts than those of higher plants. Bateman and Millar (6) reviewed the literature on this group of enzymes and their significance in phytopathology. Although pectic enzymes are produced by numerous pathogens, this is not the only property determining pathogenicity. Bateman (58) and Bateman and Lumsden (59) proposed a role for host pectinesterase in limiting lesion size in *Rhizoctonia* infections of bean hypocotyls. They suggested that calcium and perhaps other cations

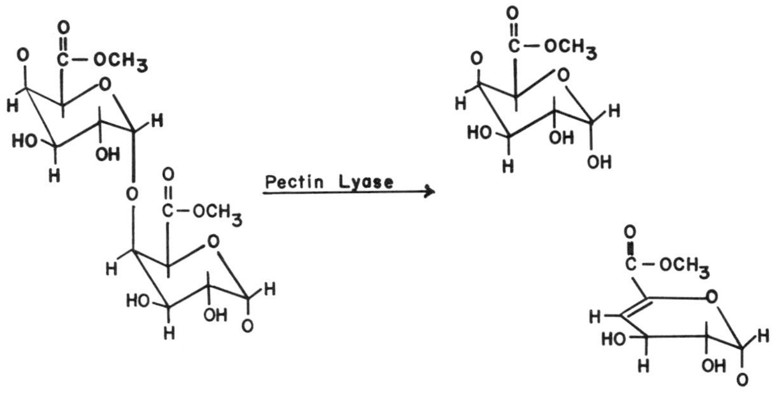

Figure 3.   Reaction catalyzed by pectin lyase

Table I.   Pectic

Name	EC Number[a]	Primary Substrate[b]
Pectinesterase	3.1.1.11	pectin
Endopolygalacturonase	3.2.1.15	polygalacturonate
Exopolygalacturonase		polygalacturonate
Oligogalacturonate hydrolase		trigalacturonate[e]
Unsaturated oligogalacturonate hydrolase		unsaturated digalacturonate[f]
Endopolymethylgalacturonase		pectin
Endopolygalacturonate lyase	4.2.2.2	polygalacturonate
Exopolygalacturonate lyase	4.2.2.9	polygalacturonate
Oligogalacturonate lyase		unsaturated digalacturonate[j]
Endopectin lyase	4.2.2.10	pectin

[a] EC numbers obtained by personal communication from Waldo E. Cohn.
[b] Primary substrate is that substrate usually identified with the enzyme. In the case of polymer splitting enzyme there is frequently slower attack on oligomers; the reverse is true with oligomer splitting enzymes. Some exo polymer splitting enzymes may have as rapid attack on oligomers as on polymers.
[c] After exhaustive degradation products are usually mono- and digalacturonates with traces of trimer.
[d] One enzyme produces dimer and not monomer.
[e] In one report (bacteria) higher oligogalacturonates were attacked at progressively slower rates, and the dimer was not attacked. In another report (fungus) the dimer was attacked.

accumulate around the point of initial infection and release and activate pectinesterase associated with host cell walls. The deesterified pectins produced by the esterase action form insoluble calcium polygalacturonates. These are resistant to hydrolysis by the polygalacturonases of the pathogens, thus limiting the infection.

Pectinesterases are found in many fruits and vegetables, with exceptionally large amounts occurring in citrus and tomato fruits (12, 20, 60, 61, 62, 63). The softening of fruit and vegetable tissue during ripening or storage is thought to be associated with deesterification by pectinesterase prior to any appreciable degradation by polygalacturonase (11, 12).

Multiple forms of pectinesterase have been reported. Reid (64) separated two active pectinesterase fractions from a fungal source and suggested that pectinesterases from various organisms differ in their initial rates and final percent deesterification of pectin. Glasziou and Inglis (65) found two pectinesterases present in tobacco pith; one of which was firmly bound to the cell walls. Hulton and Levine (66) observed three molecular forms of pectinesterase in the pulp of the

Enzymes

Mechanism	Products	Known Sources
hydrolysis	polygalacturonate and methanol	plants, fungi, bacteria
hydrolysis	oligogalacturonates[c]	plants, fungi, bacteria
hydrolysis	monogalacturonate[d]	plants, fungi, bacteria, insects
hydrolysis	monogalacturonate	bacteria, fungi
hydrolysis	unsaturated monogalacturonate and saturated $(n-1)$mer	bacteria
hydrolysis	methyl oligogalacturonates[g]	fungi
transelimination	unsaturated oligogalacturonates[h]	bacteria, fungi, protozoa
transelimination	unsaturated digalacturonate[i]	bacteria
transelimination	unsaturated monogalacturonate	bacteria
transelimination	unsaturated methyl oligogalacturonates[k]	fungi

[f] In one report higher unsaturated oligogalacturonates were attacked at progressively slower rates and saturated oligomers were not attacked.

[g] Monomer to pentamer.

[h] Final products reported after exhaustive degradation are usually the dimer and trimer.

[i] Traces of the saturated mono- and digalacturonates are produced from the non-reducing ends of chains.

[j] In one report saturated digalacturonate was attacked (one-fifth as fast as unsaturated digalacturonate) yielding an unsaturated monomer and saturated monomer. The trimer, tetramer, and polymer were degraded at progressively slower rates.

[k] In one report products were dimer to pentamer after exhaustive degradation.

banana which differed considerably in response to pH, temperature inactivation, and activation or inactivation by sodium dodecyl sulfate. The partially purified enzymes (67) differed in charge, response to cations, and overall reaction kinetics.

Pectinesterases are quite specific for galacturonide esters, which act on pectin at least 1000 times as fast as on non-galacturonide esters (62). A polygalacturonate chain structure is therefore essential to pectinesterase action. Orange pectinesterase, for example, will not attack the methyl ester of polymannuronic acid (62).

In the same study the relative activities of pectinesterases from alfalfa, tomato, orange, and a fungal source were compared using commercial pectin, methyl polygalacturonate (fully esterified), and ethyl polygalacturonate (50% esterified) as substrates. The latter two substrates were prepared by HCl-catalyzed esterification (68, 69). The enzymatic attack rate of the pectinesterases on the methyl ester was approximately 50% of that on pectin (except for the fungal enzyme where the rate was 80% of that on the pectin substrate). Kertesz (70) obtained a similar figure of 60% with the tomato enzyme. The ethyl esters were hydrolyzed at 6–16% of the rate seen with the methyl ester (2.8–13% the rate of pectin hydrolysis). Lineweaver and Ballou (71) found that alfalfa pectinesterase deesterified pectin at least several hundred times as fast as it does either α-methyl-D-galacturonate or α-methyl D-galacturonate methyl glycoside.

Solms and Deuel (13) found that the distribution of methoxyl groups along pectin chains markedly affected the reaction velocity of orange pectinesterase. The enzyme attacked completely esterified pectins [prepared by treating 70% esterified pectin with diazomethane (72)] slowly, but the enzymatic deesterification rate increased as more of the substrate was deesterified with alkali. The enzyme action rate increased in proportion to free carboxyl group content, reaching a maximum at 50% esterification. Pectin, partially deesterified by pectinesterase [presumably causing blockwise deesterification along the chain (73, 74)], resulted in a much slower enzymatic rate than with alkali deesterified substrates. It was therefore suggested that the fruit pectinesterase required a free carboxyl group next to an esterified group in order to act. Since pectin degradation by orange pectinesterase is incomplete and pectin partially reduced with $NaBH_4$ (converting methyl galacturonate residues to galactose) is only slowly deesterified, it is likely that the pectinesterase moves linearly down the chain until it reaches an obstruction (13).

Lee and Macmillan (75) purified tomato pectinesterase by ammonium sulfate precipitation followed by calcium phosphate gel adsorption and column chromatography on DEAE–cellulose and Sephadex G-75. The purified enzyme sedimented as a single homogenous band on ultracentrifugation with an $S_{20,w}$ value of 3.08 S. Disc gel electrophoresis at pH 9.5 or 4.3 indicated a single band that stained faintly with protein stains such as amido black or aniline black and more strongly with lipid stains such as sudan black B or oil red O. Lipid components were found by gas chromatography in extracts of alkaline hydrolysates of the purified enzyme, and it was tentatively concluded that tomato pectinesterase is a lipoprotein. With pectin NF (citrus pectin) as substrate, the tomato enzyme has a K_m of $4 \times 10^{-3}M$ methyl anhydrogalacturonate residues and was inhibited by polygalacturonate with a K_i of $7 \times 10^{-3}M$ anhydrogalacturonate residues. Nakagawa et al. (76) also purified tomato pectinesterase and reported a similar sedimentation coefficient (3.17 S) with a single disc gel electrophoresis band which only stained faintly with amido black.

Plant pectinesterases are activated by cations, and their pH optima shift to lower values in presence of cations (61, 77). Mayorga and Rolz (78) proposed equations predicting the activity of pectinesterases as a function of pH, cation concentration, and enzyme concentration. However their data (obtained only at pH 7.0, 7.5, and 8.0) are insufficient to support such complex equations. Their prediction that the pH optimum of tomato pectinesterase increases with increasing ionic strength is contradicted by the results of Nakagawa *et al.* (*76*), who found the pH optimum to shift downward with increasing ionic strength. At pH 7.0 tomato pectinesterase is most active in 0.05M NaCl or 0.005M CaCl$_2$ (*75*). In NaCl tomato pectinesterase is active optimally at about pH 8-8.5. In 0.05M NaCl (*75*) or 0.1M NaCl (*79*) the enzyme is active over a broad pH range. Salt affects activity particularly on the acid side. In the presence of salt half maximal activity occurred at pH 5; in the absence of salt at pH 6.5. Rate studies above pH 7.5 must be corrected for non-enzymic hydrolysis of the ester linkage. About 20 μmoles of COO$^-$/min are formed in 0.5% pectin at pH 10.0 (*75*).

Nakagawa *et al.* (*76*) reported a break in the Arrhenius plot (log tomato pectinesterase rate *vs.* 1/T) at 31°C. NaCl (0.1M) increased the activation energy and lowered the break point to 27°C.

Pressey and Avants (*80*) separated, by DEAE Sephadex A-50 chromatography, four forms of tomato pectinesterase, with molecular weights 35,500, 27,000, 23,700, and 24,300. The number and relative amounts of these components varied considerably with ripeness and variety of tomato, and no single extract contained all four. The four forms differed in heat stability and response to pH and cation concentration. The fourth form, the major species in most tomato extracts, had properties similar to those of the enzyme described by Lee and Macmillan (*75*), whose molecular weight, however, was estimated on Sephadex G-75 as 27,500 (*81*).

Lee and Macmillan (*82*) used the exo-polygalacturonate lyase from *Clostridium multifermentans* to determine how much tomato pectinesterase activity occurred at the reducing ends of chains of polymethyl polygalacturonic acid methyl glycoside [prepared by HCl-catalyzed esterification (*68, 69*)]. The exo lyase degrades polygalacturonic acid in a linear manner, removing units of Δ4:5 unsaturated digalacturonic acid. However it will not degrade glycosidic linkages in pectins which are highly esterified with methanol (*see* section on exopolygalacturonate lyase). Both lyase and esterase activities were monitored automatically in a single reaction mixture containing the highly esterified pectin. From plots of the respective activities of the two enzymes it was concluded that about half of the tomato pectinesterase activity was initiated at or near the reducing ends of highly esterified pectin molecules. The rest of the esterase activity occurred at some secondary locus or loci, perhaps next to free carboxyl groups as was suggested for orange pectinesterase (*13*). Although both esterase and lyase activties decreased as the reaction proceeded, the molar ratios of the lyase activity to esterase activity decreased gradually to about 0.5. This suggests that eventually all of the pectinesterase activity occurred at the reducing ends of molecules.

The above experiments were conducted at pH 7 although both lyase and esterase are optimally active at about pH 8.5. This was done to avoid alkaline deesterification, which according to Kohn and Furda (83, 84) is a random process. The higher esterase activities observed under alkaline conditions might be caused by the new carboxyl groups being generated by the alkaline conditions. These might act as initiation points for additional esterase attack if tomato pectinesterase acts as has been suggested for orange pectinesterase—by attacking ester linkages adjacent to free carboxyl groups. Pressey and Avants (80) have suggested that a function of one of the pectinesterase forms that they found in tomatoes may be to deesterify sites on pectin molecules which can act as the initiation point for one or more of the other pectinesterases. No experimental data were presented to support this possibility.

A partially purified pectinesterase from *Fusarium oxysporum* (MW 35,000) was assayed in the presence of *exo*-polygalacturonate lyase in experiments similar to the ones reported above with the tomato enzyme (81). The results were equivalent to those for the tomato enzyme; about half (57%) of the esterase activity was initiated at the reducing ends of highly esterified pectin chains. Added proof for this observation was obtained by showing that randomly (alkaline) deesterified pectin is not a good substrate for initial degradation by exopolygalacturonate lyase. As in the earlier experiments this lyase was used to determine the amount of esterase activity which occurred at the reducing ends of the chain.

Miller and Macmillan (85) found that the pectinesterase and exopolygalacturonate lyase from *Clostridium multifermentans* could not be separated by Sephadex G-200 chromatography, DEAE cellulose chromatography, or rate zonal centrifugation. It was concluded that the two enzymes were complexed together as a single molecular species with an apparent molecular weight of 400,000 daltons (81). After extensive dialysis in a stirred ultrafiltration apparatus (but not in cellulose tubing), active lyase of molecular weight 100,000 was observed but no active esterase (86). Esterase, but not lyase, can be inactivated by heating at 37°C (87). When the activities of the two enzymes in the complexed form are assayed simultaneously on highly esterified pectin, the esterase action rate is twice that for lyase throughout the reaction (enzyme action ceases after about 20% of the substrate is degraded). Thus the molar ratio of the two enzymes is the same since the product of the lyase is an unsaturated digalacturonic acid containing two free carboxyl groups.

Since it is known that clostridial exopolygalacturonate lyase degrades polygalacturonate linearly, beginning from the reducing ends of the substrate chains, it was concluded that the pectinesterase must hydrolyze methyl ester groups in highly esterified pectin by the same action pattern. Otherwise assuming that the lyase acts only at reducing ends where at least two galacturonates are deesterified, the 2:1 ratio of the two enzyme rates would not be possible (87). This was confirmed by our recent unpublished results in which release of terminal ^3H label from highly esterified pectin by the esterase–lyase complex was followed. The results indicated that the esterase begins at the reducing end and proceeds most of the way down to the substrate chain before moving to another molecule (*see* section on exopolygalacturonate lyase for discussion). Thus in contrast to the action of tomato and fungal pectinesterases in which only about 50% occurs at reducing ends of molecules, all of the action of the clostridial esterase takes place at that locus.

Endopolygalacturonase

Polygalacturonase was first reported by Bourquelot and Herissey in 1898 (88). They found that malt extract degraded pectic acid to low molecular weight oligomers. Jansen and MacDonnell (80) showed that polygalacturonases will cause glycosidic hydrolysis of deesterified portions of pectin molecules. Endopolygalacturonase has been found frequently in fruit and filamentous fungi and only rarely in yeast and bacteria. This group of enzymes has been well reviewed by Rombouts and Pilnik (2). The enzymes attack polygalacturonic acid molecules randomly along the chain, producing a large initial decrease in the viscosity of the solution with only a small increase in reducing power. Highly esterified pectins are not degraded in the absence of pectinesterase.

Endopolygalacturonases from *Kluyveromyces fragilis (Saccharomyces fragilis)* have been purified and well characterized by Phaff and co-workers (90–96). The yeast enzyme, unlike other fungal polygalacturonases, is formed constitutively during growth on glucose. It is produced extracellularly and is easily purified by adsorption on polygalacturonic acid gel and $(NH_4)_2SO_4$ precipitation. Only a single active protein was present in the purified preparation (91, 95). Hydrolysis of polygalacturonate occurred in stages (96): a rapid, initial stage up to 25% degradation resulted in tetra-, tri-, and digalacturonic acid; a phase that was 1/44th as fast up to 50% degradation in which tetragalacturonate was converted to tri- and monogalacturonate; and a very slow final stage, up to 70% hydrolysis, in which trigalacturonic acid was converted to di- and monogalacturonic acids. Enzymatic degradation of polygalacturonic acid was most rapid at pH 4.4, but the hydrolysis of tri- and

tetragalacturonic acid was fastest at pH 3.4. A similar drop in pH optimum was observed with polygalacturonase from *Geotrichum candidum*
(*97*).

Luh *et al.* (*98*) showed that partially purified polygalacturonase
from tomatoes had hydrolyzed 4% of the glycosidic bonds in pectic acid
at the point when viscosity had decreased by 50% which indicated an
endo mechanism. A preparation, with specific activity six times higher,
exhibited two pH optima (4.5 and 2.5), suggesting that more than one
enzyme was present (*99, 100*). Foda (*101*) and Hobson (*102*) confirmed the existence of two pH optima. Pressey and Avants (*103*) studied
the effect of substrate size on activity of tomato endopolygalacturonase
in the presence and absence of NaCl. In the absence of NaCl pectic acid
(MW $> 10^6$) was degraded optimally at pH 5, with little activity below 4.
With substrates of lower molecular weight optimal activity occurred at
lower pH, and the activity range extended lower. For example with a
substrate of molecular weight 3200 daltons the pH curve had a peak at
about 4.2 and a broad shoulder extending below pH 2.0. Monovalent
cations enhanced the activity at low pH values. Pectic acid (MW $> 10^6$)
inhibited the hydrolysis rate of low molecular weight substrates up to,
but not more than, 70%. The incomplete inhibition could be caused
by the presence of a second, noninhibited form of the enzyme. This
suggestion of two forms has been confirmed (*104*); polygalacturonase
II (MW 44,000) rapidly decreased the viscosity of high molecular weight
polygalacturonate, but levels of polygalacturonase I (MW 84,000) which
produced reducing groups at the same rate decreased the viscosity much
less, suggesting a partially exo-pattern of action. However the two
enzymes show identical decreased activity on pectic acid at low pH
which is probably the result of substrate self-association.

Recently Pressey and Avants (*105*) separated two polygalacturonases
from extracts of ripe peaches by chromatography on Sephadex G-100.
One of the enzymes decreased the viscosity of polygalacturonate 50%
by hydrolyzing only 0.03% of the glycosidic bonds and was clearly an
endopolygalacturonase. The other enzyme exhibited an exo action pattern and is discussed later. In contrast to tomato endopolygalacturonase
the peach endo enzyme was not activated by low levels of NaCl and
was inhibited by NaCl concentrations above 0.15M. This enzyme did not
respond to substrate size in the same manner as the tomato enzyme. In
this case an intermediate range substrate (110,000 daltons) was cleaved
more rapidly than either pectic acid of molecular weight greater than 10^6
or a 9,000-dalton fragment. The pH of optimal activity on pectic acid was
4.5, and, as with the tomato enzyme, the pH optimum shifted to lower
values with decreasing substrate size and in presence of NaCl.

Endopolygalacturonases from avocado were also studied, and the action pattern is similar to that of the tomato enzyme (*106, 107*). The pH optimum of 5.5 in 0.14*M* NaCl, however, is higher than that of the other fruit endopolygalacturonases and more closely resembles that of peach exopolygalacturonase (*105*). The authors suggested the possibility of contamination of their preparation with an exopolygalacturonase. As pointed out by Rombouts and Pilnik (*2*) and demonstrated by Pressey and Avants (*103*), optimum pH and pH range are profoundly affected by ionic strength, substrate properties, and temperature. Therefore it is frequently difficult and meaningless to compare literature values.

Investigations on fungal endopolygalacturonases have been recently critically reviewed by Rombouts and Pilnik (*2*). This group of enzymes is evidently widespread—[*Acrocylindrium* (*108*), *Aphanomyces euteiches* (*109, 110*), *Aspergillus niger* (*111, 112, 113, 114, 115*), *Aspergillus saito* (*116*), *Colletotrichum gloeosporoides* (*117*), *Coniothyrium diplodiella* (*118, 119, 120*), *Corticium rolfsii* (*121*), *Geotrichum candidum* (*97, 122*), *Monilia laxa* (*123*), *Penicillium expansum* (*124*), *Rhizopus tritici* (*125*), *Sclerotinia rolfsii* (*126*), *Verticillium albo-atrum* (*127*)]. They are sometimes associated with pectin lyase (*38, 111, 112, 128, 129*), exopolygalacturonase (*111, 112, 118, 119, 120, 125, 130*), polygalacturonate lyase (*81*), and pectinesterase (*81, 123, 131, 132, 133, 134*). Hydrolysis continues at the initial rate to a greater degree of degradation (25–75%) than with the yeast and fruit polygalacturonases although much variation is seen. The pH optima range from 2.5 [*Corticium rolfsii* (*121*)] to 5.5 [*Aphanomyces euteiches* (*109, 110*), *Colletotrichum gloeosporoides* (*117*)] with the majority acting optimally at pH values between 4 and 5.

The maximum amount of degradation of glycosidic bonds ranged from a low of about 55% with enzyme III from *Coniothyrium diplodiella* (*120*) and *Geotrichum candidum* (*117, 122*) to greater than 90% with *Aspergillus niger* (*114, 115*). Monogalacturonate is always an end product, usually in association with digalacturonate and sometimes trigalacturonate. The fungal endopolygalacturonases are usually more stable under acid conditions in the pH range 4–6. There are frequent exceptions with broader ranges sometimes extending into the alkaline range.

Bacterial endopolygalacturonases have been described less frequently. Nasuno and Starr (*135*) found an endopolygalacturonase produced by *Pseudomonas marginalis* along with an endopolygalacturonate lyase. The enzyme acts optimally at pH 5.2 and produces mono- to pentagalacturonic acids. These authors (*136*) purified an endopolygalacturonase from *Erwinia caratovora* which ultimately degraded polygalacturonic acid to a mixture of mono- and digalacturonic acids. The enzyme acts optimally at pH 5.2–5.4.

Recently, Horikoshi (137) isolated on an alkaline (Na_2CO_3) pectin medium a *Bacillus* species which produced a polygalacturonase which acted optimally at pH 10–10.5 (although most stable at pH 6.0). The enzyme, purified by gel filtration and ion exchange chromatography, splits polygalacturonic acid randomly, yielding mono-, di-, and higher oligogalacturonates. It is not known whether the preparation is pure as tests for homogeneity have not been done. However pectinesterase, pectin lyase, and polygalacturonate lyase were not detected.

Exopolygalacturonase

Exo-polygalacturonase has been found in plants [carrots (138, 139), peach (105)] fungi [*Aspergillus niger* (140, 141), *Coniothyrium diplodiella* (130, 142), *Rhizopus tritici* (125)], bacteria [*Erwinia aroideae* (143, 144)], and insects [*Pyrrocoris apertus* (145)]. The role of these and other exo-splitting enzymes is not clear. Bateman (58) and McClendon (146) concluded that maceration of host tissue is correlated with the action of endo-splitting enzymes but not with that of exo-splitting enzymes.

The usual end product reported is galacturonic acid except for the *Erwinia* enzyme which produces digalacturonic acid. All three enzymes studied for action pattern (138, 139, 142, 143, 144) attacked substrates from the non-reducing end. In the case of those enzymes forming monomer this was inferred by failure to attack substrates containing Δ4:5 unsaturated bonds at the non-reducing ends (139) or production of a deoxy sugar (147). The *Erwinia* enzyme presumably degraded such substrates from the non-reducing end with the formation of a molecule of digalacturonate (143, 144). The pH optimum for these enzymes was in the range of 4–5, except for the *Erwinia* enzyme which was maximally active at pH 7.5. With the carrot enzyme Hatanaka and Ozawa (138) found that the extent of hydrolysis varied considerably with the source and treatment of the pectic acid substrate, presumably as a result of blockage by side chain neutral sugars. For example peach pectic acid was degraded only 7%, citrus pectic acid 50%, and partially degraded citrus pectic acid (148) 79%.

Mill (140, 141) isolated two *exo*-polygalacturonases from mycelial extracts of *Aspergillus niger* which differed in pH optima, stability, and activation by metal ions. Exopolygalacturonase I had a pH optimum of 4.5 and was fully active only in the presence of Hg^{2+} ions. A ninefold increase in activity occurred in the presence of $1\mu M$ $HgCl_2$; the K_m was $6.2 \times 10^{-8}M$. Dialysis against 2,3-dimercaptopropanol removed all activity in absence of Hg^{2+}. Various other divalent cations and chloride ion did not act on the enzyme and addition of EDTA prevented the effect of $HgCl_2$. The second exopolygalacturonase had a pH optimum at 5.1 and was not affected by metal ions or chelating agents.

Hatanaka and Ozawa (*149*) found no effect of Hg^{2+} or EDTA on *exo*-polygalacturonases from carrot and *Coniothyrium diplodiella*. However an exopolygalacturonase from *Aspergillus niger*, with high activity on digalacturonate, was stimulated by $1\mu M$ Hg^{2+} and inhibited by $10\mu M$ (*150*). Peach exopolygalacturonase (*105*) is stimulated by Ca^{2+} and Sr^{2+} but not other divalent cations. It functions optimally at pH 5.5 and cleaves off monomer units. Substrate size does not affect the pH optimum, but the observed velocity increases with decrease of molecular weight from 10^6 to 10^5 and 3200. However these results were obtained with a constant weight concentration (0.5%). Under such conditions the concentration of substrate molecules—the number of free ends available for action by an exo enzyme—increases as the actual molecular weight decreases. The increase in activity may therefore represent the only approach to saturation of the enzyme rather than a real preference for lower molecular weight substrates. The latter can be demonstrated for exo enzymes only by observing the rate at constant molarity of substrate chains.

Such experiments were carried out by Hasegawa and Nagel (*50, 51*). They found that an oligogalacturonate hydrolase from a *Bacillus* species acts fastest on trigalacturonate at $5\mu M$. This enzyme, found intracellularly rather than in the culture broth as is the case with other bacterial and fungal enzymes, was separated from an *endo*-polygalacturonate lyase and a hydrolase acting on unsaturated digalacturonic acid. The oligogalacturonate hydrolase does not attack this latter substrate but does attack tri- and tetragalacturonic acids whose reducing group has been oxidized with hypoiodite. Results with these substrates indicated that the enzyme acts by an exo mechanism (releasing monomer) from the non-reducing end. Ca^{2+} is not required (*50*).

The same *Bacillus* species produced an intracellular enzyme which hydrolyzed short-chain unsaturated oligogalacturonates (*52*). The activity rate was maximal with unsaturated dimer, followed by unsaturated trimer, tetramer, and pentamer. This enzyme [α-1,4(4,5 dehydrogalacturonosyl) galacturonate hydrolyase] attacks specifically the glycosidic linkage adjacent to the terminal unsaturated galacturonate residue on the non-reducing ends of oligogalacturonate molecules. It releases unsaturated monomer and saturated ($n - 1$) mer which was not further attacked. The optimum pH was 6.3–6.6 for hydrolysis of unsaturated digalacturonate. The enzyme did not require a divalent cation for activity.

Endopolymethylgalacturonase

Although there are numerous reports in the literature concerning polymethylgalacturonases, the existence of such an enzyme is questionable. It is obvious now that many, perhaps all, of the reports prior to 1960, when the trans eliminase type of cleavage was discovered, were actually of pectin lyases and not polymethylgalacturonases. Furthermore workers frequently mistook impure preparations of polygalacturonase which also contained pectinesterase for this enzyme. Also it is known that polygalacturonase and polygalacturonate lyase can attack pectins to some extent in areas of the molecule which are not completely esterified. Koller and Neukom (111) described two endopolymethylgalacturonases from *Aspergillus niger* which were partially purified but still contained pectinesterase, endopolygalacturonase, and pectin lyase. Rombouts and Pilnik (2) have tried without success to isolate polymethylgalacturonases from commercial pectin enzyme preparations. These workers doubt whether this enzyme really exists.

Endopolygalacturonate Lyase

Endopolygalacturonate lyase has been described primarily from soft rot and food spoilage bacteria and related species, such as *Bacillus polymyxa* (40, 41, 151, 152), *Bacillus pumilis* (153), *Erwinia aroideae* (47, 48), *Erwinia carotovora* (42, 154, 155), *Erwinia chrysanthemi* (156), *Pseudomonas fluorescens* (157), *Pseudomonas marginalis* (135), and *Xanthomonas campestris* (158). Filamentous fungi frequently produce pectin lyase. There are few reports of polygalacturonate lyase, and they are generally concerned with *Fusarium* species (81, 159). Mah and Hungate (160) have reported polygalacturonate lyase in a protozoan, *Ophryoscolex purkynei*.

Wood, in 1951 (161), was the first to note that a bacterial enzyme capable of degrading polygalacturonate required Ca^{2+} and acted optimally in the pH range of 8.5–9.0. These properties were substantially different from those of fungal polygalacturonases and are characteristic of enzymes exhibiting a trans eliminative attack on polygalacturonate. Endopolygalacturonate lyases act optimally at slightly alkaline pH values (8–9.8; many are optimal at 8.5). A divalent cation (Ca^{2+}, or, less effectively, Sr^{2+}, Co^{2+}, or Mg^{2+}) is required ($10^{-3}M$) for activity. Cleavage of 1–3% of the glycosidic bonds results in 50% reduction of viscosity of polygalacturonate solutions. The end products encountered are usually unsaturated di- and trigalacturonic acids, but occasionally unsaturated mono-, tetra-, and pentagalacturonic acids are encountered. Molecular weights of the enzyme from 30,000 to 36,000 have been reported (156).

Nagel and Vaughn (*40, 41 162*) found that *Bacillus polymyxa* produced extracellular endopolygalacturonate lyase activity. Both the crude enzyme and four fractions separated by cellulose chromatography were able to degrade trigalacturonate to unsaturated digalacturonate and monogalacturonate. Tetramer yielded unsaturated trigalacturonate and monomer and, at a slower rate, unsaturated digalacturonate and diglacturonate. The unsaturated digalacturonic acid was purified. On the basis of ultraviolet absorption, bromine consumption, lead tetraacetate oxidation, paper chromatography, thiobarbituric acid reaction, and ozonization it was shown that the structure was *O*-(4-deoxy-β-L-*threo*-hexopyranos-4-enyluronic acid-(1,4)-D-galacturonic acid (*163*). Chromatography on carboxymethylcellulose of extracellular enzyme preparations from *B. polymyxa* yielded four fractions with differing properties (*152*). For example the pH optima were 8.4, 8.8, 9.3, and 9.5. All four enzymes appeared to be of the endo-type by comparison of K_m data and viscosity data. However paper chromatographic analysis of products formed on degradation of oligogalacturonates (trimer to octamer) indicated that two of the enzymes were more endo in character than the other two. Although two of the enzymes could have been mixtures of exo and endo enzymes, the authors concluded that this is probably not so although they present no data proving homogeneity.

McNicol and Baker (*164*) discovered that an enzyme produced by *Bacillus sphaericus* degraded the V_i antigen of *Escherichia intermedia*. This surface antigen is a polysaccharide containing α-1,4-linked 2-*N*-acetyl-3-*O*-acetyl-2-deoxy-2-amino-D-galacturonic acid. The enzymic products were a series of oligomers containing Δ4:5 double bonds at their non-reducing ends. This enzyme and the endopolygalacturonate lyase from *Bacillus polymyxa* degraded V_i antigen, *O*-deacetylated V_i antigen, pectin, and polygalacturonic acid. A commercial pectinase, probably containing polygalacturonase, was not active on the antigen. Apparently substitutions in the 2 and 3 positions do not inhibit the activity of these polygalacturonate lyases.

Starr and Moran (*42*) and Moran *et al.* (*154*) found that *Erwinia carotovora* produced an extracellular endopolygalacturonate lyase. An enzyme with similar properties was found intracellularly along with an apparent oligogalacturonate lyase. The major product of attack on citrus polygalacturonic acid was unsaturated digalacturonic acid with lesser amounts of unsaturated trigalacturonic acid and saturated mono- and digalacturonic acid. Small amounts of saturated compounds are expected from the non-reducing ends of molecules. Polygalacturonic acid (0.23%) was degraded more rapidly than oligogalacturonates ($5\mu M$). Tetragalacturonate was degraded about six times faster than trigalacturonic acid, and digalacturonic acid was not attacked. Products from tetragalacturonate were mainly unsaturated digalacturonate and digalacturonate with some monomer and unsaturated trimer. Trigalacturonic acid was cleaved into galacturonate and unsaturated digalacturonate.

The intracellular oligogalacturonate lyase from *Erwinia* was purified by Moran *et al.* (53) and degraded unsaturated digalacturonic acid into two molecules of 4-deoxy-5-hexoseulose uronic acid. This enzyme also attacks normal digalacturonic acid, producing one molecule of D-galacturonic acid and one of the deoxyketuronic acid. At equal concentrations unsaturated digalacturonic acid was degraded five times faster than normal digalacturonic acid; trimer was degraded still more slowly, and tetramer and polymer were degraded at a rate 1/400 of that of unsaturated dimer.

Hatanaka and Ozawa studied oligogalacturonate lyases from *Erwinia aroideae* (54) and a *Pseudomonas* species (55), which are apparently similar to the one described by Moran *et al.* (53). With the *Erwinia* enzyme polygalacturonic acid was degraded more slowly than oligogalacturonate, and unsaturated oligogalacturonates were attacked the most rapidly. The unsaturated dimer forms two molecules of 4-deoxy-5-keto-D-fructuronic acid. The process by which the unsaturated monomer is converted to this product is not known. Ca^{2+} was not required for activity.

The *Pseudomonas* enzyme differed from the *Erwinia* enzyme in that: saturated oligogalacturonides were degraded as rapidly as the corresponding unsaturated uronides; activity was maximal with tetramers, follower by trimer, dimer, and then polymer; and Ca^{2+} stimulated activity. With a purified enzyme preparation the reaction product from unsaturated digalacturonic acid was 4-deoxy-5-keto-D-glucuronic acid; with crude extract a mixture of this acid and 4-deoxy-5-keto-D-fructuronic acid resulted.

Exopolygalacturonate Lyase

So far exopolygalacturonate lyase has been found only as an extracellular enzyme from bacterial sources [*Clostridium multifermentans* (43, 44, 45, 85, 87), *Erwinia aroideae* (46, 49), and an *Erwinia* species (165)]. Macmillan and Vaughn (43, 44) partially purified the clostridial enzyme from culture broth and freed it of accompanying pectinesterase activity. In the absence of esterase the enzyme attacked polygalacturonic acid optimally at pH 8.5, but highly esterified pectin (prepared by HCl-catalyzed esterification) was not attacked. Citrus pectin was also attacked to a limited extent, presumably on unesterified portions of the molecules. The enzyme required divalent cations; Ca^{2+} was best, followed by Sr^{2+}, Mn^{2+}, Mg^{2+}, and Ba^{2+}. Although commercial citrus polygalacturonic acid was partially degraded without addition of divalent cations, EDTA completely inhibited activity, and polygalacturonate, which was passed through a column of Dowex 50 Na^+ to remove divalent cations, was only slightly degraded. With the latter substrate there was maximum activity in the presence of 0.5mM $CaCl_2$, and the reaction went to completion. With 0.5% substrate the initial degradation rate is the same in the presence of 0.5 and 2mM $CaCl_2$ up to about 40% degradation (one hour). After this point the rate of degradation in 2mM $CaCl_2$ was considerably reduced, and only 65% degradation occurred in 20 hours.

The main degradation product was unsaturated digalacturonic acid with small amounts of the normal monomer and dimer. The molar extinction coefficient of the purified dimer was 4800 at pH 3.7. Comparison of the number of reducing groups produced with the number of unsaturated bonds formed during degradation of polygalacturonic acid showed that the partially purified enzyme preparation did not contain polygalacturonase activity (at pH 8.0).

A total of 22% of a 0.5% polygalacturonic acid substrate was converted to unsaturated dimer when viscosity was reduced by 50%. This, with chromatographic analysis of the products on tetragalacturonic acid and trigalacturonic acid, indicated that the enzyme was the exo-splitting type. Tetragalacturonate is completely degraded to unsaturated digalacturonic acid and digalacturonic acid. Trigalacturonic acid gives unsaturated digalacturonic and galacturonic acid. (The rate of attack on tri-, tetra-, and dodecagalacturonic acids is about the same.) Digalacturonic acid is not attacked. Unsaturated digalacturonic acid was separated from the longer, as yet undegraded, chains in the reaction mixture, and the chains did not have unsaturated bonds at their non-reducing ends. This indicated that the enzyme degraded the chain from the reducing end. Normal digalacturonic acid or monogalacturonic acid was formed from the non-reducing ends of chains depending on whether there were an odd or an even number of galacturonate residues in the original chain.

Okamoto *et al.* (*46, 47, 48, 49*) isolated a polygalacturonate lyase from *Erwinia aroideae*. Since an acid soluble pectic acid was degraded to unsaturated digalacturonic acid and continually shortening chains, these authors also concluded that the attack was from the reducing end.

Recently, Hatanaka and Ozawa (*165*) found an exopolygalacturonate lyase in a species of *Erwinia* with different properties. It was activated by Co^{2+} and Mn^{2+}, but Ca^{2+} had no effect. Action on polygalacturonate resulted in only 44% degradation. The authors concluded that exo enzymes have a limited attack on polygalacturonic acid because of neutral sugars in the chains. They believe that reports of 100% degradation by such enzymes are caused by contaminating activities. Another explanation, however, could be that divalent cations present in the substrate form cross linkages between chains which the exo enzymes cannot pass. As stated above Macmillan and Vaughn (*44*) found that addition of Ca^{2+} to divalent cation-free substrate inhibited the extent of degradation.

The exopolygalacturonate lyase from *Clostridium multifermentans* is apparently a complex with pectinesterase also present in a molecule with a molecular weight of 4×10^5 (*87*). A polygalacturonic acid substrate (degree of polymerization 17) labelled at the reducing end with NaB^3H_4 was subjected to degradation by the enzyme and release of tritiated product and total unsaturated dimer were measured (*166*). The percentage release of tritiated product was closely proportional to the total amount of product release throughout the reaction. It was concluded that the substrate was degraded terminally by a single chain action pattern. A similar conclusion was reached for the pectinesterase action also present in the substrate. Evidently the two types of active sites act in a coordinated manner on highly esterified pectin; the polysaccharide chain passing from the esterase site to the lyase site without intermediate dissociation and rebinding. Thus the two sites constitute a molecular disassembly line—a type of system which may be of general significance in synthesis and degradation of biological polymers.

Pectin Lyase

Neukom and Deuel (*34*) studied the alkaline degradation of pectin and speculated that cleavage of glycosidic linkages occurred by a trans elimination mechanism rather than by hydrolysis. Albersheim, Neukom, and Deuel (*37*) demonstrated that the glycosidic bonds in pectin are very susceptible to this type of degradation. Slow trans elimination occurs even at pH 6.8 at elevated temperatures. These authors then reported the first example of enzyme-mediated trans elimination and discovered pectin lyase activity in a commercial enzyme preparation, Pectinol R-10, from *Aspergillus niger*. So far all pectin lyases studied have been from filamentous fungi and of the endo type. It is not known whether other organisms can produce endo pectin lyases or whether the exo type exists.

Albersheim and Killias (*167*) purified the enzyme ninefold from Pectinol R-10. The optimal pH for activity was 5.1–5.2, and its isoelectric point was between 3 and 4. Pectin lyase was active on 65% esterified citrus pectin but not on polygalacturonic acid. The enzyme was more active in citrate and phosphate buffers than in acetate. The addition of $CaCl_2$ to reaction mixtures buffered with citrate or phosphate inhibited the reaction. Product inhibition was markedly increased by addition of plant auxins (*168*). Bull (*169*) suggested that the auxin effects were artifacts in spectrophometrically dense reaction mixtures.

Edstrom and Phaff (*38, 39*) purified pectin lyase from the culture broth of *Aspergillus fonsecaeus* and removed contaminating pectinesterase, polygalacturonase, and cellulase. The enzyme did not attack polygalacturonic acid, but citrus pectin pretreated with yeast endopolygalacturonase was a good substrate. With this substrate both Ca^{2+} and Na^- ions stimulated activity. The enzyme was optimally active at pH 5.2, but in the presence of Ca^{2+}, a second peak of activity appeared at pH 8.5. The endo nature of the enzyme was demonstrated by chromatographic analysis on paper of the products found from 95% esterified pectin (HCl catalyzed esterification in methanol). At 27.5% degradation the products were a series of unsaturated methyl oligogalacturonates with chain length of 2–8 residues. With exhaustive treatment with a higher enzyme concentration the main end products were the methyl esters of unsaturated tri-, tetra-, and pentagalacturonates. With chemically esterified methyl esters of oligogalacturonates (mono- to hexa-) pectin lyase attacked only tetramers and higher oligomers. The hexamer was rapidly split at the central bond, yielding trimer and unsaturated trimer. A slower reaction yielded dimethyl digalacturonate and unsaturated tetramer. Pentamer yielded unsaturated trimer and normal dimer, and the products from tetramer were unsaturated trimer and (saturated) methylgalacturonate. It was concluded that endo pectin lyase cannot cleave the two bonds nearest the reducing end of highly esterified pectins and has limited ability to split bonds nearest the non-reducing end.

Ishii and Yokotsuka recently used pectin lyases from *Aspergillus sojae* (*170, 171, 172*) and *Aspergillus japonicus* (*173*) for the enzymatic clarification of fruit juices. This is discussed later in this review. Bush and Codner (*174*) compared the pectin lyases produced by *Penicillium digitatum* and *Penicillium italicum* and found them remarkably similar. Both were endo acting, had optimal activities at pH 5.5, and had the same K_m. The pectin lyase from *P. italicum*, however, was less stable than that of *P. digitatum*. Spalding and Abdul-Baki (*175*) reported the production of a pectin lyase by *Penicillium expansum* growing on apple tissue or a pectin–polypectate mixture.

A second group of endo pectin lyases is characterized by optimum activity at slightly basic pH and by the ability to degrade polygalacturonic acid (even though pectin is still favored). Sherwood (*129*) isolated this type of pectin lyase from *Rhizoctonia solani*. The enzyme had an optimal activity at pH 7.2 and degraded pectin faster than pectic acid. Bateman (*128*) found two pectin lyases from this microorganism which, when purified, were stimulated by cations and inhibited by EDTA. Optimal activity was between pH 8.0 and 9.0.

Voragen (*176*) observed the activity of three pectin lyases upon pectin with differing degrees of esterification. V_{max} for all three enzymes was optimal at pH 5.8–6.5 and was unaffected by the degree of esterification (in the range of 74–95% and above pH 5.0). However the K_m increased with decreasing percent esterification but decreased with decreasing pH. Thus at low concentrations of less esterified pectin an apparently lower pH optimum was observed. Substrates with blocks of esterified and unesterified carboxyls were preferred to those with free carboxyls distributed randomly. The pectin lyases have no absolute requirement for cations, except for the enzyme produced by *Fusarium solani* which requires calcium (*177*).

Applications

Through the centuries man has used pectic enzymes produced by microorganisms in natural fermentations to modify plant tissue to produce more useful products. One of the oldest fermentations is the process of retting, in which textile fibers, such as flax, hemp, and jute are loosened from their plant stems. *Clostridium felsineum*, one of the organisms involved in anaerobic retting, produced endopolygalacturonase and endopolygalacturonate lyase but not pectinesterase (*178, 179*).

In the production of coffee beans the outer skin and fibrous pulp of the cherry are removed mechanically. The removal of the residual mucilaginous coating surrounding the bean is accomplished by a natural fermentation, taking 24–48 hours and involving the action of pectic enzymes and other hydrolases. Alternatively commercial pectic enzymes have been used to liquify the mucilage in 1–10 hours, depending on enzyme concentration (*7*). The curing or fermentation of cocoa, tea, and tobacco also can involve pectic enzymes (*17*).

Naturally occurring pectic enzymes in fruit can be either advantageous or deleterious in commercial processes. Canned fruit may soften during storage, and suspending syrups become more viscous unless the fruit enzymes are inhibited or inactivated during processing (*180*). Enzymatic degradation of pectin in overripe fruit for jams or jellies results in loss of gel forming properties. In preparing cloudy juices such as tomato and citrus it is essential that pectic substances are not degraded. However it is impossible to make tomato juice concentrates unless pectins are first degraded, and this can be accomplished by action of tomato polygalacturonase and pectinesterase. Citrus fruits contain considerable amounts of pectinesterase which can deesterify pectin and cause the formation of undesirable calcium pectate gels in juices. According to Fogarty and Ward (*180*) the measurement of pectinesterase and polygalacturonase activities in fruit products has provided clues to several industrial problems involving gelling, viscosity changes, and cloud instability.

Commercial sources of fungal pectic enzymes have been used since the 1930's (*181, 182, 183*) for clarifying fruit juices and disintegrating plant pulps to increase juice yields. Pectic enzymes are widely used in producing apple juice, grape juice, and wine (7). Fungal pectinesterases are used to prepare pectins of low methyl ester content which can form stable gels in the presence of divalent cations at sugar concentrations much lower than those required for normal pectin jellies. Pectinesterase, free of polygalacturonase, can be used to produce these gels directly in fruit juices in the presence of calcium ions (7).

Freshly pressed apple juice contains about 0.15% pectin (actual amounts vary considerably, depending on variety, ripeness, etc.) in colloidal suspension. This pectin is closely associated with insoluble particles and causes them to remain suspended in the juice. The haze is difficult to remove by conventional filtration or centrifugation. If an appropriate amount of a commercial pectic enzyme preparation is added to freshly pressed juice, the haze begins to flocculate within a few minutes, and after several hours the floc settles out, leaving a relatively clear supernatant liquid. The juice is finished by centrifugation or filtration through diatomaceous earth (7, 180).

Much early work which tried to relate the activity of specific pectic enzymes to the clarification reaction is now of limited value because many of the investigators were not aware of the many types of pectic enzymes and the different substrates which they degrade. Usually crude enzyme preparations containing a variety of activities in addition to pectic enzymes were used, and the results were confusing. Furthermore the colloidal particles contain other materials beside pectin. For example Yamasaki *et al.* (184) reported up to 36% protein in the precipitate obtained by ultracentrifugation of apple juice. The most definitive work relating clarification to a specific enzyme was by Endo (185). He purified three endopolygalacturonases, one exopolygalacturonase, and two pectinesterases from preparations derived from *Coniothyrium diplodiella*. When added separately no single enzyme fraction was capable of clarifying apple juice satisfactorily.

Mixtures of the polygalacturonases or of the pectinesterases were also ineffectual. Only the combined action of at least one of the polygalacturonase fractions with one of the pectinesterase fractions was effective in clarification. Purified endopolygalacturonase alone could decrease the viscosity of a solution of citrus pectin (64% esterification), but it was completely ineffectual when apple pectin was used [90% esterified (186)].

Although pectic enzymes are used universally by the apple juice industry, the actual clarification mechanism has not been completely elucidated. Endo showed (186, 187) that hydrolysis of pectin, accompanied by a decrease in viscosity of the juice, is the principal step during clarification. Endo separated the clarification process into three phases: a pectin fraction is solubilized; viscosity decreases; and previously colloidally suspended particles flocculate. Only a small percentage of the total glycosidic linkages present needed to be split to cause flocculation and resultant clarification.

Yamasaki et al. (184) reported that apple pectin is suspended as colloidal particles with surfaces negatively charged at pH 3.5. Positively charged colloids (gelatin) enhanced the clarification reaction while negatively charged colloids (i.e., sodium alginate) inhibited the clarification reaction. It was postulated [Yamasaki et al. (184, 188)] that the initial depolymerization of pectin is followed by a nonenzymatic, physical (electrostatic neutralization) coagulation step. Negatively charged colloidal pectin is thought to surround a positively charged protein core. As pectic enzymes attack this outer coating, sections of the positively charged protein become exposed. Presumably these positively charged sections on one colloidal particle react with undegraded, negatively charged sections on a neighboring particle, and aggregates are formed which precipitate.

Okada et al. (189) reported an unknown enzyme from Aspergillus niger which was not a polygalacturonase nor a pectinesterase but which was a potent clarifier of citrus juices. This factor had no effect by itself, but it strongly accelerated juice clarification when added with polygalacturonase. In subsequent papers (190, 191) the enzyme was reported purified and found to be a hemicellulase.

Recently Ishii and Yokotsuka (170, 171, 172) reported that a purified pectin lyase from Aspergillus sojae initiated the clarification reaction in apple juice. This enzyme has a molecular weight of 32,000 and a pH optimum of 5.5 for 68% esterified pectin (pH 7.0 for 98% esterified pectin). Divalent cations stimulated activity and shifted the pH optimum towards neutrality. The enzyme was stable between pH 4 and 7. One mg of purified pectin lyase (76.5 units of activity) could clarify 30–40 liters of apple juice within one hour at 40°C [0.018–0.024 units needed to clarify 10 ml (172)]. At the point of complete clarification 50% of the pectin in the juice had been converted to a form soluble in 75% ethanol.

According to Reed (7) the use of pectic enzymes to clarify grape juice represents an anomaly with Concord varieties since a juice of good body and high viscosity is desired. The juice, however, is difficult to express from this variety because of its slimy consistency. In recent years the desire for high yields has led to widespread use of pectic enzymes. Crushed grapes are treated with enzyme at 60–65°C for about 30 minutes, and most of the juice is then easily removed by conventional mechanical methods. Enzyme treatment is designed to give high yields without serious loss in viscosity of the final juice product. Ishii and Yokotsuka (172, 192) found that grape juice was more resistant to clarification by pectin lyases than apple juice. One reason for this is that apple pectin is more highly esterified and a better substrate for pectin lyase than grape pectin [87–90% esterified *vs.* 45–50% esterified (*184*)]. A mixture of pectin lyase and endopolygalacturonase was more effective in clarifying grape juice than either alone (*193*).

The use of pectic enzymes to clarify wine was first proposed by Cruess and Besone (*194*). A discussion of the literature can be found in Reed (*7*). Better results are obtained by adding enzyme directly to musts before fermentations. In extracting juice for winemaking it is necessary to control the amount of pectin degradation to leave sufficient body in the wine (*7*). Pectic enzymes are particularly recommended for red wines which are fermented with their skins (*195*). Berg and Marsh (*196*) compared the effects of enzymatic *vs.* nonenzymatic treatment of grapes for production of California wine. With certain grapes enzyme use was important for clarification. On dry red wines stored for 24 months, no difference could be detected between enzyme-treated and control wines in color intensity. However there was color loss in the enzyme-treated wines after 54 months. Other wines gained color.

Other novel uses of pectic enzymes have been proposed. Fogarty and Ward (*180*) suggested that commercial softwoods such as Sitka and Norway spruce be treated with pectic enzyme preparations or specific bacteria which produce them. This treatment was effective in making these woods more permeable to preservatives. Miller *et al.* (*197*) proposed that pectic enzymes could be immobilized in a support matrix for possible use in industrial apparatus for continuous clarification of apple juice. The pectin lyase from *Aspergillus japonicus* was partially purified and immobilized on collagen films according to the procedure of Wang and Vieth (*198*). In this procedure a concentrated solution of the enzyme is mixed with a 1% suspension of collagen (cow tendon). Under acid conditions the collagen enzyme dispersion forms a gel which is spread onto a sheet of Mylar with a motor driven knife, forming a membrane of uniform thickness. The films are allowed to dry and then peeled off the Mylar support.

A reactor module incorporating a piece of the dried membrane coiled around a glass rod with a Vexor mesh spacer was constructed. This module was used to test kinetics of the immobilized pectin lyase and to determine whether it could clarify apple juice on a continuous basis. A reaction mixture containing 0.5% pectin was passed through the module, and the amount of product formed was continuously monitored spectrophotometrically at 235 nm. The reactor was stabilized at a constant activity level by passing sufficient reaction mixture through the module to wash away loosely bound enzyme located near the surface of the membrane. Activity stabilized at approximately 20% of the initial activity (representing less than 1% of the free enzyme originally added to the collagen). This reactor was used intermittently for several weeks without further loss of activity. The lyase was more stable to temperature in the immobilized form than in the nonimmobilized form. Optimum activity of the free enzyme was observed at pH 6.0. With the collagen-bound lyase, however, the pH optimum shifted to around 4.5. Shifts of this magnitude are not unusual, however, when bound enzymes are compared with their free counterparts.

The small experimental reactor was able to clarify apple juice. However its life was severely limited when used for this purpose. Decline in activity probably resulted from precipitated solids clogging the reactive sites in the module or through nonspecific inactivation of the bound lyase by tannins in the circulating juice (199).

Literature Cited

1. Whitaker, J. R., "Principles of Enzymology for the Food Sciences," pp. 469-479, Marcel Dekker, New York, 1972.
2. Rombouts, F. M., Pilnik, W., CRC Crit. Rev. Food Technol. (1972) 3, 1.
3. Codner, R. C., J. Appl. Bacteriol. (1971) 34, 147.
4. Voragen, A. G. J., Pilnik, W., Z. Lebensm. Unters. Forsch. (1970) 142, 346.
5. Pilnik, W., Voragen, A. G. J., in "The Biochemistry of Fruits and their Products," A. C. Hulme, Ed., pp. 53-87, Academic, New York, 1970.
6. Bateman, D. F., Millar, R. L., Annu. Rev. Phytopathol. (1966) 4, 119.
7. Reed, G., "Enzymes in Food Processing," pp. 73-88, Academic, New York, 1966.
8. Neukom, H., Schweiz. Landw. Forsch. (1963) 2, 112.
9. Joslyn, M. A., Advan. Food Res. (1962) 11, 1.
10. Wood, R. K. S., Annu. Rev. Plant Physiol. (1960) 11, 299.
11. Deuel, H., Stutz, E., Advan. Enzymol. (1958) 20, 341.
12. Demain, A. L., Phaff, H. J., Wallerstein Lab. Commun. (1957) 20, 119.
13. Solms, J., Deuel, H., Helv. Chim. Acta (1955) 38, 321.
14. Whistler, R. L., Smart, C. L., "Polysaccharide Chemistry," Academic, New York, 1953.
15. Sumner, J. B., Somers, G. F., "Chemistry and Methods of Enzymes," p. 83, Academic, New York, 1953.
16. Lineweaver, H., Jansen, E. F., Advan. Enzymol. (1951) 11, 267.
17. Kertesz, Z., "The Pectic Substances," Interscience, New York, 1951.
18. Kertesz, Z., McColloch, R. J., Advan. Carbohyd. Chem. (1950) 5, 79.
19. Joslyn, M. A., Phaff, H. J., Wallerstein Lab. Commun. (1947) 10, 39.
20. Phaff, H. J., Joslyn, M. A., Wallerstein Lab. Commun. (1947) 10, 133.
21. Hirst, E. L., Jones, J. K. N., Advan. Carbohyd. Chem. (1946) 2, 235.
22. Hobson, G. E., Phytochemistry (1967) 6, 1337.
23. Spencer, M., in "Plant Biochemistry," J. Bonner, J. E. Varner, Eds., pp. 793-825, Academic, New York, 1965.
24. LaMotte, C. E., Gochnauer, C., LaMotte, L., Mathur, J. R., Davies, L., Plant Physiol. (1969) 44, 21.
25. Lund, B. M., J. Appl. Bacteriol. (1971) 34, 9.
26. Vaughn, R. H., King, A. D., Nagel, C. W., Ng, H., Levin, R. E., Macmillan, J. D., York II, G. K., J. Food Sci. (1969) 34, 224.

27. Vaughn, R. H., Jacubczyk, T., Macmillan, J. D., Higgins, T. E., Davé, B. A., Crampton, V. M., *Appl. Microbiol.* (1969) **18**, 771.
28. McCready, R. M., Gee, M., *J. Agr. Food Chem.* (1960) **8**, 510.
29. Rombouts, F. M., Thesis, Landbouwhogeschool, Wageningen, Netherlands, 1972.
30. Palmer, K. J., Merrill, R. C., Owens, H. S., Ballantyne, M., *J. Phys. Colloid Chem.* (1947) **51**, 710.
31. Palmer, K. J., Lotzkar, H., *J. Amer. Chem. Soc.* (1945) **67**, 883.
32. Palmer, K. J., Ballantyne, M., *J. Amer. Chem. Soc.* (1950) **72**, 736.
33. Merrill, R., Weeks, M., *J. Amer. Chem. Soc.* (1945) **67**, 2244.
34. Neukom, H., Deuel, H., *Chem. Ind.* (1958) 683.
35. Baker, G. L., *Advan. Food Res.* (1948) pp. 395-427.
36. Roboz, E., Barratt, R. W., Tatum, E. L., *Biochem. J.* (1952) **48**, 459.
37. Albersheim, P., Neukom, H., Deuel, H., *Arch. Biochem. Biophys.* (1960) **90**, 46.
38. Edstrom, R. D., Phaff, H. J., *J. Biol. Chem.* (1964) **239**, 2403.
39. Edstrom, R. D., Phaff, H. J., *J. Biol. Chem.* (1964) **239**, 2409.
40. Nagel, C. W., Vaughn, R. H., *Arch. Biochem. Biophys.* (1961) **93**, 344.
41. Nagel, C. W., Vaughn, R. H., *Arch. Biochem. Biophys.* (1961) **94**, 328.
42. Starr, M. P., Moran, F., *Science* (1962) **135**, 920.
43. Macmillan, J. D., Phaff, H. J., *Methods Enzymol.* (1966) **8**, 632.
44. Macmillan, J. D., Vaughn, R. H., *Biochemistry* (1964) **3**, 564.
45. Macmillan, J. D., Phaff, H. J., Vaughn, R. H., *Biochemistry* (1964) **3**, 572.
46. Okamoto, K., Hatanaka, C., Ozawa, J., *Agr. Biol. Chem.* (1963) **27**, 596.
47. Okamoto, K., Hatanaka, C., Ozawa, J., *Agr. Biol. Chem.* (1964) **28**, 331.
48. Okamoto, K., Hatanaka, C., Ozawa, J., *Ber. Ohara Inst. Landwirt. Biol. Okayama Univ.* (1964) **12**, 107.
49. Okamoto, K., Hatanaka, C., Ozawa, J., *Ber. Ohara Inst. Landwirt. Biol. Okayama Univ.* (1964) **12**, 115.
50. Hasegawa, S., Nagel, C. W., *Nature* (1967) **213**, 207.
51. Hasegawa, S., Nagel, C. W., *Arch. Biochem. Biophys.* (1968) **124**, 513.
52. Nagel, C. W., Hasegawa, S., *J. Food Sci.* (1968) **33**, 378.
53. Moran, F., Nasuno, S., Starr, M., *Arch. Biochem. Biophys.* (1968) **125**, 734.
54. Hatanaka, C., Ozawa, J., *Agr. Biol. Chem.* (1970) **34**, 1618.
55. Hatanaka, C., Ozawa, J., *Agr. Biol. Chem.* (1971) **35**, 1617.
56. Fremy, E., *Ann. Chim. Phys.* (1848) **24**, 5.
57. International Union of Biochemistry, "Enzyme Nomenclature," Elsevier, New York, 1965.
58. Bateman, D. F., *Phytopathology* (1964) **54**, 438.
59. Bateman, D. F., Lumsden, R. D., *Phytopathology* (1965) **55**, 734.
60. Kertesz, Z., *Food Res.* (1938) **3**, 481.
61. MacDonnell, L. R., Jansen, E. F., Lineweaver, H., *Arch. Biochem.* (1945) **6**, 389.
62. MacDonnell, L. R., Jang, R., Jansen, E. F., Lineweaver, H., *Arch. Biochem.* (1950) **28**, 260.
63. Bell, T., Etchells, J., Jones, I., *Arch. Biochem. Biophys.* (1951) **31**, 431.
64. Reid, W. W., *Nature* (1950) **166**, 569.
65. Glasziou, K., Inglis, S., *Aust. J. Biol. Sci.* (1958) **11**, 127.
66. Hulton, H. O., Levine, A. S., *Arch. Biochem. Biophys.* (1963) **101**, 396.
67. Hulton, H. O., Sun, B., Bulger, J., *J. Food Sci.* (1966) **31**, 320.
68. Morell, S., Link, K. P., *J. Biol. Chem.* (1933) **100**, 385.
69. Morell, S., Baur, L., Link, K. P., *J. Biol. Chem.* (1934) **105**, 1.
70. Kertesz, Z., *J. Biol. Chem.* (1937) **121**, 589.
71. Lineweaver, H., Ballou, G., *Fed. Proc.* (1943) **2**, 66.
72. Deuel, H., Huber, G., Leuenberger, R., *Helv. Chim. Acta* (1950) **33**, 1226.

73. Schultz, T. H., Lotzkar, H., Owens, H. S., Maclay, W. D., *J. Phys. Colloid Chem.* (1945) **49**, 554.
74. Speiser, R., Copley, M. J., Nutting, G. C., *J. Phys. Colloid Chem.* (1947) **51**, 117.
75. Lee, M., Macmillan, J. D., *Biochemistry* (1968) **7**, 4005.
76. Nakagawa, H., Yanagawa, Y., Takehana, H., *Agr. Biol. Chem.* (1970) **34**, 991.
77. Lineweaver, H., Ballou, G., *Arch. Biochem.* (1945) **6**, 373.
78. Mayorga, H., Rolz, C., *J. Agr. Food Chem.* (1971) **19**, 179.
79. Nakagawa, H., Yamagawa, Y., Takehana, H., *Agr. Biol. Chem.* (1970) **34**, 998.
80. Pressey, R., Avants, J. K., *Phytochemistry* (1972) **11**, 3139.
81. Miller, L., Macmillan, J. D., *Biochemistry* (1971) **10**, 570.
82. Lee, M., Macmillan, J. D., *Biochemistry* (1970) **9**, 1930.
83. Kohn, R., Furda, I., *Collect. Czech. Chem. Commun.* (1967) **32**, 1925.
84. Kohn, R., Furda, I., *Collect. Czech. Chem. Commun.* (1967) **32**, 4470.
85. Miller, L., Macmillan, J. D., *J. Bacteriol.* (1970) **102**, 72.
86. Sheiman, M., Macmillan, J. D., Miller, L., *Bacteriol. Proc.* (1972) p. 167.
87. Lee, M., Miller, L., Macmillan, J. D., *J. Bacteriol.* (1970) **103**, 595.
88. Bourquelot, E., Herissey, H., *Compt. Rend.* (1898) **127**, 191.
89. Jansen, E. F., MacDonnell, L. R., *Arch. Biochem.* (1945) **8**, 97.
90. Luh, B. S., Phaff, H. J., *Arch. Biochem. Biophys.* (1951) **33**, 212.
91. Phaff, H. J., *Methods Enzymol.* (1966) **8**, 636.
92. Demain, A. L., Phaff, H. J., *Nature* (1954) **174**, 515.
93. Luh, B. S., Phaff, H. J., *Arch. Biochem. Biophys.* (1954) **48**, 23.
94. Luh, B. S., Phaff, H. J., *Arch. Biochem. Biophys.* (1954) **51**, 102.
95. Phaff, H. J., Demain, A. L., *J. Biol. Chem.* (1956) **218**, 875.
96. Demain, A. L., Phaff, H. J., *J. Biol. Chem.* (1954) **210**, 381.
97. Barash, I., *Phytopathology* (1968) **58**, 1364.
98. Luh, B. S., Leonard, S. J., Phaff, H. J., *Food Res.* (1956) **21**, 448.
99. Patel, D. S., Phaff, H. J., *Food Res.* (1960) **25**, 37.
100. Patel, D. S., Phaff, H. J., *Food Res.* (1960) **25**, 47.
101. Foda, Y. H., Ph.D. Thesis, University of Illinois, 1957.
102. Hobson, G. E., *Biochem. J.* (1964) **92**, 324.
103. Pressey, R., Avants, J. K., *J. Food Sci.* (1971) **36**, 486.
104. Pressey, R., Avants, J. K., *Biochem. Biophys. Acta* (1973) **309**, 363.
105. Pressey, R., Avants, J. K., *Plant Physiol.* (1973) **52**, 252.
106. McCready, R. M., McComb, E. A., Jansen, E. F., *Food Res.* (1955) **20**, 186.
107. Reymond, D., Phaff, H. J., *J. Food Sci.* (1965) **30**, 266.
108. Uchino, F., Kurono, Y., Doi, S., *Agr. Biol. Chem.* (1966) **30**, 1066.
109. Ayers, W. A., Papavizas, G. C., *Phytopathology* (1965) **55**, 249.
110. Ayers, W. A., Papavizas, G. C., Lumsden, R. D., *Phytopathology* (1969) **59**, 925.
111. Koller, A., Neukom, H., *Mitt. Geb. Lebensmittelunters. Hyg.* (1967) **58**, 512.
112. Koller, A., Neukom, H., *Eur. J. Biochem.* (1969) **7**, 485.
113. Mill, P. J., Tuttobello, R., *Biochem. J.* (1961) **79**, 57.
114. Rexová-Benková, L., *Collect. Czech. Chem. Commun.* (1967) **32**, 4504.
115. Rexová-Benková, L., Slezarik, A., *Collect. Czech. Chem. Commun.* (1966) **31**, 122.
116. Yamasaki, M., Yasui, T., Arima, K., *Agr. Biol. Chem.* (1966) **30**, 1119.
117. Barash, I., Khazzam, S., *Phytochemistry* (1970) **9**, 1189.
118. Endo, A., *Agr. Biol. Chem.* (1964) **28**, 535.
119. Endo, A., *Agr. Biol. Chem.* (1964) **28**, 543.
120. Endo, A., *Agr. Biol. Chem.* (1964) **28**, 551.
121. Kaji, A., Okada, T., *Arch. Biochem. Biophys.* (1969) **131**, 203.

122. Barash, I., Eyal, Z., *Phytopathology* (1970) **60**, 27.
123. Slezarik, A., Rexová-Benková, L., *Bologia* (Bratislava) (1967) **22**, 407.
124. Swinburne, T. R., Corden, M. E., *J. Gen. Microbiol.* (1969) **55**, 75.
125. McClendon, J. H., Kreisher, J. H., *Anal. Chem.* (1963) **5**, 295.
126. Bateman, D. F., *Neth. J. Plant Pathol.* (1968) **74** (Suppl. 1), 67.
127. Mussell, H. W., *Phytopathology* (1973) **63**, 62.
128. Bateman, D. F., *Proc., Conf. Gamagori, Japan* (1967) 58.
129. Sherwood, R. T., *Phytopathology* (1966) **56**, 279.
130. Endo, A., *Agr. Biol. Chem.* (1964) **28**, 639.
131. Keen, N. T., Horton, J. C., *Phytopathology* (1966) **56**, 603.
132. Reid, W. W., *Biochem. J.* (1952) **50**, 289.
133. Lyr, H., *Arch. Mikrobiol.* (1963) **45**, 198.
134. Tani, T., *Proc., Conf. Gamagori, Japan* (1967) 40.
135. Nasuno, S., Starr, M. P., *Phytopathology* (1966) **56**, 1414.
136. Nasuno, S., Starr, M. P., *J. Biol. Chem.* (1966) **241**, 5298.
137. Horikoshi, K., *Agr. Biol. Chem.* (1972) **36**, 285.
138. Hatanaka, C., Ozawa, J., *Agr. Biol. Chem.* (1964) **28**, 627.
139. Hatanaka, C., Ozawa, J., *Ber. Ohara Inst. Landwirt. Biol. Okayama Univ.* (1966) **13**, 161.
140. Mill, P. J., *Biochem. J.* (1966) **99**, 557.
141. Mill, P. J., *Biochem. J.* (1966) **99**, 562.
142. Hatanaka, C., Ozawa, J., *Ber. Ohara Inst. Landwirt. Biol. Okayama Univ.* (1966) **13**, 175.
143. Hatanaka, C., Ozawa, J., *Agr. Biol. Chem.* (1969) **33**, 116.
144. Hatanaka, C., Ozawa, J., *Nippon Nogei Kagaku Kaishi* (1969) **43**, 764.
145. Courtois, J. E., Percheron, F., Foglietti, M. J., *C. R. Acad. Sci. Paris, Serie D* (1968) **266**, 164.
146. McClendon, J. H., *Amer. J. Bot.* (1964) **51**, 628.
147. Hatanaka, C., Ozawa, J., *Nippon Nogei Kagaku Kaishi* (1969) **43**, 139.
148. Ozawa, C., *Nogaku Kenkyu.* (1955) **42**, 157.
149. Hatanaka, C., Ozawa, J., *Ber. Ohara Inst. Landwirt. Biol. Okayamu Univ.* (1969) **14**, 197.
150. Hatanaka, C., Ozawa, J., *Ber. Ohara Inst. Landwirt. Biol. Okayamu Univ.* (1969) **15**, 15.
151. Nagel, C. W., Anderson, M. M., *Arch. Biochem. Biophys.* (1965) **112**, 322.
152. Nagel, C. W., Wilson, T. M., *Appl. Microbiol.* (1970) **20**, 374.
153. Davé, B., Vaughn, R. H., *J. Bacteriol.* (1971) **108**, 166.
154. Moran, F., Nasuno, S., Starr, M. P., *Arch. Biochem. Biophys.* (1968) **123**, 298.
155. Mount, M. S., Bateman, D. F., Grant Basham, H., *Phytopathology* (1970) **60**, 924.
156. Garibaldi, A., Bateman, D. F., *Physiol. Plant Pathol.* (1971) **1**, 25.
157. Fuchs, A., *Antonie van Leeuwenhoek, J. Microbiol. Serol.* (1965) **31**, 323.
158. Nasuno, S., Starr, M. P., *Biochem. J.* (1967) **104**, 178.
159. Hancock, J. G., *Phytopathology* (1968) **58**, 62.
160. Mah, R. A., Hungate, R. E., *J. Protozool.* (1965) **2**, 131.
161. Wood, R. K. S., *Nature* (1951) **167**, 771.
162. Nagel, C. W., Vaughn, R. H., *J. Bacteriol.* (1962) **83**, 1.
163. Hasegawa, S., Nagel, C. W., *J. Biol. Chem.* (1962) **237**, 619.
164. McNicol, L., Baker, E., *Biochemistry* (1970) **9**, 1017.
165. Hatanaka, C., Ozawa, J., *Agr. Biol. Chem.* (1972) **36**, 2307.
166. Sheiman, M. I., Ph.D. Thesis, Rutgers University, 1974.
167. Albersheim, P., Killias, U., *Arch. Biochem. Biophys.* (1962) **97**, 107.
168. Albersheim, P., *Plant Physiol.* (1963) **38**, 426.
169. Bull, T., *Phytochemistry* (1968) **7**, 209.
170. Ishii, S., Yokotsuka, T., *J. Agr. Food Chem.* (1971) **19**, 958.

171. Ishii, S., Yokotsuka, T., *Agr. Biol. Chem.* (1972) **36**, 146.
172. Ishii, S., Yokotsuka, T., *J. Agr. Food Chem.* (1972) **20**, 787.
173. Ishii, S., Yokotsuka, T., *J. Agr. Food Chem.* (1973) **21**, 269.
174. Bush, D. A., Codner, R. C., *Phytochemistry* (1970) **9**, 87.
175. Spalding, D. H., Abdul-Baki, A. A., *Phytopathology* (1973) **63**, 213.
176. Voragen, A. G., Thesis, Landbouwhogeschool, Wageningen, Netherlands, 1972.
177. Bateman, D. F., *Phytopathology* (1966) **56**, 238.
178. Kapitonova, L. S., Rodionova, N. A., Feniksova, R. V., *Prikl. Biokhim. Mikrobiol.* (1972) **8**, 539.
179. Rodionova, N. A., Kapitnova, L. S., Feniksova, R. V., *Dokl. Akad. Nauk. SSSR* (1972) **207**, 466.
180. Fogarty, W. M., Ward, O. P., *Process Biochem.* (August, 1972) 13.
181. Kertesz, Z., *NY State Agr. Exp. Sta. (Geneva, NY)* (1931) Bull. **589.**
182. Willaman, J. J., Kertesz, Z. I., *NY State Agr. Exp. Sta. (Geneva, N. Y.)* (1931) Bull. **178.**
183. Mehlitz, *Biochem. Z.* (1930) **221**, 217.
184. Yamasaki, M., Yasui, T., Arima, K., *Agr. Biol. Chem.* (1964) **28**, 779.
185. Endo, A., *Agr. Biol. Chem.* (1965) **29**, 129.
186. Endo, A., *Agr. Biol. Chem.* (1965) **29**, 137.
187. Endo, A., *Agr. Biol. Chem.* (1965) **29**, 229.
188. Yamasaki, M., Kato, A., Chu, S., Arima, K., *Agr. Biol. Chem.* (1967) **31**, 552.
189. Okada, S., Inque, M., Fukumoto, J., *Nippon Nogei Kagaku Kaishi* (1969) **43**, 99.
190. Inque, M., Okada, S., Fukumoto, J., *Nippon Nogei Kagaku Kaishi* (1970) **44**, 1.
191. Inque, M., Okada, S., Fukumoto, J., *Nippon Nogei Kagaku Kaishi* (1970) **44**, 8.
192. Ishii, S., Yokotsuka, T., *J. Food Sci. Technol.* (1972) **19**, 151.
193. Ishii, S., Yokotsuka, T., *J. Agr. Food Chem.* (1973) **21**, 269.
194. Cruess, W. V., Besone, J., *J. Fruit Prod.* (1941) **20**, 365.
195. Boulin, J., Barthe, J. C., *Ind. Aliment. Agr. (Paris)* (1963) **80**, 116.
196. Berg, H. W., Marsh, G. L., *Food Technol.* (1956) **10**, 4.
197. Miller, L., Sheiman, M., Macmillan, J. D., *Bacteriol. Proc.* (1973) 148.
198. Wang, S. S., Vieth, W. R., *Biotechnol. Bioeng.* (1973) **15**, 93.
199. Goldstein, J., Swain, T., *Phytochemistry* (1965) **4**, 185.

RECEIVED November 5, 1973.

5

Lipolytic Enzymes

H. BROCKERHOFF[1]

Fisheries Research Board of Canada, Halifax Laboratory,
Halifax, Nova Scotia, Canada

A major group of lipolytic enzymes typified by pancreatic lipase consists of nonspecific esterases probably of the serine-histidine type. Another group, the phospholipases 2 of exocrine glands, are calcium activated and have exacting stereospecific substrate requirements. The enzymes of both groups hydrolyze water-insoluble esters. They must not only adsorb to oil–water or micelle–water interfaces but also must position their active sites toward the matrix (oil droplet, micelle, or membrane) in which the substrate molecules are imbedded, the "supersubstrate." It is postulated that lipolytic enzymes are hydrolases that have developed supersubstrate binding sites for attachment and orientation toward lipids. Such a binding site, which is topographically distinct from the substrate binding sites of the reactive center, may have hydrophobic or electrostatic character.

A large part of the living matter on earth consists of lipids, and lipolytic enzymes play an important role in their biological turnover. Lipids can be classified into four major groups: (1) neutral esters of glycerol, especially triglycerides, (2) cholesterol and its relatives, (3) phospholipids, and (4) glycolipids. (The fourth group is not considered in this review). Table I shows the classification of some major lipolytic enzymes according to these categories. The enzymes are important because none of the fatty acid esters—triglycerides, cholesterol esters, or phospholipids—can be used for energy generation or in other metabolic reactions without prior enzymatic hydrolysis. Furthermore, the lipids cannot pass through the biological food chains except after hydrolysis; animals must hydrolyze fats to digest them. This fact makes lipolytic

---

[1] Present address: N. Y. State Institute for Research in Mental Retardation, 1050 Forest Hill Rd., Staten Island, N. Y. 10314.

Table I. Groups of Lipids and Major Lipolytic Enzymes

Triglycerides:	*Cholesterol Esters*
Lipases:	Cholesterol esterase
Microbial	*Phospholipids*
Plants	Phospholipase 2
Milk	Phospholipase 1
Pancreatic	Lysophospholipase
Lipoprotein	
Hormone-sensitive	

enzymes interesting for the nutritionist. In addition, many practical problems in food processing involve lipolytic enzymes.

The enzymes that hydrolyze triglycerides are lipases (Table I). Many microorganisms such as molds and bacteria produce these enzymes. Lipases may cause spoilage of food but may also be beneficial. The aroma and flavor of Roquefort cheese, for example, is partly a result of the hydrolysis of milk fat by the lipase of a mold. The characteristic flavor of many Italian cheeses, on the other hand, is produced by pregastric lipase which is added to the milk as an extract from the calf stomach. Plant lipases are found in the seeds. They are needed during germination when the seed oils are degraded to provide energy for rapid growth. They cause hydrolysis of an oil during extraction, and this requires additional production steps.

Milk contains at least two lipases. One of them is activated by foaming which can be caused by air leaks in the pipelines that carry fresh, raw milk. The result is a rancid flavor caused by free short-chain fatty acids such as butyric acid.

Three lipases are important to man and higher animals:

(a) pancreatic lipase, which digests triglycerides to fatty acids and monoglycerides, which can then pass through the intestinal wall,

(b) lipoprotein lipases, which degrade the triglycerides that circulate in the blood; there are at least two different enzymes of this kind, one produced in the liver and one in other tissues (1),

(c) hormone-sensitive lipase, which is triggered by adrenalin to hydrolyze and mobilize the depot fat of adipose tissue.

Lipases can be divided into two groups according to their positional specificity. Some lipases hydrolyze only the ester bonds in positions 1 and 3 of glycerol, *i.e.*, the primary esters. This is true for pancreatic lipase (Table II). Other lipases, such as many microbial lipases, hydrolyze all three ester bonds. In this case, the primary esters are probably hydrolyzed faster than the secondary ester.

The best-known cholesterol esterase (Table I) is also a pancreatic enzyme. Cholesterol esterases not only hydrolyze cholesterol esters but also synthesize them from cholesterol and fatty acid in a reversible reaction. This is important in the digestion of cholesterol itself because cholesterol is esterified in the intestinal wall before it is incorporated into chylomicrons and delivered into the blood. Cholesterol esterase is a very unspecific enzyme which will also hydrolyze monoglycerides and water-soluble esters (*2*).

Many of the phospholipases, on the other hand, are very selective enzymes. The pancreatic phospholipase 2 (or phospholipase A_2) hydrolyzes only esters in position 2 of *sn*-3 phospholipids; *i.e.*, the enzyme is stereospecific (Table II). Phospholipases I (or A_1) attack only the 1-position. Lysophospholipases remove the remaining fatty acid, either in position 1 or 2, from the lysophospholipid. Phospholipases play a role in the post-mortem degradation of meat and fish.

In the remainder of the paper, I concentrate on the two best-known lipolytic enzymes—pancreatic lipase and phospholipase 2—and then speculate on the special nature of lipolytic reactions and how lipolytic enzymes differ from hydrolases.

Table II.   Positional Specificity of Pancreatic Enzymes

sn-position

Pancreatic Lipase

The most thoroughly investigated pancreatic lipase is from the pig. Most of our knowledge of the structure and properties of this enzyme comes from P. Desnuelle and his co-workers in Marseille (*3*) (Table III). The lipase occurs in two similar forms, isoenzymes Lipase A and Lipase B. There is no known proenzyme (zymogen). The enzyme does not react with diisopropylfluorophosphate (DFP), the standard inhibitor

Table III.   Molecular Properties of Porcine Pancreatic Lipase

Molecular weight 48,000
No proenzyme
Amino acid sequence: leu-*ser*-gly-his
Reactive residues: ser, his
Carbohydrates: mannose$_4$, N-acetylglucosamine$_3$

of esterases that have serine as a reactive amino acid. It does react with a similar inhibitor, diethyl-*p*-nitrophenyl phosphate (DNP) (4). Beside serine, a histidine is a reactive residue. The lipase is probably a serine–histidine enzyme like many other esterases and proteinases such as chymotrypsin or elastase. The amino acid sequence around the reactive serine is remarkable. Where other enzymes have glycine or an acidic residue such as aspartic acid, lipase has leucine. The total amino acid sequence of the lipases is not yet known. There are four or five disulfide bridges in the molecule and two free SH groups which are not involved in the catalytic mechanism (5). The carbohydrate residues are less than 2% of the molecule (6) and have no obvious function.

The substrate specificity of lipase has been investigated in my laboratory. An ester such as a glyceride is likely to be hydrolyzed by nucleophilic attack:

$$R-O-\overset{\overset{\textstyle O}{\|}}{\underset{\underset{\textstyle X}{\uparrow}}{C}}-R^1$$

The chemistry of such reactions is well understood. The attack is facilitated by withdrawal of electrons from the carbonyl carbon. This can be achieved by electrophilic substituents in R. The electron-withdrawing power is expressed as the Hammett constant (Table IV). If this constant is positive, the substituent is electron withdrawing relative to hydrogen (7). The inductive power increases with increasing electronegativity of the substituent and is quite strong in ester groups. When esters of different alcohols are hydrolyzed by pancreatic lipase (Table V) and the maximal reaction rates are compared the inductive effect of the substituent is also apparent (8). The reaction proceeds by nucleophilic attack. Triglycerides are good substrates for the lipase because each of their ester groups is activated by two other ester groups.

A second factor in nucleophilic reactions is steric hindrance. Table VI shows how the change from primary to secondary to tertiary ester influences the rate of alkaline hydrolysis of an ester (7). A parallel

Table IV.   Hammett Constants σ of Electrophillic Induction (7)

CH_3	-0.07	⬡ $+0.22$	$R-\overset{\overset{O}{\|\|}}{C}-O$ $+0.40$
OH	-0.002	F $+0.34$	CN $+0.68$
OCH_3	$+0.12$	Cl $+0.37$	$\overset{O}{\underset{O}{N}}-$ $+0.71$

relationship between steric hindrance and reaction rates is observed when different esters of oleic acid are hydrolyzed by lipase (Table VII) (8). Beta-monoglycerides (2-acyl-glycerols) are resistant to pancreatic lipase. This resistance and the corresponding specificity of the lipase for the primary positions can be explained by steric hindrance.

A third factor regulating the speed of lipolysis is the hydrophobicity of the ester. The normal substrates of lipase—the natural triglycerides—are insoluble in water, and the enzyme acts at the oil–water interface. Thus, lipases have been defined as esterases that act on insoluble substrates at such interfaces. However, triglycerides that are similar in steric and inductive effects may still react with different velocities even if all reactions take place at oil–water interfaces. For instance, emulsified tributyrin is hydrolyzed 20 times faster than emulsified triacetin. This difference is caused by the different hydrophobicity of the triglycerides not by the different chain lengths of butyric and acetic acid.

When comparing the reaction rates of a series of esters of similar structure (9) (Figure 1), it is seen that the esters of highest solubility—

Table V.   Relative Maximal Rates V of Hydrolysis of Oleic Acid Esters by Pancreatic Lipase, Compared with Triolein, V = 1.0 (7)

$HO-\bullet-\bullet-O-$	0.05	$R\overset{\overset{O}{\|\|}}{C}O-\bullet-\bullet-O-$	0.27	$Br-\bullet-\bullet-O-$	0.16
˅˄˅˄-O-	0.08	$F-\bullet-\bullet-O-$	0.30	$Cl-\bullet-\bullet-O-$	0.25
CH_3-O-	0.07	⬡$-\bullet-O-$	0.27	Cl⬡$-\bullet-O-$	1.0
˅˄˅˄O$-\bullet-\bullet-O-$	0.16	$NC-\bullet-O-$	0.80	NO_2⬡$-\bullet-O-$	>1.0

Table VI.   Relative Rates of Alkaline Hydrolysis of Methyl, Ethyl, Isopropyl, and *tert*-Butyl Acetate (7)

Table VII.   Relative Rates V of Hydrolysis of Sterically Hindered Oleic Acid Esters by Pancreatic Lipase, Compared with Triolein, V = 1.0 (8)

i.e., higher hydrophilicity—are very poor substrates although they are offered to the enzyme as emulsions and the reaction takes place at the interface. It appears then that the lipase prefers hydrophobic substrates.

The chain-length specificity of pancreatic lipase has been a subject of much controversy. Since lipase hydrolyzes only primary esters, it can be used to determine the distribution of fatty acids in positions 1 plus 3 as opposed to position 2 of triglycerides. If this method is applied to milk fat, *i.e.,* butter, the lipase hydrolyzes more butyric ester in relation to other acids than it should. A comparison of monoester sequences of different fatty acid chain lengths (Figure 2) (*10*) shows that the enzyme has some preference for butyric acid. A more interesting feature of Figure 2 is the activity shown at chain length 1 in the uppermost graph (the second cross refers to a cinnamoyl ester). It shows that fatty acid esters with no chain, namely formic esters, also make good substrates. We come to the paradoxical conclusion that an enzyme which is specific for a lipid need not engage in any lipophilic binding with its substrate, *i.e.,* hydrophobic binding of an aliphatic fatty acid chain. I discuss the resolution of this paradox later.

From the available information it can be concluded that the catalytic apparatus of lipases is similar to that of chymotrypsin (*11*) and other proteinases (Figure 3) (*12*). A serine oxygen attacks the substrate. The nucleophilicity of this oxygen is enhanced by hydrogen bonding of the hydroxyl proton to histidine. In contrast to the proteinases the lipase has no additional sites that would fix the substrate on the surface of the enzyme. However, there is evidence that the substrate has two-dimensional orientation on the enzyme (*13*). This is a result probably not of attachment to a binding site but of the necessity of the leaving alkoxy group—which is itself a strong nucleophile—to react with the proton between serine and histidine. This can only happen if the substrate is oriented as shown in Figure 3.

A more complete but more speculative picture of the reactive site of the lipase (Figure 4) (*12*) shows the hydrophobic leucine next to the reactive serine. I suggest that the leucine is not buried inside the enzyme but exposed on the surface and held in place by steric restriction or hydrophobic binding by another amino acid. The leucine could then contribute not only to the hydrophobicity of the reactive site but also to its sterically hindered nature. An aspartic acid residue in chymotrypsin assists in the activation of the nucleophilic serine hydroxyl through a charge relay system. Lipase may have a similar system, and aspartic acid is therefore included in the model (Figure 4).

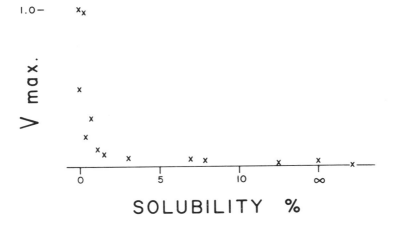

Figure 1.   Relative rates of hydrolysis of emulsions of esters of different solubilities by pancreatic lipase (9)

Microbial Lipases

Pancreatic lipase can be considered a model for all other lipases. Most if not all of these enzymes seem to be nonspecific carboxyl ester hydrolases of the serine histidine type. Their specificity consists, by definition, in their ability to hydrolyze insoluble substrates, but apart

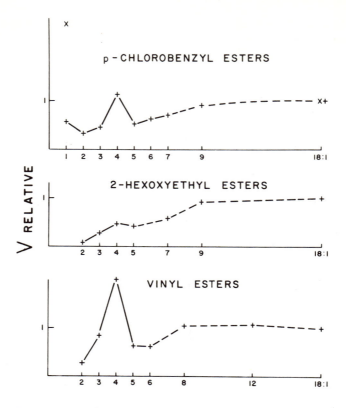

*Figure 2. Relative rates of hydrolysis of fatty acid esters
of different chain lengths by pancreatic lipase (10)*

from this they have no specific requirements for the structure of their substrates, except that they may be more or less sensitive to steric hindrance in the substrate. The particular nature of the fatty acid that is to be split or the structure of the rest of the molecule is important only as far as is steric, inductive, or hydrophobic properties are concerned.

The only known lipase that is nonspecific is the enzyme of a mold, *Geotrichum candidum*. It has a decided preference for fatty acids with a cis-Δ9 double bond such as oleic acid (*14*). There is no explanation for this specificity. The lipases of most other microorganisms are very similar to pancreatic lipase. For example, the lipase of the mold *Rhizopus arrhizus* is an exoenzyme which contains carbohydrates that are not essential for activity. It is not inhibited by DFP, it is specific for primary ester groups, and it does not distinguish between different fatty acids (*15*). Other microbial lipases are similar in these respects but may differ in their recognition of steric hindrance. The lipase of *Stapyhlococcus aureus* for example, hydrolyzes secondary as well as primary esters showing no positional specificity towards triglycerides (*16*).

Hormone-Sensitive Lipase

The mobilization of the depot fat of adipose tissues requires hydrolysis of the triglycerides. Adrenalin or other lipolytic hormones trigger a

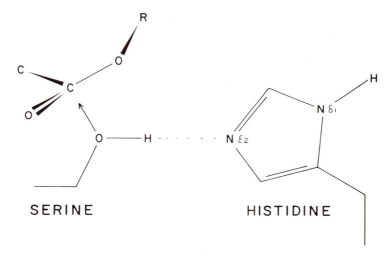

Chemistry and Physics of Lipids

Figure 3.   Hypothetical alignment of substrate and amino acids residues of the catalytic center of lipase (12)

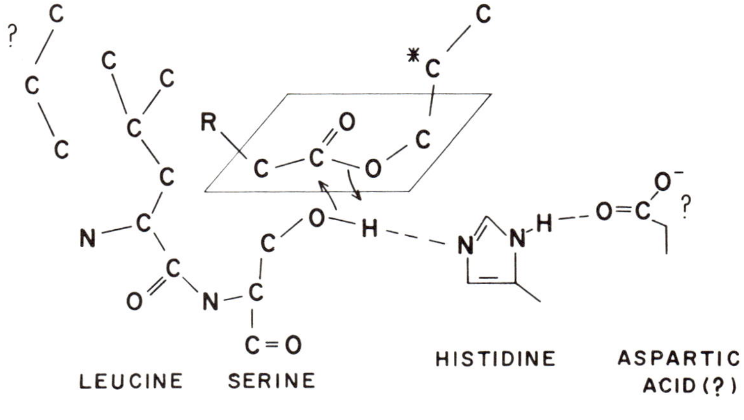

Chemistry and Physics of Lipids

Figure 4.   Hypothetical catalytic center of pancreatic lipase (12)

chain of reactions which ends in the activation of the hormone-sensitive lipase. This activation is probably a phosphorylation of an inactive form of the enzyme (17).

Lipoprotein Lipases

The lipoprotein lipases of higher animals hydrolyze the triglycerides of the blood which circulate in the form of lipoproteins, *i.e.*, multimolecular aggregates of triglycerides, cholesterol, phospholipid, and protein. The largest lipoproteins are the chylomicrons that appear in the blood after ingestion of fat. The lipoprotein lipases can be released from tissues by the administration of heparin (an acidic polysaccharide), either through injection or *in vitro* by incubation of the tissue with heparin. The lipoprotein lipases from extrahepatic tissues cannot hydrolyze triglycerides that are simply emulsified because they need activators that are present in blood. These activators have been identified as serum proteins, in particular, apolipoproteins, especially those with glutamic acids as a C-terminal residue (18). Phospholipids can amplify the activating effect if the primary activator is present. It is likely that the activators act on the substrate, not on the enzyme. They probably modify the matrix in which the substrate is imbedded, *e.g.*, the surface of chylomicrons, and make the substrate molecules available to the enzyme. The lipoprotein lipase of the liver is also released by heparin, but it does not need activation by serum proteins (19).

Milk Lipases

The milk lipase that is activated by foaming and causes the rancidity of milk is a glycoprotein or a family of glycoproteins. It is inhibited by DFP and is specific for primary ester bonds (14). The physiological function of the lipase is mysterious since new-born animals already possess their own digestive lipases. Milk also contains a lipoprotein lipase which has the properties typical for such an enzyme; it is sensitive to heparin and activated by serum proteins. This enzyme is probably serum lipoprotein lipase that has leaked into the milk (14).

Pancreatic Phospholipase

Our knowledge of porcine pancreatic phospholipase comes from the laboratory of G. H. de Haas and his co-workers. The enzyme is quite different from pancreatic lipase (Table VIII). Its molecular weight is quite small, it is a metalloenzyme that requires Ca^{2+} ion as a cofactor, and it is excreted from the pancreas as a proenzyme which is then activated by trypsin with removal of a heptapeptide. The molecule has six

Table VIII.   Molecular Properties of Pancreatic Phospholipase

Molecular weight 13,700
Cofactor: Ca^{2+}
Proenzyme: heptapeptide-enzyme
Amino acid sequence known
Six S—S bridges

S–S bridges and is very stable, even at 100°C. Figure 5 shows the complete primary structure of the proenzyme with the point of attack of trypsin at residue 7 and with the S-S bridges (*20*).

Like the lipase, the phospholipase hydrolyzes its substrates at an interface; in this case the interface of micelles and water. However pure phospholipid micelles are not digested, and even micelles in which the substrate molecules are spaced apart by inclusion of solvents such as ether or by cationic detergents are not attacked. Micelles containing anionic detergents such as bile salts are attacked. The enzyme requires a negative surface charge on the micelle even though the substrate molecule itself may be an electrically neutral lipid such as phosphatidyl choline.

The sensitivity against the surface charge of the aggregated substrate is a general property of phospholipases. The charge required by a phos-

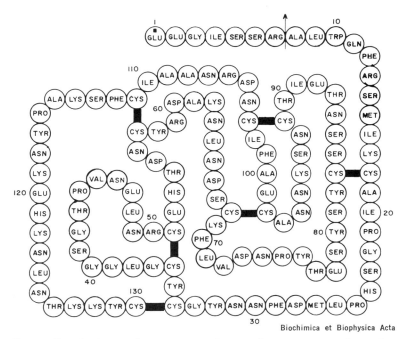

Biochimica et Biophysica Acta

*Figure 5.   Amino acid sequence and secondary structure of porcine
pancreatic prophospholipase* (*20*)

pholipase is independent of the charge of the substrate molecules that are hydrolyzed (21).

An important observation was made in the de Haas laboratory in Utrecht, where the activities of the pancreatic phospholipase and its proenzyme on the same substrate have been tested (22). The phospholipase hydrolyzes substrates that are part of a micelle, but the prophospholipase is completely inactive against micelles. The phospholipase can also hydrolyze short chain phospholipids that are in monomolecular solution although at much slower rates. The prophospholipase is almost as active as the enzyme against such substrates. It must be concluded that the reactive site of the phospholipase is already completely assembled and active in the proenzyme. Tryptic activation forms a new site that binds to micelles. This site is called the anchoring site or recognition site (22).

Other Phospholipases

Phospholipases 2 are found in high concentrations in the venom of reptiles and invertebrates. These enzymes are probably similar to the pancreatic phospholipase 2 in structure and catalytic mechanism. They break down cellular membranes in the victim's tissues by hydrolysis of the phospholipids and thus assist in spreading the toxin.

Phospholipases are found in most cell compartments. Mitochondria have a heat stable and Ca^{2+}-dependent phospholipase 2 in their outer membranes (23). Phospholipases 1 and 2 occur in microsomal and lysosomal cell fractions. Very little is known about these enzymes. The lysosomal enzymes may be responsible for the post-mortem degradation of phospholipids in meat and fish.

Although bacteria may contain phospholipase 2, phospholipase 1 is more common. This enzyme has been isolated from the outer membrane of *E. coli* (24). It proved to depend on Ca^{2+} and be very stereospecific since only the ester bond in position 1 of the natural *sn*-3 phospholipids was hydrolyzed.

Enzymes that hydrolyze lysophospholipids are found in nearly all tissues and organisms. They seem to be non-specific esterases of the serine-histidine type (25) and hardly deserve the name lysophospholipase because they also hydrolyze esters other than phospholipids. They should probably be considered together with such enzymes as cholesterol esterases and monoglyceride lipases as amphiphilic carboxyl ester hydrolases. These non-specific esterases have a preference for amphiphilic (hydrophilic-lipophilic) substrates. Such an enzyme may perhaps hydrolyze lysophospholipis, monoglycerides, diglycerides, and cholesterol esters.

Enzymic Reactions at Interfaces: Substrate and Supersubstrate

The characteristic that sets lipolytic enzymes apart from other hydrolases is their ability to hydrolyze insoluble substrates at lipid–water interfaces at very high rates. These substrates are imbedded in the matrix of a surface or a membrane that is much larger than the enzyme. They cannot leave this matrix and can only move laterally in the surface. I shall refer to the matrix as the "supersubstrate." Much confusion concerning enzymic interfacial reactions has resulted from the ambiguous use of the word "substrate." It has not only been used for the individual substrate molecule but also for the oil droplet or micelle of which this molecule is a part. This confusion is the cause of the paradox concerning lipase: the enzyme binds to its "substrate," the oil, as it obviously must, but there is no lipophilic binding in the enzyme–substrate complex. If we call the substrate molecule "substrate" and call the matrix in which the molecule is embedded "supersubstrate," the paradox is easily resolved. The enzyme binds hydrophobically to the supersubstrate but not to the substrate. Figure 6 shows how this clarification can be translated into a

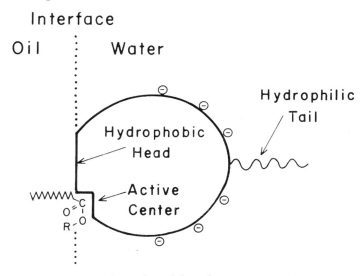

Figure 6.   *Hypothetical model of the orientation of pancreatic lipase at the oil–water interface*

model of the pancreatic lipase at the interface. The enzyme has a supersubstrate binding site, the hydrophobic head, which is separate from the reactive site. Both sites must be near each other because the hydrophobic head serves not only to attach the enzyme to the supersubstrate but also to bring the reactive site into reach of the substrate molecules which cannot leave the supersubstrate. Hydrophilic amino acids and the carbohydrate residues of the lipase may provide a tail for the enzyme for further stabilization of the correct orientation.

Phospholipases must have a similar mechanism of orientation at the interface because their substrates are also immobilized (except for lateral movements) in a supersubstrate structure. The enzyme must attach itself to the supersubstrate in the correct orientation in order to place the reactive center toward the substrate (Figure 7). In pancreatic phospholipase and many other phospholipases the binding to the supersubstrate may be electrostatic rather than hydrophobic so that the enzyme has an electrostatic head. Supersubstrate binding sites of mixed type or hydrogen bonding sites may also exist.

The definition of lipolytic enzymes as esterases that act at interfaces (26) can now be further specified. Lipolytic enzymes are hydrolases that have developed special lipophilic supersubstrate binding sites which enable them to align their reactive centers towards their substrates.

The distinction between substrate binding and supersubstrate binding as it has developed from studies on lipolytic enzymes has far-reaching implications. Many of the substrates of interacellular metabolism are partially immobilized, i.e., they are parts of biological membranes. The enzymes acting on such substrates are much larger than the substrate molecules, and the reactive center of the enzyme occupies only a small part of the surface of the enzymic protein. The rest of the protein must interact with the environment and in particular with the matrix in which

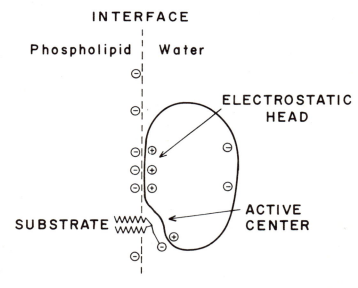

Figure 7.   Hypothetical model of a phospholipase at the phospholipid–water interface

the substrate is embedded. It is not sufficient that enzyme and substrate are ready to react; they must be able to come together. Even though the substrate molecule may be in the correct ionic state as required for the reaction, if the enzyme has an electronegative head and the supersubstrate has a net negative charge, no approach between substrate and reactive site can take place. I suggest that supersubstrate binding sites and orientation mechanisms must be very common among intracellular enzymes and that they play a decisive regulatory role in intracellular metabolism.

Literature Cited

1. La Rosa, J. C., Levy, R. I., Windmueller, H. G., Fredrickson, D. S., *J. Clin. Invest.* (1970) **49**, 55a.
2. Erlanson, C., Borgström, B., *Scand. J. Gastroenterol.* (1970) **5**, 395.
3. Desnuelle, P., in "The Enzymes," P. D. Boyer, Ed., 3rd ed., Vol. VII, p. 575, Academic, New York and London, 1972.
4. Maylié, M. F., Charles, M., Desnuelle, P., *Biochim. Biophys. Acta* (1972) **270**, 162.
5. Verger, R., Sarda, L., Desnuelle, P., *Biochim. Biophys. Acta* (1971) **242**, 580.
6. Garner, C. W., Jr., Smith, L. C., *J. Biol. Chem.* (1972) **247**, 561.
7. Hine, J., "Physical Organic Chemistry," McGraw-Hill, New York, 1956.
8. Brockerhoff, H., *Biochim. Biophys. Acta* (1968) **159**, 296.
9. Brockerhoff, H., *Arch. Biochem. Biophys.* (1969) **134**, 366.
10. Brockerhoff, H., *Lipids* (1971) **6**, 942.
11. Blow, D. M., Birktoft, J. J., Hartley, B. S., *Nature* (1969) **221**, 337.
12. Brockerhoff, H., *Chem. Phys. Lipids* (1973) **10**, 215.
13. Brockerhoff, H., *Biochim. Biophys. Acta* (1970) **212**, 92.
14. Jensen, R. G., *Progr. Chem. Fats Other Lipids* (1971) **11**, 347.
15. Benzonana, G., *Lipids*, in press.
16. Alford, J. A., Pierce, D. A., Suggs, F. G., *J. Lipid Res.* (1964) **5**, 390.
17. Steinberg, D., in "Pharmacological Control of Lipid Metabolism," W. L. Holmes, R. Paoletti, D. Kritchevsky, Eds., p. 77, Plenum, New York, 1972.
18. Fredrickson, D. S., Levy, R. I., in "The Metabolic Basis of Inherited Disease," J. B. Stanbury, J. B. Wyngaarden, D. S. Fredrickson, Eds., 3rd ed., p. 545, McGraw-Hill, New York, 1972.
19. Assmann, G., Krauss, R. M., Fredrickson, D. S., Levy, R. I., *J. Biol. Chem.* (1973) **248**, 1992.
20. De Haas, G. H., Slotboom, A. J., Bonsen, P. P. M., Nieuwenhuizen, W., van Deenen, L. L. M., Maroux, S., Dlouha, V., Desnuelle, P., *Biochim. Biophys. Acta* (1970) **221**, 54.
21. Dawson, R. M. C., *Colloq. Ges. Physiol. Chem.* (1965) **16**, 29.
22. Pieterson, W. A., Vidal, J. C., Volwerk, J. J., de Haas, G. H., unpublished data.
23. Scherphof, G. L., Waite, M., van Deenen, L. L. M., *Biochim. Biophys. Acta* (1966) **125**, 406.
24. Scandella, C. J., Kornberg, A., *Biochemistry* (1971) **10**, 4447.
25. Van den Bosch, H., Aarsman, A. J., de Jong, J. G. N., van Deenen, L. L. M., *Biochim. Biophys. Acta* (1973) **296**, 94.
26. Desnuelle, P., *Adv. Enzymol.* (1961) **23**, 129.

RECEIVED September 17, 1973.

6

Structure, Function, and Evolution of Acid Proteases

THEO HOFMANN

Department of Biochemistry, University of Toronto,
Toronto, Canada. M5S 1A8

Acid proteases are inactivated by active-site specific re-
agents, diazoacetylnorleucine ethyl ester and other diazo
compounds, and epoxy (p-nitrophenoxy)propane. Cova-
lently labelled aspartic acid peptides have been isolated
from pepsin, chymosin (= rennin), and penicillopepsin. The
peptides labelled with the diazo compounds have similar
sequences and differ from the epoxy (p-nitrophenoxy)pro-
pane labelled peptides. These results indicate two aspartic
acids at the active site and suggest homology between the
enzymes. The latter is confirmed by a comparison of the
sequence data. Studies of the action of porcine pepsin and
penicillopepsin on some dipeptides with free N-terminal
groups show transpeptidation involving a covalent acyl
intermediate. It is proposed that there are differences in
the mechanism of action of pepsin which are determined
by the nature of the substrate.

From a historical perspective, the acid proteases are without a doubt the most interesting group of enzymes. This claim is easily justified since one member of the group—chymosin, in the form of rennet—has been used as an isolated, although impure, enzyme preparation for thousands of years and was the first enzyme to be isolated by man. (The name "chymosin" will be used for the pure enzyme as suggested by Foltmann (*1*) instead of "rennin." "Rennin" is often confused with "renin," a proteolytic enzyme from the kidney. "Chymosin" is a name originally suggested over 100 years ago for the milk-clotting principle. "Rennet" will be retained to describe the calf-stomach extract used for cheese making.) According to a number of sources, the first use of chymosin in cheese making goes back 8000 or 9000 years. References to cheese as an important food are found on Sumerian cuneiform tablets dating from 4000 B.C. From then on frequent references appear in Egyptian, Judaen, and Greek writings. Many legends tell how travellers accidentally discovered rennet when carrying milk in a sheep stomach. After a hot journey, a pleasantly sweet tasting coagulated product was formed. Acid proteases have been used for a long time in Asiatic countries as part of fungal fermentation processes. The use of fermented products of the soy bean is particularly well known. Soy beans were used as food in Eastern Asia long before written records existed and were first mentioned in a book by Emperor Shang-Nung (2) of China in 2838 B.C. When the use of fermented products developed is not certain, but Komiya (3) suggests that shoyu (soy sauce) originated during the Chau dynasty (1134-246 B.C.).

Pepsin, another member of the family of acid proteases was the object of the first scientific enzymatic study. In 1783 Spallanzani [cited in (4)] conducted experiments on digestion mechanism. He questioned the prevalent theory that food was mechanically ground down inside the body to a state where it could be absorbed. He fed lumps of meat encased in metal cages to hawks. When the indigestible cages were regurgitated empty, it proved that the meat had been digested by means other than mechanical grinding. Subsequently he showed that a cell-free solution of gastric juice contained an agent which could digest meat completely. In modern scientific investigations pepsin is also of interest. Although it was the second protein to be crystallized (5), it was the first enzyme whose crystals were examined by x-ray crystallography by Bernal (6). The x-ray diffraction patterns resembled those of small molecule crystals and showed that proteins have regular structures when crystalline.

It is surprising that, as of this date, there are not completed x-ray crystallographic analyses of any of the acid proteases. At least three are at an advanced stage of investigation while work on others has been started. These enzymes are pepsin (7), chymosin(8), penicillopepsin (9), *Rhizopus* acid protease (10), *Endothia parasitica* acid protease (11), and Mucor-rennin (12). Furthermore, at the time of writing no complete amino acid sequences have been published. Tang's laboratory has, however, completed the sequence of porcine pepsin (13), and Moravek and Kostka (14) have published a nearly complete sequence. Large segments of the sequences of chymosin (15), and penicillopepsin (16, 17, and unpublished) are also known.

Many acid proteases, however, have been isolated, especially in recent years, and have been obtained in highly purified form. They are listed in Table I along with their occurrence, function where known, and possible use.

Table I.   Acid

Name	*Occurrence*
Vertebrates	
pepsins and gastricsin	vertebrate stomachs
chymosin = rennin	abomasum (young ruminants)
Cathepsins, D, E	many vertebrate organs
Fungi	
Penicillopepsin	extracellular, *P. janthinellum*
Aspergillopeptidase A	extracellular, *A. saitoi*
Aspergillus oryzae acid	
protease[a]	extracellular, *A. oryzae*
Rhizopus-pepsin	extracellular, *R. chinensis*
Mucor rennins	extracellular, *Mucor pusillus*
	extracellular, *Mucor miehei*
Endothia acid protease	extracellular, *Endothia parasitica*
Other Plants	
Lotus acid protease	seeds
Nepenthes acid protease	pitcher plant, secretion
Protozoan	
Tetrahymena	intracellular, *T. pyriformis*

[a] Davison, R., Hofmann, T., Gertler, A., unpublished.

Acid proteases have been found in other organisms and have been the subject of recent reviews (*21, 29*). The widespread occurrence of these enzymes is surprising since they function optimally at pH values which may be 5 to 6 units below a "physiological" pH. The necessity for a low pH optimum for pepsin and chymosin is understandable, but why do cathepsins require a low pH for optimal activity? Their role is at present unclear, but at least one enzyme—an acidic cathepsin from polymorphonuclear leucocytes—has been implicated in responses to inflammatory processes (*30*). It is not clear why an acidic enzyme would be required in such a situation unless local pockets of low pH are created intracellularly.

Primitive fungi do occur in acidic media. Penicillopepsin is formed by *Penicillium janthinellum* if the pH of the medium remains below 4.5 (*31*). On a synthetic medium the enzyme appears only at a development stage where mycelial growth has ceased and sporulation has started (*32*). The exact role in the sporulation process is, however, still obscure. No information is available about a possible role in a physiological environment.

Acid Proteases in Food Processing

Acid proteases play an important role in two major areas of food processing—cheese making and the production of fermented foods by molds from soy beans, rice, and other cereals.

Proteases

Function	Use	References
digestion	—	(18)
digestion	cheese making	(1, 19)
?	—	(20)
in sporulation	—	(21)
?	soya fermentation	(22)
		a
?		(21)
?	cheese making	(23)
?	cheese making	(24)
?	cheese making	(25)
?	—	(26)
digestion	—	(27)
digestion	—	(28)

Chymosin. Chymosin is the essential milk-clotting component of rennet, the crude extract of the abomasum of calves. Rennet is used in the production of several hundred cheeses in almost all parts of the world.

Chymosin is secreted by the abomasum in the form of an intially inactive precursor, prochymosin, which is converted autocatalytically to chymosin (*see* below, p. 176). Many reviews on the properties of the zymogen and the enzyme have been published. One of the most recent, by Foltmann (1), also contains references to earlier reviews. The action of chymosin on κ-casein is primarily responsible for the milk-clotting process. The numerous studies of this reaction have been reviewed recently by Mackinlay and Wake (32). Because of the number of different milk proteins and the complexity of their interactions, especially in the casein micelle, the full details of the coagulation process are not yet understood.

Two phases of the clotting process, however, can be distinguished readily (33). In the first phase a low-molecular weight peptide material is released (34) from the κ-casein fraction (35). It is a glycopeptide and represents the C-terminal 64 amino acids of this protein in the case of the bovine species (36). This peptide is produced by a chymosin-catalyzed cleavage of the peptide bond between methionine-105 and phenylalanine-106. In the second phase the casein-complex becomes insoluble and coagulates because of destabilization of the micelles. Calcium ions are necessary for coagulation (32). The cleavage by chymosin, limited as it is to the Met-Phe bond, is not caused by a high specificity of chymo-

Table II.   Food Fermentations Involving Organisms

Product	Organism
Shoyu (Chiang-yu, or soy sauce)	*Aspergillus oryzae* (or *Asp. soyae*) *lactobacillus* and yeast
Miso (Chiang)	*Aspergillus oryzae* *Saccharomyces rouxii*
Ketjap	*Aspergillus oryzae*
Hamanatto	*Aspergillus oryzae*
Tempeh	*Rhizopus oligosporus* and other *Rhizopus* species
Sufu (beancake)	mainly *Actinomucor elegans* (Also other *Mucor* species)
Ragi	*Mucor* and *Rhizopus*

sin but by the fact that this bond is probably particularly labile and exposed on the surface of the micelle. This would explain why many different proteolytic enzymes have pronounced milk-clotting ability. It has been shown that pepsin and chymotrypsin degrade κ-casein in essentially the same way as chymosin (32). Another factor which supports the concept of an exceptionally labile bond is the fact that chymosin action in milk occurs rapidly at pH 6.6-6.8. As a general proteolytic enzyme, chymosin has a pH optimum of 3.5-4.0 (with denatured hemoglobin as substrate) and shows virtually no activity at pH 5.0 (37). This also explains why chymosin exerts little if any proteolytic action on the milk clot (cheese).

Rennet Substitutes. Because of the ever increasing shortage of calf-rennet created by the increasing demand for both beef and cheese, many laboratories have been searching for substitutes. Some 800 strains of fungi and microorganisms have been investigated in the last 10 years (23). Enzyme preparations from at least three organisms are used now in many countries for the manufacture of a variety of cheeses. The preparations are "Sure-Curd" from *Endothia parasitica* (Charles Pfizer and Co.) (25, 38), "Rennilase" from *Mucor Miehei* (Novo Industri A/S) (39, 41), and Mucor-Rennin from *Mucor pusillus* (23). Although many enzyme preparations can clot milk, most are not suitable for cheese production because their proteolytic activity does not end with the cleavage of the glycopeptide from κ-casein. Further degradation of caseins occurs during cheese maturation and leads to products with unacceptable texture or bitter tastes.

It is interesting to know that the plant yellow bedstraw (*Galium verum*) has been used formerly in Britain to curdle milk although it never became a useful rennet substitute.

Acid Proteases in Fungal Fermentations. Soy beans and cereals are fermented by a variety of organisms to give different products used

Which Produce Acid Proteases (42)

Starting Material	*Major users*
soybeans and wheat	China, Japan, Philippines
soybeans sometimes with rice and other cereals.	China, Japan
black soybeans	Indonesia
soybeans	Japan
soybeans	Indonesia
soybeans	China, Taiwan
rice	Indonesia, China

mainly in East Asia. The most important of these products are listed in Table II (*see* Table I for purified enzymes). Although many enzymes play a role in fermentation, acid proteases are undoubtedly among the most important. The initial stages of the fermentation usually involve incubation of suspensions of crushed and ground soy beans and/or cereals with cultures of fungi, such as Aspergilli, Rhizopus, and others. During this stage these organisms produce acid proteases which break down the proteins in the substrates. Cell-free enzyme preparations have not been used to any extent for the digestion of soy beans, but some attempts have been made to produce "chemical shoyu" by acid hydrolysis (42).

Acid proteases probably also play a role in the breakdown of cheese proteins by species of *Penicillia* used to produce blue cheeses (Roquefort, Stilton, Danish Blue) and soft cheeses (Camembert, Brie, etc.). The curds are inoculated with spore preparations of the appropriate mold. The growing mold then converts the curd into the desired cheese through the action of different enzymes.

General Properties of Acid Proteases

pH Optimum. In 1960, Hartley (43) proposed that proteolytic enzymes be divided into four classes. Three of these—serine proteases, metallo-proteases, and sulfhydryl proteases— were defined according to the nature of a specific group at the catalytic site. The fourth class was acid proteases, the endopeptidases with a pH optimum in the acidic pH range of 1.5 to 5.0. The optimum pH is not only characteristic of the enzyme but is also greatly influenced by the substrate and, in the case of proteins, the conformational state. Thus, pepsin acts optimally on hemoglobin and serum albumin near pH 2, but after the protein substrates have been denatured, the optimum shifts to about pH 3.5 (44). The ionic strength of the medium may also affect the optimum pH values. All the enzymes listed in Table I have pH optima in the acid range.

Molecular Weight. Detailed information on the molecular weights of pepsin from several species, chymosin, and fungal enzymes is given by Fruton (46) and Matsubara and Feder (29). They are in the range of 30,000-40,000 for the mammalian gastric proteins and microbial enzymes. A protease from *Saccharomyces cerevisiae* with a molecular weight of 60,00 appears to be an exception (48). However, a wide range of molecular weights for acid cathepsins has been reported. The molecular weight of thyroid acid protease is 47,000-52,000 according to Kress *et al.* (48) while McQuillan and Trikojus (49) report a value of 21,000. Spleen cathepsin D has a molecular weight of 50,000-58,000 (20). The only highly purified plant enzyme, lotus seed acid protease, has one of about 36,000 (26).

Table III.   Acid Proteases Inhibited by Diazoacetyl Norleucine Methyl Ester (DAN)

Enzyme	*DAN-Labelled Peptide Isolated*	*Reference*
Pepsin, porcine	Ile-Val-Asp[*][a]-Thr-Gly-Thr-Ser	(50, 51)
Pepsin C, porcine	Ile-Val-Asp[*][a]-Thr	(52)
Pepsin, bovine	Ile-Val-Asp[*][a]-Thr-Gly-Thr-Ser	(53)
Gastricsin, human[b]		(54)
Chymosin, bovine	Asp[*][a]-Thr-Gly-Thr-Ser-Leu[e]	(55)[c]
Cathepsin D, spleen		(56, 57)[d]
Cathepsin E, bone marrow		(58)
Thyroid acid protease		(59)
Penicillopepsin (*Penicillium janthinellum*)	Ile-Ala-Asp[*][a]-Thr-Gly-Thr-Thr-Leu	(60)
Awamorin (*Aspergillus awamori*)	Ile-Ala-Asp[*][a]	(61)[d]
Rhizopus-pepsin (*Rhizopus chinensis*)		(17, 55)
Aspergillopeptidase A (*Aspergillus saitoi*)		(55)
Mucor-rennin (*Mucor pusillus*)		(56)
Endothia parasitica acid protease		[c]
Rhodotorula glutinis acid protease		(62)

[a] The asterisk (*) indicates the labelled residue, aspartic acid.
[b] Diazoacetylglycine ethyl ester was used as inhibitor.
[c] Mains, G., Hofmann, T., unpublished observation.
[d] Diazoacetyl-2,4-dinitrophenylethylene diamine was used as inhibitor.
[e] Chang, W.-J., personal communication.

Amino Acid Compositions. Amino acid compositions of mammalian pepsinogen and pepsin, gastricsin, and prochymosin and chymosin have been collected by Fruton (46). Those of five fungal enzymes have been collected by Matsubara and Feder (29).

The amino acid compositions vary widely and do not indicate any relationship among the enzymes. However, there is one unusual feature. In four proteins the basic amino acid content is exceptionally low. Human gastricsin and human pepsin have no lysine and contain only one histidine and three arginine residues per molecule. Porcine pepsin has one lysine, one histidine, and two arginine residues. Penicillopepsin has five lysines, three histidines, and no arginine. The significance of this unusual feature, which is not shared by other acid proteases, is not clear.

Reaction with Diazo Reagents. The classification of the acid proteases on the basis of their pH optima as suggested by Hartley (43) was useful in the absence of other information on the nature of active site residues. Since Hartley's proposal, however, it has been discovered—first with porcine pepsin—that in the presence of Cu^{2+} ions these enzymes can be specifically inactivated with diazotized dipeptide esters. An ester linkage is formed between one specific aspartic acid side chain and the inhibitor (50).

The most commonly used inhibitor is diazoacetylnorleucine methyl ester (DAN) because it allows the determination of the label incorporation by amino acid analysis after complete hydrolysis of the protein. (Abbreviations used are: DAN, diazoacetylnorleucine methyl ester; EPNP, 1,2-epoxy 3-(nitrophenoxy)propane; Z-, carbobenzoxy; Phe(NO)$_2$, p-nitrophenylalanyl.) The reaction proceeds as follows:

$$-CH_2COOH \ + \ N_2CH_2-CONH-\overset{\displaystyle CH_3}{\underset{\displaystyle (CH_2)_3}{\overset{\displaystyle |}{\underset{\displaystyle |}{CH}}}}-COOCH_3 \ \longrightarrow$$

$$-CH_2COOCH_2-CONH-\overset{\displaystyle CH_3}{\underset{\displaystyle (CH_2)_3}{\overset{\displaystyle |}{\underset{\displaystyle |}{CH}}}}-COOCH_3$$

Enzymes which have been specifically inhibited by DAN are listed in Table III. Since the active site aspartic acid of the enzymes reacts specifically with DAN and since only active but not denatured forms of the enzyme react, it is reasonable to conclude that the reaction is caused by the unique nature of the active site and reflects an aspect of the mechanism. However, the relation between the DAN reaction and the catalytic mechanism is still obscure.

This contrasts with the serine proteases. These enzymes are inhibited by diisopropylfluorophosphonate which reacts with the active site serine. The inhibitor is considered a pseudo-substrate which forms a stable acyl-intermediate which cannot deacylate (63). The reaction of DAN with acid proteases is unlikely to be analogous since DAN does not resemble a substrate. Nevertheless, the specificity of the DAN reaction suggests that these enzymes share a common mechanism. Furthermore, a comparison of the aspartic acid peptide sequences which show five identical and two very similar residues indicates evolutionary homology, at least between penicillopepsin and the mammalian pepsins. This makes it probable that the acid proteases belong to one family of enzymes related through a common ancestor although it is quite possible that not all the enzymes classified as acid proteases will show this relationship. The partial sequences compared in the section on page 173 provide much stronger evidence for homology between penicillopepsin and the mammalian enzymes.

Specificity of Acid Proteases

Traditionally the specificity of proteases has been defined in terms of the amino acid side chains surrounding the peptide bond which is being cleaved. This is illustrated by the exclusive specificity of trypsin, thrombin, and plasmin for bonds involving the carboxyl group of arginine and lysine residues (64–66), by thermolysin which has a preference for bonds involving the amino group of hydrophobic amino acid residues with bulky side chains (67), and by other enzymes. However, amino acids not involved with the cleaved bond can make substantial contributions to binding and to catalytic efficiency. The work of Berger's laboratory on papain (68), carboxypeptidase A (69), and elastase (70), as well as evidence from many other laboratories show the necessity for considering the specificity and action of proteolytic enzymes, not only in terms of the amino acids involved in the bond that is cleaved, but also in terms of secondary effects. This becomes particularly evident with pepsin and other acid proteases.

Early studies on pepsin going back to 1938 (71) showed a preference for aromatic amino acids contributing the NH-group to the sensitive bond. Subsequently Baker (72) showed that bonds with aromatic side chains on both sides of the sensitive bond are preferred in substituted dipeptides. However, cleavage of even the best small substrates is slow and not comparable with the hydrolysis rate of proteins and polypeptides. In contrast to studies with small peptides, the use of pepsin in sequence work indicated that it had a broad side-chain specificity. In order to obtain a clearer picture, Fruton initiated studies on the action of pepsin on a large number of synthetic peptides with two to seven amino acids and a variety of blocking groups [*see* Ref. 73 for a detailed review]. The most important finding was that increases in the peptide chain length caused large increases in the hydrolysis rate. The increases were almost entirely caused by increases in k_{cat} while K_M was not much affected (k_{cat} represents the overall rate constant of the hydrolytic reaction and was calculated from the maximum velocity by dividing V_M by the molar enzyme concentration). Values for k_{cat} for four representative substrates are given in Table IV. The best substrate found so far is Z-Ala-Ala-Phe-

Table IV.   Kinetic Constants for Various Substrates at 37° and pH 3.4–3.5

Substrate	Porcine Pepsin		Penicillopepsin	
	K_M (*mM*)	$k_{cat}(sec^{-1})$	K_M (*mM*)	$k_{cat}(sec^{-1})$
Z-Glu-Tyr (at 31.6°, pH 4.0)	1.89[a]	0.001[a]	—	not detected[b]
Ac-Phe-Tyr	4.4[c]	0.016[c]		
Z-His-Phe-Trp ethyl ester	0.23[c]	0.51[c]	—	not detected[b]
Z-Gly-Gly-Phe-Phe-3-(4-pyridyl)-propyl ester	0.42[c]	71.8[c]	1.2	0.02[b]
Serum albumin	0.038	0.53[b]	0.067	5.05[b]

[a] (75).
[b] (76).
[c] (73).

Phe-3-(4-pyridyl)propyl ester which has K_M = 0.04 mM and k_{cat} = 260 sec^{-1} (77). The large differences in the catalytic parameters are interpreted best in terms of a secondary binding site. Binding of the extended peptide chain probably causes conformational changes at the catalytic site which lead to an increase in catalytic efficiency (73). It has been suggested that the secondary binding site has hydrophobic character (73). So far no direct evidence for a conformational change associated with binding in the secondary binding site has been obtained. It is interesting to note that binding of the best dipeptide substrate, acetylphenylalanyl-3,5-diiodotyrosine, causes a marked change in the 260-290 nm region of the circular dichroism spectrum (74).

As far as the fungal proteases are concerned, studies in Fruton's laboratory (75) and our own laboratory (76) have provided evidence that secondary binding sites are equally or more important for increased

Table V.   Number of Peptide Bond Cleavages
Other Proteases of Low

	Cleavages by	
	Penicillopepsin and Enzyme in Column 1	*Penicillopepsin, not by Enzyme in Column 1*
Acidic proteases		
Pepsin	10	6
Pepsin C	5	11
Chymosin	9	7
Gastricsin[c]	5	6
Spleen cathepsin D	5	11
Adrenal cathepsin D	6	10
Thyroid acid protease	6	10
Rhizopus chinensis	11	5
Mucor miehei	5	11
Mucor pusillus	8	8
Endothia parasitica	7	9
Other proteases of low specificity		
Subtilisin	5	11
Subtilisin[c]	5	6
Pseudomonas alkaline protease	8	8
Aspergillus flavus Alkaline protease	7	9
Serratia sp. E15 protease	8	8
Aspergillus oryzae Alkaline protease	9	7
Streptomyces griseus protease A	7	9
Other proteases with restricted specificity		
Pancreatic elastase	4	12
Ficin	4	12
Myxobacter 495 α-Lytic protease	2	14
Myxobacter AL-1 protease	2	14
Aspergillopeptidase C	2	14
Chymotrypsin	2	14
Thermolysin	6	10

[a] (74).
[b] With oxidized β-chain of insulin, as substrate, except where noted.

catalytic efficiency with these enzymes than with porcine pepsin. Table IV shows that N-substituted dipeptide substrates are not hydrolyzed by penicillopepsin, that Z-Gly-Gly-Phe-Phe-3-(4-pyridyl)propyl ester is hydrolyzed much more slowly by penicillopepsin than by pepsin but that the initial rate of hydrolysis of a protein substrate, bovine serum albumin, is higher than that by pepsin. Note, however, that some dipep-

**Catalyzed by Acid Proteases Compared with
Side Chain Specificity**[a, b]

	Cleavages by

Enzyme in Column 1 not by Penicillopepsin	*Ref.*
0	(79,80)
0	(81)
0	(82)
1	(79)
0	(83)
0	(84)
0	(81)
1	(86)
0	(83)
0	(84)
1	(89)
3	(90)
5	(90)
6	(91)
4	(92)
4	(93)
5	(94)
6	(95)
3	(94)
2	(95)
4	(96)
1	(97)
3	(98)
1	(99)
0	(100)

[c]With glucagon, as substrate.

tides with free N-terminal groups are hydrolyzed by penicillopepsin (*see* p. 164). Other fungal acid proteases also fail to hydrolyze dipeptide substrates [aspergillopeptidase A (22) and *Endothia* acid protease (25)] but readily act on proteins. Voynick and Fruton (78) have compared the action on synthetic substrates of *Rhizopus chinensis* acid protease, mucor-rennin (from *Mucor pusillus*), and chymosin with that of pepsin.

Table VI. Cleavage Sites on S-Sulfo B-Chain of Insulin by Acidic Proteases [a]

Penicillopepsin	Pepsin	B-chain	Chymosin	Rhizopus
		NH$_2$		
			Phe		
- - →	→		←	←
		Val		
- - →				
		Asn		
		Gln		
→	→			
		His		
→				←
		Leu		
		CyS.SO$_3$		
- - →				
		Gly		
		Ser		
		His		
→				←
		Leu		
→	→		← - -	←
		Val		
		Glu		
→	→		←	
		Ala		
→	→		←	←
		Leu		
→	→		←	←
		Tyr		
→	→		←	←
		Leu		
→			←	
		Val		
		CyS.SO$_3$		
		Gly		
→				←
		Glu		
		Arg		
		Gly		
→	→			←

Table VI. Continued

Penicillopepsin	Pepsin	B-chain	Chymosin	Rhizopus
		$\mid$		
		Phe		
$\longrightarrow$	$\longrightarrow$	$\mid$	$\longleftarrow$	$\longleftarrow$
		Phe		
$\longrightarrow$	$\longrightarrow$	$\mid$	$\longleftarrow$	$\longleftarrow$
		Tyr		
		$\mid$		
		Thr		
		$\mid$		
		Pro		
		$\mid$		
		Lys		
		$\mid$		
		Ala		
		$\mid$		
		COOH		

[a] Penicillopepsin, (76); *Rhizopus chinensis*, (86); chymosin, (82); pepsin, (80).

They found evidence for a secondary binding site requirement although the catalytic efficiency of chymosin and the fungal enzymes on the same substrates was lower than that of pepsin, in analogy to our findings for penicillopepsin.

A qualitative comparison of the specificities of acid proteases can also be made on the basis of their effects on the B-chain of insulin. The hydrolysis of this substrate has been investigated by many workers (for references *see* Table V). The sites of observed cleavages for four representative enzymes show a similar pattern and are presented in Table VI. An analysis was made to find whether the similarities in the action of acid proteases reflected a common property, which may have its origins in an evolutionary relationship. Since a number of the bonds cleaved by acid proteases (Table VI) are also hydrolyzed by clearly unrelated enzymes of low specificity, the possibility had to be considered that an apparent similarity in the action of the acid proteases might be fortuitous and not indicative of a relationship between them. Therefore, a comparison was made of a variety of different proteases whose action on the B-chain of insulin had been investigated. This is shown in Table V. A comparison between penicillopepsin and other enzymes was made in terms of the number of identical bonds and of different bonds cleaved. The second column of Table V lists those cleavages which are common to penicillopepsin and the enzymes in the first column. The third column lists the cleavages observed for penicillopepsin but not for the enzymes in the first column. The fourth column lists cleavages shown by the

**Table VII.   Penicillopepsin: Nature of Amino Acids at Cleavage Points
in Glucagon and B-Chain of Insulin**[a]

Nature of Side Chain	No. of Residues in Position							
	P_4[b]	P_3	P_2	P_1	P_1'	P_2'	P_3'	P_4'
Total hydrophobic	14	13	11	16	23	14	13	17
Total hydrophilic	13	14	17	13	6	15	16	11

Bond
Cleavage

[a] (76).
[b] Symbols "Pn" indicate position of amino acid relative to bond cleaved, as defined by Berger and Schechter, (103).

enzymes in column one but not by penicillopepsin and shows that penicillopepsin cleaves virtually all the bonds cleaved by other acid proteases.

This is also expressed in the high ratios of column two to column four, the lowest being 5:1. On the other hand, proteases of low specificity which do not belong to the acid proteases hydrolyze many bonds not hydrolyzed by penicillopepsin. For this group of enzymes the ratios of column two to column four range between 1.2 and 2. Even the enzymes with restricted specificity listed in the third group show low ratios of columns two over column four. The only exception is thermolysin, an enzyme with a high specificity for hydrophobic side chains. This enzyme splits six bonds which are also hydrolyzed by penicillopepsin and other acid proteases. However, the apparent similarity is probably accidental for two reasons: (1) only three of the thermolytic splits are major splits for penicillopepsin, and (2) spleen, thyroid and adrenal cathepsin, and pepsin C hydrolyze only three of the bonds hydrolyzed by thermolysin while all the bonds split by the cathepsins are major cleavages for penicillopepsin.

Although this kind of comparison has obvious limitations, it does give a definite indication of a relationship among the acid proteases. The basis of this relationship may be the secondary binding site requirement.

In an attempt to find out more about the nature of the secondary binding site in penicillopepsin, Mains et al. (76) analysed the nature of the amino acid side chains at positions removed from the sensitive peptide bond. An abbreviated summary of the results is given in Table VII. The number of hydrophobic and hydrophilic side chains respectively are listed for four amino acids on either side of every peptide bond broken in the B-chain of insulin and in glucagon. The positions are numbered P_1, P_2, P_1', P_2' etc. as defined by Berger and Schechter (103). The choice of four positions was taken from the Fruton's work (73) on the specificity of pepsin and the effect of chain length on catalytic efficiency. The largest effects were observed with substrates having three to four amino acids on either side of the sensitive bond. As Table VII shows, there is a similar number of hydrophilic and hydrophobic side chains at every position except P_1'. (Amino acids defined as hydrophobic are: Gly, Ala, Val, Leu, Ile, Met, Phe, Tyr, Trp, and Pro. All other amino acids are considered hydrophilic.)

There is a high preference for hydrophobic side chains, particularly large aliphatic and aromatic side chains in position P_1'. There is no evidence for a specific side chain requirement in any of the other positions, some of which are required for interaction in the secondary binding site. The main effect in the secondary binding site could be induced by binding of the protein backbone. This contrasts with the evidence for a hydrophobic nature of the secondary binding site obtained by Fruton (73). It is hoped that future studies on specificity and the x-ray analysis of the enzymes will characterize the binding site in more detail and lead to a better understanding of its function.

Nature of Groups Involved in Pepsin Catalysis

The sections of this review which deal with aspects of the mechanism of acid proteases will be concerned with porcine pepsin and penicillopepsin only since little information is available on other acid proteases.

Work from many laboratories involving a variety of small substrates has indicated that the activity of porcine pepsin is controlled by two ionizing groups, one with pK_a about 1 required in its deprotonated form and the other with pK_a about 4 in its protonated form [for references *see* review by Fruton (46)]. The identification of one of these groups as a carboxyl group is based on specific inactivation by DAN and isolation of a labelled peptide (*see* p. 151). More recently Tang (104) found that another carboxyl-group reagent, 1,2-epoxy 3-(p-nitrophenoxy)propane (EPNP), also caused irreversible inactivation. Two p-nitrophenoxypropyl groups were incorporated, one of these was esterified to a β-carboxyl group of an aspartic acid and the other to a methionine. Subsequently Hartsuck and Tang (105) showed that only the reaction with aspartic acid was responsible for the inactivation, and Chen and Tang (106) isolated a labelled peptide. The reaction with EPNP and the sequence of the isolated peptide are as follows:

Reaction of porcine pepsin with EPNP

Active Site peptide isolated (106)

Ile-Val*-Asp-Thr-Gly-Ser-Ser-Asn

* The correct amino acid in this position is Phe as reported in the complete sequence (*13*).

In the complete sequence (*13*), the aspartic acid occupies position 32 from the N-terminal of pepsin. Other acid proteases are also inactivated by EPNP. They are listed in Table VIII, together with the number of incorporated moieties of the label. In the case of penicillopepsin, essentially complete inactivation was obtained when only one residue reacted. The linkage was tentatively identified as an ester to an aspartic acid residue. It is also interesting to note that the partial sequences available for chymosin (*15*) and penicillopepsin (Kurosky, A., Rao, L., Hofmann, T., to be published) show peptides which align with the EPNP-labelled peptide from pepsin (Table IX). The similarity of the three sequences strongly indicates similar or identical functional roles and suggests that it is the aspartic acid in this peptide which reacts with EPNP in chymosin and pencillopepsin.

The pH dependence of the reaction between EPNP and pepsin suggests that the reactive aspartic acid-32 has a pK_a below pH 3 (*105*).

Table VIII.   Acid Proteases Inhibited by 1,2-Epoxy 3-(*p*-Nitrophenoxy)propane

Enzyme	EPNP-Labelled Peptide	No. of EPNP Residues Incorporated	Reference
Pepsin, porcine	Ile-Phe-Asp-Thr-Gly-Ser-Ser-Asn [*a] [*]	2	(*104,106*)
Pepsin, human		2.02	(*104*)
Gastricsin, human		1.97	(*104*)
Chymosin, bovine	Asp-Thr-Gly-Ser-Ser [*a] [*]	4.2	(*104*)
Chymosin, bovine		2.2	(*107*)
Rhizopus-pepsin (*Rhizopus chinensis*)		1.87	(*107*)
Aspergillopeptidase A (*Aspergillus saitoi*)		1.7	(*107*)
Acid protease (*Trametes sanguinea*)		1.76	(*107*)
Penicillopepsin		1.1	c
Endothia parasitica acid protease		2	c

[a] The asterisk (*) indicates the labelled residue.
[b] Chang, W.-J., personal communication.
[c] Mains, G., Hofmann, T., to be published.

It has been postulated that this is the residue which is required as a carboxylate ion in the active site of pepsin. The DAN-reactive aspartic acid-215 may be required in its protonated form. Thus, the two reactive aspartic acids are probably those which are responsible for the pH-dependence of k_{cat}.

Residues with ionizable side chains other than carboxyl groups could in principle be responsible for the pH dependence of the catalytic reaction. The group with pK_a about 4.5 could be an α-amino group or a histidine, but there is evidence from chemical modification studies which makes this highly unlikely. Knowles (*108*) has discussed this in detail.

Table IX.   Alignment of Peptide Segments of Chymosin (*15*) and Penicillopepsin with EPNP-reactive Peptide from Porcine Pepsin (*13*)[a]

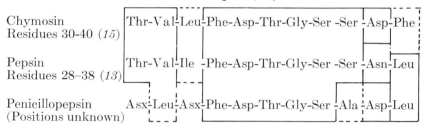

Chymosin Residues 30-40 (*15*)	Thr-Val-Leu-Phe-Asp-Thr-Gly-Ser -Ser -Asp-Phe
Pepsin Residues 28–38 (*13*)	Thr-Val-Ile -Phe-Asp-Thr-Gly-Ser -Ser -Asn-Leu
Penicillopepsin (Positions unknown)	Asx-Leu-Asx-Phe-Asp-Thr-Gly-Ser -Ala -Asp-Leu

[a] ———— surrounds residues which are identical between two or more enzymes, ------ surrounds residues which are chemically similar.

Additional groups have been implicated in the active site of pepsin although their modification does not abolish activity completely. Chemical modifications of pepsin have been carried out in many laboratories since the first studies of Herriott and Northrop in 1934 (*109*). They have been reviewed by Fruton (*46*), and only the most important findings are discussed briefly here. The early workers (*109*) who used ketene as an acetylating reagent concluded that modification of tyrosines reduced activity towards hemoglobin. When acetyl imidazole was used to acetylate tyrosines, Lokshina and Orekhovich (*110*) found an increase in peptidase activity while proteinase activity was inhibited. Fruton (*46*) concludes that acetylation of tyrosine residues occurs in the secondary binding site and causes a conformational change which leads to rate enhancement of small substrates except those with extended binding sites. Iodination of tyrosines, on the other hand, leads to loss of activity towards all substrates (*111*).

Tyrosine-9 and tyrosine-175 can be iodinated with a 75% activity loss (*112*). Whether one or both of these residues is responsible for loss of activity is not clear.

p-Bromophenacylbromide reacts in porcine pepsin with an aspartic acid and causes a partial loss of activity towards hemoglobin (*113*). The composition of the isolated labelled peptide shows, however, that this aspartic acid is different from the DAN- or the EPNP-reactive residues. Penicillopepsin is not inactivated by *p*-bromophenacylbromide (*114*).

Since the review of Fruton (*46*) the modification of arginine is the most important which has been carried out. Using phenylglyoxal, Kitson and Knowles (*115*) modified one of the two arginines of pepsin and found a loss of activity and a shift of the lower arm of the pH dependence to higher values. Huang and Tang (*116*) modified an arginine residue [now known to be in position 316 in the complete sequence, (*13*)], with biacetyl with an 85% activity loss. Both groups concluded that an arginine residue contributes to the catalytic apparatus by interacting with one of the catalytic carboxyl groups.

In this laboratory it has been observed (Mains, G., Hofmann, T., unpublished) that modification of one of the five lysine residues of penicillopepsin causes an effect similar to that observed when arginine is modified in pepsin (*115*). If a positively charged residue analogous to the arginine in pepsin is indeed required for penicillopepsin activity, it would have to be a lysine since penicillopepsin contains no arginine.

Mechanism of Pepsin- and Penicillopepsin-Catalyzed Reactions

Four reviews dealing with the mechanism of action of pepsin have been published in recent years (*46, 73, 108, 117*). Other recent publications deal with various aspects of this mechanism (*118–120*). In this section, therefore, the main emphasis will be placed on the significance of studies on hitherto unobserved pepsin- and penicillopepsin-catalyzed transpeptidation reactions, especially as they affect the mechanisms proposed by various authors. The question we are concerned with is the role of the two carboxyl groups which are involved in the catalytic action. We shall not further consider the role of other functional groups which have been discussed in the previous section.

One of the essential questions relating to the mechanism is whether the catalytic process involves any covalent intermediates between the substrate moieties and the enzyme. In the case of pepsin two possible intermediates could be formed: the amino intermediate which involves the transfer of the amino moiety of the substrate to one of the carboxyl groups on the enzyme and the acyl intermediate which involves the transfer of the acyl moiety as shown on page 165. The examples used are substrates involved in transpeptidation reactions in which the respective intermediate has been demonstrated (*see* below).

Amino intermediate

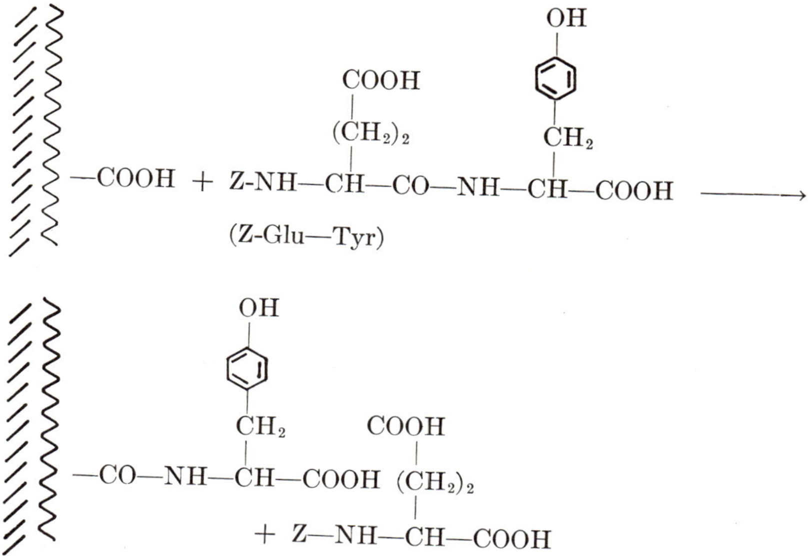

$$\text{—COOH} + \text{Z-NH—CH—CO—NH—CH—COOH} \longrightarrow$$

with side chains:
- On first CH: $(CH_2)_2$ with COOH
- On second CH: CH_2 with a phenol ring bearing OH

(Z-Glu—Tyr)

$$\text{—CO—NH—CH—COOH} \quad \text{(CH}_2)_2$$
$$+ \text{ Z—NH—CH—COOH}$$

with side chains:
- On CH: CH_2 with a phenol ring bearing OH
- COOH group and $(CH_2)_2$

Acyl intermediate

$$\text{—COOH} + \text{NH}_2 \text{ —CH—CO—NH—CH—CONH}_2 \longrightarrow$$

with side chains:
- On first CH: C_4H_9
- On second CH: CH_2 with a phenol ring bearing OH

(Leu—Tyr amide)

$$\text{—CO—O—CO—CH—NH}_2 + \text{NH}_2\text{—CH—CONH}_2$$

with side chains:
- On first CH: C_4H_9
- On second CH: CH_2 with a phenol ring bearing OH

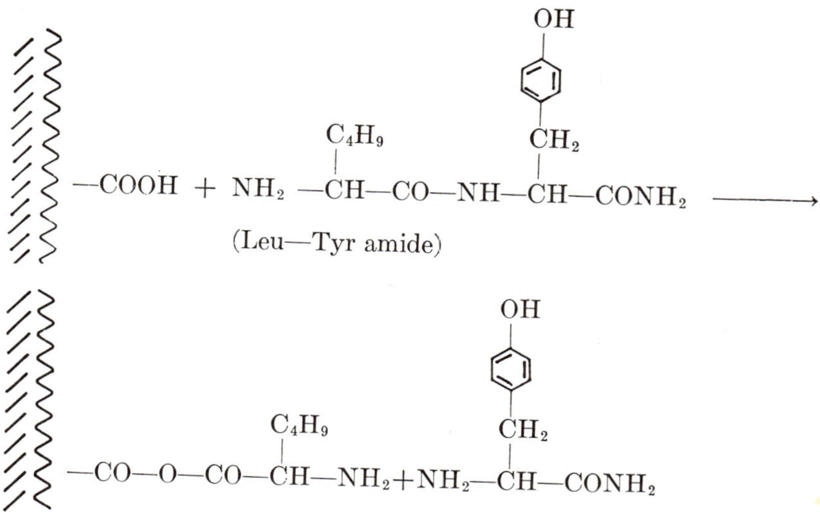

Evidence for the formation of an amino intermediate comes from the work of Neumann *et al.* (*121*) and Fruton *et al.* (*122*). Neumann *et al.* (*121*) showed that when substrates such as Z-Glu-Tyr or Z-Phe-Tyr are incubated with pepsin, detectable amounts of Tyr-Tyr are formed. Fur-

Table X.   Transpeptidation of Leu-Tyr-Leu by Penicillopepsin[a]

Products	Yield (nmoles)	Recovery, %
Tyr-Leu	722	22.3
Leu-Leu	537	16.8
Leu	247	7.7
Leu-Tyr	194	6.0
Tyr	161	5.0

[a] Penicillopepsin (0.2 mg) was incubated with Leu-Tyr-Leu (3.2 μmoles) in 1 ml pyridine-acetate buffer (0.05M in acetate) pH 3.6 at 35° C for 24 hr. The products were separated by high-voltage electrophoresis at pH 3.1 and pH 1.9. Ninhydrin positive bands were eluted, hydrolyzed, and quantitated by amino acid analysis. Identification was made by sequence analysis by standard procedures (details to be published by Takahashi and Hofmann).

ther, Fruton et al. (122) showed that a covalent amino intermediate is involved in this reaction since radioactive labelled free tyrosine was not incorporated into the transpeptidation product. The inhibition kinetics by products and product analogs (123, 124) are most easily interpreted if an amino enzyme is involved in the hydrolytic reaction since the kinetic studies suggest a mechanism involving an ordered release of the product with the acyl moiety released first. Thus the evidence for a covalent amino intermediate is substantial but by no means unambiguous. First, as Fruton (46) points out, the transpeptidation reactions have been observed only with substrates which are relatively resistant to the action of pepsin. Also, the yield of the transpeptidation product from the amino transfer reaction is very low, and according to Neumann and Sharon (125) activation of pepsinogen at pH 3 yields a pepsin preparation that does not catalyze the transpeptidation reaction.

The evidence for involvement of an acyl intermediate comes from the fact that pepsin catalyzes the exchange of [18]O between free carboxyl groups of acyl amino acids (products) and water (126, 127). This is fully discussed by Knowles (108) and Fruton (46). Attempts to trap putative acyl enzymes have failed so far. Furthermore, Shkarenkova et al. (128) showed that [18]O is rapidly incorporated from H_2 [18]O into an active site carboxyl group of pepsin in the absence of an acyl amino acid, and that the rate of [18]O loss from [18]O labelled pepsin is similar to the pepsin-catalyzed rate of [18]O exchange with acetylphenylalanine. The ^{18}O-exchange experiments, therefore, do not require the formation of an acyl intermediate. With this in mind Knowles proposed a mechanism for pepsin-catalyzed reactions (108) which involves a covalent amino intermediate but not an acyl intermediate.

However, since the time these mechanisms were proposed, we have obtained more direct evidence for an acyl intermediate in pepsin- and penicillopepsin-catalyzed reactions from transpeptidation reactions which only proceed via an acyl transfer (129). Some of the experimental evidence for the acyl transfer reaction is presented here although a full account of the experiments is being prepared for publication.

When penicillopepsin was incubated with a tripeptide Leu-Tyr-Leu under the conditions given in Table X, a substantial amount of Leu-Leu was formed. In these experiments the isolation and identification of the reaction products were made by high voltage electrophoresis (Table X). The separated bands were eluted and identified by sequence analysis and composition. Quantitation was made on the amino acid analyzer using a system developed for the separation of dipeptides (*130*). Because of considerable losses during the isolation procedure, recovery of the products was not quantitative, but ratios of the various products obtained are strongly indicative of their yield. It is apparent that the yield of the transpeptidation product is considerable. The relatively high yield of the dipeptide Tyr-Leu as compared with the dipeptide Leu-Tyr suggests that the leucine involved in the transpeptidation comes from the N-terminal position although the possibility of an involvement of the C-terminal leucine is not excluded.

Table XI.   Transpeptidation of Phe-Tyr-Thr-Pro-Lys-Ala by Penicillopepsin[a]

Products	Yield (nmoles)	Recovery, %
Tyr-Thr-Pro-Lys-Ala	950	35.0
Phe	635	23.5
Thr-Pro-Lys-Ala	144	5.3
Pro-Lys-Ala	99	3.7
Phe-Phe	57	4.2
Tyr	51	1.9

[a] The substrate (2.7 μmoles) was incubated with 0.3 mg penicillopepsin in 2 ml pyridine-acetate, pH 3.6, for 24 hr at 35° C. Isolation and identification of products were as described in footnote to Table X.

This ambiguity was resolved in the experiment shown in Table XI. The peptide in this case represents the C-terminal sequence of the B-chain of insulin. The major products were the penta-peptide lacking the N-terminal phenylalanine and free phenylalanine indicating that hydrolysis was the major reaction. In addition, however, there was a significant amount of Phe-Phe. This dipeptide could only come from a transpeptidation involving an acyl transfer. The acyl transfer presumably proceeds *via* a covalent acyl intermediate. The evidence for this comes from the experiment shown in Table XII where Leu-Tyr-Leu was incubated with both porcine pepsin and penicillopepsin in the presence of a 10-fold excess of ^{14}C-leucine over Leu-Tyr-Leu. The products, leucine and leucylleucine, were separated by high voltage electrophoresis and analyzed for their specific radioactivity. At most, only traces of radio-

Table XII.   Transpeptidation of Leu-Tyr-Leu in Presence of ¹⁴C-Leucine[a]

Enzyme	Electroph. Band	nmoles	cpm	cpm/nmole
Porcine pepsin	leucine	477	4620	9.7
	Leu-Leu	14	(1.5)	(0.1)
Penicillopepsin	leucine	356	3210	9.0
	Leu-Leu	13	(3)	(0.23)

[a] Penicillopepsin (0.3 mg) *or* porcine pepsin (0.5 mg) was incubated with 1.2 μmoles Leu-Tyr-Leu and 7.6 μmoles ¹⁴C-leucine in 0.52 ml pyridine-acetate buffer (0.05M in acetate) pH 3.6 for 24 hr. at 35°C. The products were separated by high-voltage electrophoresis. The bands of leucine and Leu-Leu were eluted, counted, and quantitated by amino acid analysis.

activity had been incorporated into the transpeptidation product. This experiment also showed that the transpeptidation reaction involving an acyl transfer is catalyzed not only by penicillopepsin with which enzyme it was discovered but also by porcine pepsin.

Dr. Wang subsequently discovered substrates which were more suited to quantitative study of the transpeptidation reaction and gave

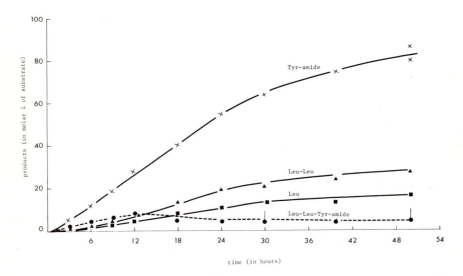

Figure 1.   Time course of action of penicillopepsin on Leu-Tyr-amide at pH 3.4, 37°C. Penicillopepsin (0.6 mg/ml) was incubated with Leu-Tyr amide (2 μmoles/ml) in 0.05M citrate. Products were analyzed on a Beckman 120 B amino acid analyzer using a column of UR 30 resin (20 × 0.9 cm) with two sodium citrate buffers (0.2M, pH 4.3 and 5.25) at 56°C. Tyr amide was determined on a column of PA 35 resin (3 × 0.9 cm) with sodium citrate (0.35M, pH 5.25) at 66°C.

unambiguous evidence for the acyl transfer reaction with both penicillo-pepsin and porcine pepsin. One of the substrates used for this purpose was leucyltyrosine amide. The results obtained with this substrate and both enzymes are shown in Figures 1 and 2. The figures present progress curves of the action of the enzymes on the substrate. The products were analyzed by modified amino acid analyzer methods. Figure 1 represents the reaction with penicillopepsin and shows that as with the previous substrate used, the formation of the dipeptide, leucylleucine, predominates over hydrolysis. Furthermore Figure 1 also shows that a tripeptide amide, Leu-Leu-Tyr amide, is probably an intermediate in the reaction.

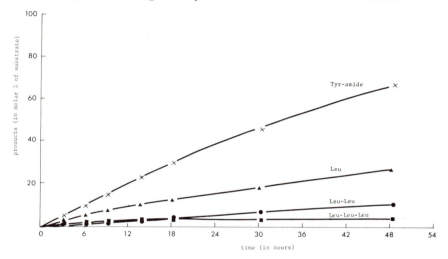

Figure 2.   Time course of action of porcine pepsin on Leu-Tyr amide at pH 3.4, 37°C. The experimental conditions were as described for Figure 1 with 0.6 mg/ml pepsin.

While the reaction with porcine pepsin (Figure 2) is qualitatively similar to that with penicillopepsin, there are obvious differences. First, hydrolysis predominated over transpeptidation and second, another trans-peptidation product, tri-leucine, was formed. The putative intermediate Leu-Leu-Tyr has, however, been detected. A detailed discussion of the significance of these experiments will be presented elsewhere. For the purpose of this review suffice it to say that they provide strong evidence for the involvement of an compulsory acyl intermediate in some pepsin and penicillopepsin catalyzed reactions. Hence, mechanistic proposals which do not involve acyl intermediates are insufficient for a complete description of pepsin catalyzed reactions.

The question now arises as to which is the pathway of *hydrolytic* reactions. There are five possibilities.

(1) Only the amino intermediate exists as proposed by Fruton (73), Knowles (108), and others while acyl intermediates are involved only in transpeptidation reactions of the type observed in our laboratory.

(2) Only the acyl intermediate exists in hydrolytic reactions while amino intermediates are involved only in transpeptidation reactions.

(3) Appropriate covalent intermediates are necessary only for transpeptidation. Hydrolytic reactions proceed without forming covalent intermediates.

(4) Pepsin-catalyzed reactions involve both an amino and an acyl intermediate as proposed by Bender and Kezdy (131). The order of release would be determined by the nature of the substrate.

(5) Any reaction can proceed via one intermediate or the other (but not both). The nature of the substrate would determine which intermediate is on the pathway of any particular reaction.

A detailed discussion of these possibilities is beyond the scope of this review, but we give the major reasons why we think the fifth mechanism is the most satisfactory one although we would not rule out the third possibility at this time.

(1,2) The evidence for the amino intermediate comes from the transpeptidation reaction and from the kinetic studies of Greenwell et al. (123). In both types of studies poor substrates (N-substituted dipeptides) were used. No evidence for an amino intermediate has been obtained as yet from good substrates (substrates with high k_{cat}). Similarly, the evidence for the acyl intermediate comes from transpeptidation studies as discussed. So far these reactions too have been observed using poor substrates. Thus, there is no a priori reason for preferring one of the intermediates over the other for the hydrolytic reaction.

(3) The possibility that hydrolysis proceeds without the formation of covalent intermediates has been mentioned by Fruton (73) and must be given serious attention in future studies. Because of the very large effects of the secondary binding sites on the catalytic efficiency, it is essential that these studies be carried out with good substrates. Silver and Stoddard (120) suggest that hydrolysis of small substrates proceeds without an amino intermediate. However, they did not consider an acyl intermediate.

(4) There is also the possibility that both acyl and amino intermediates are required for all the pepsin-catalyzed reactions. Bender and Kezdy (131) suggested that two carboxylic acid groups on the enzyme could reversibly form an anhydride. It could then undergo an exchange reaction with the peptide substrate to form an amino intermediate from

the amino moiety and an acyl intermediate from the acyl moiety of the substrate. This double intermediate, in which the acyl group would be linked to the carboxyl group of the enzyme through a carboxylic anhydride and the amino group through an amide bond, would readily react with water to liberate the carboxylic acid moiety. At this point an amino intermediate would be left which could be hydrolyzed through internal catalysis by the neighboring carboxylic acid group on the enzyme to give the hydrolytic reaction. A model compound for an internally catalyzed hydrolysis of an amide bond is phthalamic acid in which a carboxyl group in the ortho position to the carboxamide group catalyzes the hydrolysis of the amide (*132*). Alternatively, it could react with suitable acceptors to give transpeptidation of the amino transfer type. While this mechanism could conceivably account for the latter and for hydrolysis

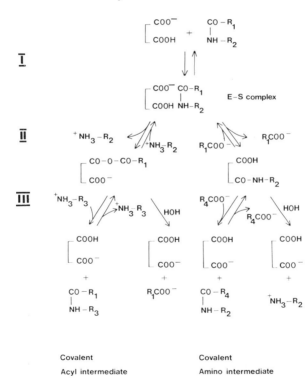

Figure 3. Proposed mechanisms for pepsin-catalyzed reactions (see text for details). The orientation of the peptide bond does not imply any particular mechanism of the intermediate formation. The positions of the protonated and unprotonated carboxyl groups in the E-S complex could be reversed, depending on the mechanism of interaction.

involving an amino intermediate, it is difficult to see how it could account for the transpeptidation reactions of the acyl transfer type. The action of pepsin and penicillopepsin on leucyltyrosine amide give a high yield of the transpeptidation product, leucylleucine. This suggests that the acyl intermediate is highly stabilized. In order to convert the double intermediate proposed by Bender into the acyl intermediate, it would be necessary to invoke an additional catalytic mechanism which would cleave the amide bond of the amino intermediate and leave the acyl intermediate intact. At present there is no evidence for such a catalytic system.

(5) We therefore propose that not all pepsin-catalyzed reactions proceed *via* the same mechanism. We suggest that the nature of the substrate will determine the type of covalent intermediate formed on the pathway of the reaction. The proposed pathways are summarized in Figure 3. The first step would, of course, be the formation of an enzyme substrate complex (Reaction I). Reaction II presents the alternative formation of an acyl intermediate or an amino intermediate, depending on the substrate.

The basis of this proposal is the fact that the catalytic center of the the enzyme in the native state is in a nearly inactive configuration as shown by the low catalytic efficiency towards small substrates. As has been stressed, a secondary interaction is required which induces a conformational change probably involving the catalytic apparatus to generate high catalytic efficiency. The evidence from Fruton's extensive studies with peptides of increasing chain length shows that secondary interactions take place on both sides of the scissile bond and lead to increase in k_{cat} without significant changes in K_M [Table XIII, taken from Tables III and IV of (73)]. Examples (1) and (2) show a 100-fold increase in k_{cat} when the peptide chain is lengthened on the C-terminal side. Examples (3) and (4) show a similar effect on the N-terminal side. This strongly suggests that there are two secondary binding sites, A and B, where A is the site which would be occupied by residues P_2-P_n, and B is the site occupied by residues P_2'-P_n'. It is quite possible, if not prob-

Table XIII. Action of Porcine Pepsin on Synthetic Peptides[a]

Substrate	$k_{cat}(sec^{-1})$	$K_m(mM)$
Z-His-Phe(NO$_2$)-Phe-OMe[b]	0.29	0.46
Z-His-Phe(NO$_2$)-Phe-Ala-Ala-OMe[b]	28.0	0.12
Z-Phe-Phe-3-(4-pyridyl)-propyl ester[c]	0.49	0.7
Z-Gly-Gly-Phe-Phe-3-(4-pyridyl)-propyl ester[c]	56.5	0.8

[a] From Tables III and IV (73) where further details are given.
[b] pH 4.0, 37° C.
[c] pH 2.0, 37° C.

able, that the conformational changes induced by binding in A differ from those induced by binding in B although both increase catalytic efficiency. Those substrates with which amino transpeptidation has been observed can bind only on site A or on the primary binding site occupied by P_1 next to A since so far only transfer of free C-terminal amino acids has been observed. Conversely, substrates which give acyl transpeptidation can bind only on site B or on the primary site occupied by P_1' next to B because only free N-terminal amino acids are transferred. We postulate therefore, that the changes induced by binding in site A lead to a configuration of the catalytic apparatus which favors formation of the amino intermediate while binding in site B favors formation of the acyl intermediate. Substrates which bind efficiently in both sites could proceed *via* one intermediate or the other or induce a configuration which would not require a covalent intermediate for hydrolysis. In the case of transpeptidation, Reactions III (Figure 3) are the reverse of Reactions II. They also present the hydrolysis of the intermediate which could proceed by mechanisms such as that proposed by Knowles (*108*) in the case of the amino intermediate or by a general acid catalysis in the case of the acyl intermediate.

From this discussion and the evidence and arguments presented, it is clear that the mechanism of action of pepsin and penicillopepsin, let alone other acid proteases, is little understood, and a great deal of additional work is required before detailed mechanisms can be formulated.

Sequence Studies in Acid Proteases

It was pointed out in the first part of this article that at the time of writing no complete sequences of acid proteases had been published although the sequence of porcine pepsin is completed (*13*). Nevertheless sufficient information is available from partial sequences to allow a meaningful comparison between a number of acid proteases, porcine, bovine, and human pepsins, human gastricsin, chymosin, and penicillopepsin. Table XIV gives an alignment of the N-terminal sequences of two mammalian pepsins, chymosin, and two fungal pepsins. The two mammalian enzymes are clearly homologous. However, there are difficulties with the alignment of rhizopus pepsin penicillopepsin with the mammalian enzymes. While they show evidence for homology with each other, the evidence for their homology with the mammalian enzymes from the first 20 residues is weak. Thus, in order to obtain a minimal alignment with penicillopepsin four insertions are necessary. A good alignment, however, is easily obtainable with aspartic acid-32 which is the residue that reacts with EPNP, as was shown on page 161. In chy-

Table XIV.   Alignment of N-terminal Sequences

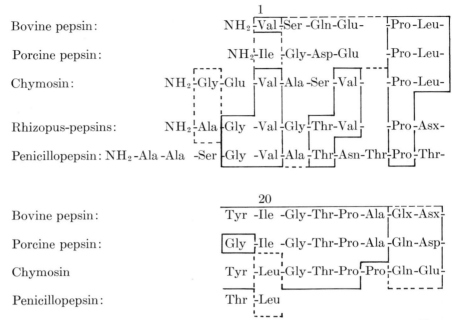

Bovine pepsin: NH₂-Val-Ser -Gln-Glu-   -Pro-Leu-

Porcine pepsin: NH₂-Ile -Gly-Asp-Glu   -Pro-Leu-

Chymosin: NH₂-Gly-Glu -Val-Ala -Ser -Val-   -Pro-Leu-

Rhizopus-pepsins: NH₂-Ala -Gly -Val-Gly-Thr-Val-   -Pro-Asx-

Penicillopepsin: NH₂-Ala -Ala   -Ser -Gly -Val-Ala -Thr-Asn-Thr-Pro-Thr-

20

Bovine pepsin: Tyr -Ile -Gly-Thr-Pro -Ala -Glx-Asx-

Porcine pepsin: Gly -Ile -Gly-Thr-Pro -Ala -Gln-Asp-

Chymosin   Tyr -Leu-Gly-Thr-Pro-Pro-Gln-Glu-

Penicillopepsin: Thr -Leu

[a] Taken in part from (15). Porcine pepsin sequences are numbered according to Tang et al. (13), bovine pepsin sequences are from Pederson and Foltmann (133) and Foltmann

mosin and bovine pepsin, this residue is in an equivalent position to porcine pepsin, but there is no direct evidence for its involvement in catalysis.

Table XV shows two fragments each of chymosin and penicillopepsin which align with sections of the interior of the porcine pepsin molecule. The first fragment which shows extensive similarity between all three enzymes centers around aspartic acid-171. The second fragment centers around aspartic acid-215, the aspartic acid which reacts with DAN (page 151). Table XVI shows an alignment of the known C-terminal sequences of five acid proteases. Again there is extensive similarity between the sequences. Taking into account all five sequences, well over 50% of the residues are either identical or highly conservative replacements (such as lysine for arginine or isoleucine for valine). The sequences in Table XVI allow the conclusions that (1) the mammalian enzyme chymosin, the pepsins, and gastricsin are homologous, and (2) that the fungal penicillopepsin is homologous with the mammalian enzymes. More extensive comparisons between chymosin and porcine and bovine pepsin are given by Foltmann et al. (15).

of Acid Proteases (Active Forms)[a]

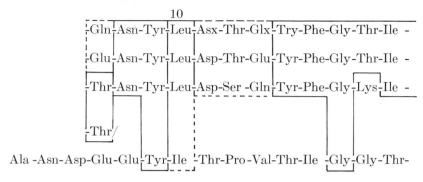

(personal communication); rhizopus-pepsin (17) and penicillopepsin (16). ————— and
- - - - - - - - - are explained in Table IX.

In our laboratory we have accounted for all the amino acids of penicillopepsin in 20 unique sequence fragments. Now that the complete sequence of porcine pepsin is available, we have attempted to align these fragments with those of porcine pepsin. It is interesting to note that apart from the sequences shown in Tables XIV to XVI no other unambiguous alignments are possible at present. This suggests that the sequences shown in these tables are those which are the most highly conserved while other sections of the sequence have undergone mutations which make alignment difficult. It is very interesting that the best alignment obtained is around aspartic acid residues: first, around those two which react with specific active site inhibitors (DAN and EPNP), and second, around aspartic acid-171, aspartic acid-304, and aspartic acid-315. Whether or not these latter aspartic acid residues or residues near them are essential for the enzymic function remains to be seen. The suggestion has been made that at least three carboxyl groups are involved in the catalytic site of pepsin (105). As has also been mentioned, Huang and Tang (116) have implicated arginine-316 in the active site of pepsin. It is noteworthy that this arginine has no equivalent residue in penicillo-

Table XV.   Alignment of Some Central Fragments of Chymosin

Fragment at Asp-171

Porcine pepsin	. . . . Leu-Leu-Gly-Gly-Ile 170
Chymosin[b]	/Gly-Ala-Ile
Penicillopepsin[c]	/Asp-Phe-Gly-Phe-Ile

Fragment at Asp-215

Porcine pepsin	. . . . Cys-Gln-Ala -Ile 210
Chymosin[d]	/Gln-Ile
Penicillopepsin[c],[e]	/Ser -Gly-Ile

[a] Numbered from complete pepsin sequence (*13*).
[b] Sequence from (*134*).
[c] Harris, C.I., Kurosky, A., Rao, L., Hofmann, T., to be published.

pepsin. However, arginine-308 has an equivalent lysine residue in penicillopepsin. In view of the similarity in the mechanism between penicillopepsin and porcine pepsin and in view of the paucity of basic amino acids in these enzymes, one wonders whether or not the basic residue in position-308 also has an important function in the mechanism of pepsins.

Evolution of Zymogens

It is now well known that the mammalian enzymes pepsin, gastricsin, and chymosin are secreted by the respective organisms as zymogens which are activated by the removal of over 40 residues from the N-terminal end.

The activation of zymogens has been studied in many laboratories and was reviewed by Ottesen (*139*). Until fairly recently it had been generally accepted that zymogens were the completely inactive precursors and were converted into active enzymes by limited proteolysis. However, Foltmann (*140*) first suggested on the basis of kinetic studies that prochymosin had inherent activation power, *i.e.* was a protein with some proteolytic activity. Since then other zymogens have been found to possess weak inherent activity. Evidence for this has been found in zymogens belonging to the four major classes of proteases: acid proteases, serine proteases, sulfhydryl proteases, and metalloenzymes. This aspect was reviewed recently by Kassell and Kay (*141*). As far as the acid proteases are concerned, it appears that both prochymosin and pepsinogen

and Penicillopepsin with the Porcine Pepsin Sequence[a]

180

-Asp-Ser -Ser -Tyr-Tyr-Thr-Gly-Ser -Leu-Asn-Trp . . . .

-Asp-Pro -Ser -Tyr-Tyr/

-Asp-Ser -Ser -Lys-Tyr-Thr-Gly-Ser -Leu-Thr-Tyr/

220

-Val -Asp-Thr-Gly-Thr-Ser -Leu- . . . .

-Leu-Asp-Thr-Gly-Thr-Ser/

-Ala -Asp-Thr-Gly-Thr-Thr-Leu/

[a] (15).
[e] Sequence 213-220 is from (60).
——— and - - - - - - are expalined in Table IX.

can form active zymogens which then are converted into active enzymes. It is not clear, however, whether this conversion is intramolecular or inter-molecular (141). Evidence for both mechanisms has been presented by various authors [see (141)].

While vertebrate pepsins, chymosin, and gastricsins have been shown to originate from zymogens, there is no evidence for zymogen forms of the fungal enzymes. We made many attempts to find evidence for or against a zymogen of penicillopepsin but were unsuccessful in detecting any kind of precursor. This enzyme is formed only at a late stage in the development of the fungus. The possibility exists that it is stored in an inactive form during the logarithmic phase of growth of the organism and secreted as an active enzyme only when the stationary phase is reached and sporulation sets in. We have, therefore, examined cultures of *Penicillium janthinellum* at various stages of development (Sodek and Hofmann, unpublished) for the presence of a precursor. The mycelium as well as the medium was subjected to various procedures which are known to activate other zymogens, including trypsinogen. They were incubated at various pH values with and without active trypsin and active pepsin, enterokinase, calcium, and other ions. No detectable activity was produced in either mycelial extract or in the medium during the logarithmic phase, and activity did not increase once secretion of penicillopepsin had begun. It is, therefore, highly probable that there is no precursor of penicillopepsin. On a previous occasion (142), it was

Table XVI.   Alignment of C-Terminal

300

Human gastricsin	
Human pepsin	/Ile -Leu-Gly-Asp-Val -Phe
Porcine pepsin	. . . . Trp-Ile -Leu-Gly-Asp-Val -Phe
Pencillopepsin	/Ser -Ile -Gly-Asp-Ile -Phe
Chymosin	. . . . Trp-Ile -Leu-Gly-Asp-Val -Phe
Bovine pepsin	

320

Human gastricsin	Phe-Asp-Arg-Ala -Asn-Gln-Lys
Human pepsin	Phe-Asp-Arg-Ala -Asn-Asn-Gln
Porcine pepsin	Phe-Asp-Arg-Ala -Asn-Asn-Lys
Pencillopepsin	Phe-Asp-Ser -Asp-Gly-Pro -Gln
Chymosin	Phe-Asp-Arg-Ala -Asn-Asn-Leu
Bovine	Phe-Asp-Arg-Gly-Asn-Asn-Gln

a Numbered according to (*13*). Human gastricsin and human pepsin are from (*135*), porcine pepsin (*136,137*), penicillopepsin (*16*), chymosin (*15*), bovine pepsin (*138*).

proposed that zymogens of serine proteases and acid proteases evolved relatively late and were introduced through a mutational process leading to an extension of the peptide chain at the N-terminal end. This proposal was based on the well-known fact that activation of zymogens of proteolytic enzymes is caused by a peptide bond cleavage and removal of an N-terminal fragment. In the case of the serine proteases, this fragment may range from only six amino acids [bovine trypsinogen, (*143*)] to many hundreds [bovine prothrombin, (*65*)]. While the sequence of the B-chain of active thrombin shows conclusively that it is homologous with the trypsin-family of serine proteases, the activation peptide is probably not homologous. Though there are three bacterial proteases, α-lytic protease from Myxobacter 495 (*144*), and two enzymes from *Streptomyces griseus* (*145*), for which there is sufficient evidence for homology with the trypsins, so far no evidence has been presented for the presence of zymogens. Also, the homology extends over the active enzyme only

Sequences of Acid Proteases[a]

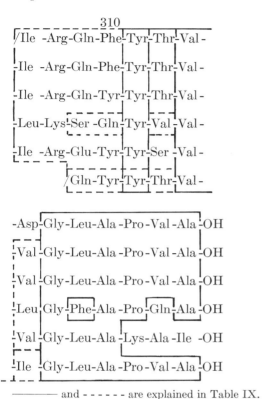

```
                      310
/Ile  -Arg-Gln-Phe-Tyr-Thr-Val -

-Ile  -Arg-Gln-Phe-Tyr-Thr-Val -

-Ile  -Arg-Gln-Tyr-Tyr-Thr-Val -

-Leu-Lys-Ser -Gln-Tyr-Val -Val -

-Ile  -Arg-Glu-Tyr-Tyr-Ser -Val -

        /Gln-Tyr-Tyr-Thr-Val -
```

```
-Asp-Gly-Leu-Ala -Pro-Val -Ala -OH

-Val -Gly-Leu-Ala -Pro-Val -Ala -OH

-Val -Gly-Leu-Ala -Pro-Val -Ala -OH

-Leu-Gly-Phe-Ala -Pro-Gln-Ala -OH

-Val -Gly-Leu-Ala -Lys-Ala -Ile -OH

-Ile  -Gly-Leu-Ala -Pro-Val -Ala -OH
```

———— and - - - - - - are explained in Table IX.

and does not include activation peptides. This could indicate that zymogens evolved relatively late. The evolution of the zymogens of acid proteases is indicated schematically in Figure 4. It is suggested that the common ancestor was an active enzyme and was converted to a zymogen by an extension of the peptide chain as indicated in Figure 5. Penicillo-

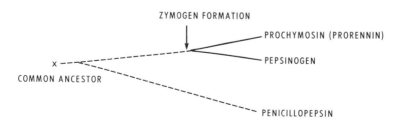

Figure 4.   Proposed scheme for the evolution of acid proteases

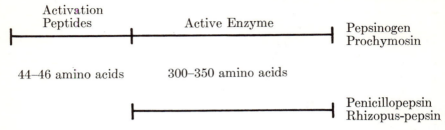

Figure 5. *Probable sequence alignments of acid proteinases and zymogens*

Table XVII.   Alignment of N-Terminal

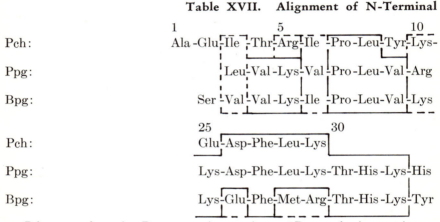

a Pch = prochymosin; Ppg = porcine pepsinogen; Bpg = bovine pepsinogen. Sequences of Ppg are taken largely from Ong and Perlmann (*147*), and Pedersen and Foltmann (*148*), and those of Pch and Bpg from (*15*) and (*149*).

pepsin and rhizopus-pepsins are aligned with the active enzyme because there is some evidence for homology in this region as shown in Table XIV. In contrast we have found no evidence for homology between 28 N-terminal amino acids of penicillopepsin and the activation peptides of pepsinogen and prochymosin using the diagram method of Gibbs and McIntyre (*146*). The activation peptides of porcine and bovine pepsinogen and prochymosin on the other hand are clearly homologous as the alignment in Table XVII shows. Therefore the introduction of the zymogen would have occurred before the gene duplication which led to the formation of pepsinogen and prochymosin, respectively.

The nature of the mutational event which could lead to a peptide chain extension can only be guessed at. A mutation in an initiation codon or in the recognition site in the intercistronic region could have been sufficient to form the first zymogen. Alternatively a partial genetic cross-over could lead to such an extension.

Conclusion

It is clear that a great deal of work is still required in the field of acid proteases before we reach the level of understanding attained for other groups of proteolytic enzymes. Fortunately the amino acid sequence work on at least three enzymes—pepsin, chymosin, and penicillopepsin— is well advanced, and complete three-dimensional structures should become available in the near future. A tentative structure for rhizopus-pepsin has been obtained, but in the absence of sufficient sequence information interpretation of the electron density maps is difficult. We can

Sequences of Acid Protease Zymogens[a]

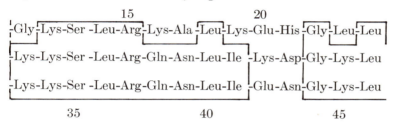

15   20

-Gly-Lys-Ser -Leu-Arg-Lys-Ala -Leu-Lys-Glu-His -Gly-Leu-Leu

-Lys-Lys-Ser -Leu-Arg-Gln-Asn-Leu-Ile -Lys-Asp-Gly-Lys-Leu

-Lys-Lys-Ser -Leu-Arg-Gln-Asn-Leu-Ile -Glu-Asn-Gly-Lys-Leu

35   40   45

-Asn-Pro-Ala -Ser -Lys-Tyr-Phe-Pro-Glu-Ala -Ala -Ala -Leu[b]

-Asn-Leu-Gly-Ser -Lys-Tyr-Ile -Arg-Glu-Ala -Ala -Thr-Leu[b]

[b] Leu in position 46 is involved in the peptide bond which is cleaved during the activation process.
———— and - - - - - - - - are explained in Table IX.

now look forward to rapid advances in our understanding of acid proteases.

Acknowledgments

I am deeply indebted to my collaborators John Graham, Ian Harris, Anne Hui, Stephen Jones, Alec Kurosky, Geoff Mains, Letty Rao, Jaro Sodek, Miho Takahashi, and Tusn-Tien Wang for their hard work and the many stimulating ideas which they contributed to the studies on acid proteases carried out in this laboratory. The work was generously supported by the Medical Research Council of Canada. I am grateful to B. Foltmann, B. Kassell, J. Tang, and W.-J. Chang for their permission to include material unpublished at the time of writing.

Literature Cited

1. Foltmann, B., in "Milk Proteins: Chemistry and Molecular Biology," Vol. 2, McKenzie, H. A., Ed., p. 217–254, Academic, New York, 1971.
2. Smith, A. K., Circle, S. J., in "Soybeans: Chemistry and Technology," Vol. 1, Smith, A. K., Circle, S. J., Eds., p. 1–26, Avi, Westport, Conn., 1972.
3. Komiya, A., *Soybean Digest* (1964) **24**, 43.
4. Gillespie, A. L., "The Natural History of Digestion," p. 16, London, 1898.
5. Northrop, J. H., *J. Gen. Physiol.* (1930) **13**, 739–766.
6. Bernal, J. D., Crowfoot, D., *Nature* (1934) **133**, 794–795.
7. Borisov, V. V., Melik-Adamyan, B. R., Shutskeva, N. E., *Cryst.* (1972) **17**, 309–315.
8. Bunn, C. W., Moews, P. C., Baumber, M. E., *Proc. Roy. Soc. Lond.* (1971) **B178**, 245–258.
9. Camerman, N., Hofmann, T., Jones, S. R., Nyburg, S. C., *J. Mol. Biol.* (1969) **44**, 569–572.
10. Swan, I. D. A., *J. Mol. Biol.* (1971) **60**, 405–407.
11. Moews, P. C., Bunn, C. W., *J. Mol. Biol.* (1970) **54**, 395–397.
12. Moews, P. C., Bunn, C. W., *J. Mol. Biol.* (1972) **68**, 389–390.
13. Tang, J., Sepulveda, P., Marciniszyn, Jr., J. P., Chen, K. C. S., Huang, W.-Y., Tao, N., Lin, D., Lanier, J. P., *Proc. Natl. Acad. Sci.* (USA) (1973) **78**, 3437.
14. Moravek, L., Kostka, V., *FEBS Letters* (1973) **35**, 276–278.
15. Foltmann, B., Kauffman, D., Parl, M., Andersen, P. M., *Netherlands Milk Dairy J.* (1973) **27**, 165–175.
16. Harris, C. I., Kurosky, A., Rao, L., Hofmann, T., *Biochem. J.* (1972) **127**, 34P–35P.
17. Graham, J. E. S., Sodek, J., Hofmann, T., *Can. J. Biochem.* (1973) **51**, 789–796.
18. *Methods Enzymol.* (1971) **19**, Chap. 20–24, 27.
19. Foltmann, B., *Methods Enzymol.* (1971) **19**, 421–436.
20. Mycek, M. J., *Methods Enzymol.* (1971) **19**, 285–314.
21. Sodek, J., Hofmann, T., *Methods Enzymol.* (1971) **19**, 372–396.
22. Ichishima, E., *Methods Enzymol.* (1971) **19**, 397–405.
23. Arima, K., Yu, J., Iwasaki, S., *Methods Enzymol.* (1971) **19**, 446–459.
24. Ottesen, M., Rickert, W., *Methods Enzymol.* (1971) **19**, 459–460.
25. Whitaker, J. R., *Methods Enzymol.* (1971) **19**, 436–445.
26. Shinano, S., Fukushima, K., *Agr. Biol. Chem.* (1971) **35**, 1488–1494.
27. Amagase, S., *J. Biochem. (Tokyo)* (1972) **72**, 73–81.
28. Dickie, N., Liener, I. E., *Biochim. Biophys. Acta* (1962) **64**, 41–51.
29. Matsubara, H., Feder, J., in "The Enzymes," Vol. 3, 3rd ed., Boyer, P. D., Ed., p. 723–744, Academic, New York, 1971.
30. Movat, H. Z., Uruibara, T., Takeuchi, Y., McMorine, D. R. L., *Int. Arch. Allerg. Appl. Immunol.* (1971) **40**, 197–217.
31. Sodek, J., Hofmann, T., *Can. J. Biochem.* (1970) **48**, 425–431.
32. Mackinlay, A. G., Wake, R. G., in "Milk Proteins: Chemistry and Molecular Biology," Vol. 2, McKenzie, H. A., Ed., p. 175–215, Academic, New York, 1971.
33. Berridge, N. J., *Adv. Enzymol.* (1954) **15**, 423–448.
34. Nitschmann, H., Keller, W., *Helv. Chim. Acta* (1955) **30**, 804–815.
35. Jollès J., Alais, C., Jollès, P., *Biochim. Biophys. Acta* (1968) **168**, 591–593.
36. Mercier, J.-C., Uro, J., Ribadeau-Dumas, B., Grosclaude, F., *Eur. J. Biochem.* (1972) **27**, 535–547.
37. Berridge, N. J., *Biochem. J.* (1945) **39**, 179–186.
38. Canadian Patent No. **739,298**; Danish Patent No. **109,506**; British Patent No. **1,035,897**; U.S. Patent No. **3,275,453**.

39. Prins, J., Nielsen, T. K., *Process Biochem.* 1970, May issue.
40. Belnke, U., Siewert, R., *Milchstandard* (1969) **11**, 66–72.
41. Ramet, J. P., Alais, C., *Le Lait* (1973) **53**, 154–162.
42. Hesseltine, C. W., Wang, H. L., in "Soybeans: Chemistry and Technology," Vol. 1, Smith, A. K., Circle, S. J., Eds., p. 389–419, Avi, Westport, Conn., 1972.
43. Hartley, B. S., *Ann. Rev. Biochem.* (1960) **29**, 45–72.
44. Schlamowitz, M., Peterson, L. U., *J. Biol. Chem.* (1959) **234**, 3137–3145.
45. Schlamowitz, M., Peterson, L. U., *Biochem. Biophys. Acta* (1961) **46**, 381–383.
46. Fruton, J. S., in "The Enzymes," Vol. 3, 3rd ed., Boyer, P. D., Ed., p. 119–164, Academic, New York, 1971.
47. Hata, T., Hayashi, R., Doi, E., *Agr. Biol. Chem.* (1967) **31**, 357–367.
48. Kress, L. F., Peanasky, R. J., Klitgaard, H. M., *Biochim. Biophys. Acta* (1966) **113**, 375–389.
49. McQuillan, M. T., Trikojus, in "Tissue Proteinases," Barrett, A. J., Dingle, J. T., Eds., p. 157–166, North-Holland, Amsterdam, 1970.
50. Rajogopalan, T. G., Stein, W. H., Moore, S. J., *Biol. Chem.* (1966) **241**, 4295–4297.
51. Bayliss, R. S., Knowles, J. R., Wybrandt, G. B., *Biochem. J.* (1969) **113**, 377–386.
52. Kay, J., Ryle, A. P., *Biochem. J.* (1971) **123**, 75–82.
53. Meitner, P. A., *Biochem. J.* (1971) **124**, 673–676.
54. Tang, J. J., *Methods Enzymol.* (1970) **19**, 406–420.
55. Takahashi, K., Mizobe, F., Chang, W. J., *J. Biochem. (Tokyo)* (1972) **71**, 161–164.
56. Keilová, H., *FEBS Letters* (1970) **6**, 312–314.
57. Stepanov, V. M., Orekhovich, V. N., Lobareva, L. S., Mzhel'skaya, T. I., *Biokhimiya* (1969) **34**, 209–210.
58. Keilová, H., Lapresle, C., *FEBS Letters* (1970) **9**, 348–350.
59. Smith, G. D., Murray, M. A., Nichol, L. W., Trikojus, V. M., *Biochim. Biophys. Acta* (1969) **171**, 288–298.
60. Sodek, J., Hofmann, T., *Can. J. Biochem.* (1970) **48**, 1014–1017.
61. Kovaleva, G. G., Shimanskaya, M. P., Stepanov, V. M., *Biochem. Biophys. Res. Commun.* (1972) **49**, 1075–1081.
62. Lin, C. L., Ohtsuki, K., Hatano, H., *J. Biochem. (Tokyo)* (1973) **73**, 671–673.
63. Blow, D. M., in "The Enzymes," Vol. 3, 3rd ed., Boyer, P. D., Ed., p. 185–212, Academic, New York, 1971.
64. Keil, B., in "The Enzymes," Vol. 3, 3rd ed., Boyer, P. D., Ed., p. 250–277, Academic, New York, 1971.
65. Magnusson, S., in "The Enzymes," Vol. 3, 3rd ed., Boyer, P. D., Ed., p. 278–321, Academic, New York, 1971.
66. Robbins, K. C., Summaria, L., *Methods Enzymol.* (1971) **19**, 184–199.
67. Matsubara, H., *Methods Enzymol.* (1971) **19**, 642–651.
68. Schechter, I., Berger, A., *Biochem. Biophys. Res. Commun.* (1967) **27**, 157–162.
69. Abramowitz, N., Schechter, I., Berger, A., *Biochem. Biophys. Res. Commun.* (1967) **29**, 862–867.
70. Atlas, D., Berger, A., *Biochemistry* (1973) **12**, 2573–2577.
71. Fruton, J. S., Bergmann, M., *Science* (1938) **87**, 557–558.
72. Baker, L. E., *J. Biol. Chem.* (1951) **193**, 809–819.
73. Fruton, J. S., *Adv. Enzymol.* (1970) **33**, 401–443.
74. Perlmann, G. E., Kerwar, G. A., *Arch. Biochem. Biophys.* (1973) **157**, 145–147.
75. Casey, E. J., Laidler, K. J., *J. Amer. Chem. Soc.* (1950) **72**, 2159–2164.

76. Mains, G., Takahashi, M., Sodek, J., Hofmann, T., *Can. J. Biochem.* (1971) **49**, 1134–1149.
77. Sachdev, G. P., Fruton, J. S., *Biochemistry* (1970) **9**, 4465–4469.
78. Voynick, I. M., Fruton, J. S., *Proc. Nat. Acad. Sci.* (1971) **68**, 257–259.
79. Huang, W. Y., Tang, J., *J. Biol. Chem.* (1969) **244**, 1085–1091.
80. Sanger, F., Tuppy, H., *Biochem. J.* (1951) **49**, 481–490.
81. Ryle, A. P., Leclerc, J., Falla, F., *Biochem. J.* (1968) **110**, 4P.
82. Foltmann, B., *C.R. Trav. Lab. Carlsberg* (1964) **34**, 326–345.
83. Press, E. M., Porter, R. R., Cebra, J., *Biochem. J.* (1960) **74**, 501–514.
84. Dopheide, T. A. A., Todd, P. E. E., *Biochim. Biophys. Acta* (1964) **86**, 130–135.
85. Dopheide, T. A. A., Trikojus, V. M., *Biochim. Biophys. Acta* (1966) **118**, 435–437.
86. Tsuru, D., Hattori, A., Tsuji, H., Yamamoto, T., Fukumoto, J., *Agr. Biol. Chem.* (1969) **33**, 1419–1426.
87. Rickert, W., *C.R. Trav. Lab. Carlsberg* (1970) **38**, 1–17.
88. McCullough, J. M., Whitaker, J. R., *J. Dairy Sci.* (1971) **54**, 1575–1578.
89. Williams, D. C., Whitaker, J. R., Caldwell, P. V., *Arch. Biochem. Biophys.* (1972) **149**, 52–61.
90. Johansen, J. T., Ottesen, M., Svendsen, I., Wybrandt, G., *C.R. Trav. Lab. Carlsberg* (1968) **36**, 365–384.
91. Morihara, K., Tsuzuki, H., *Arch. Biochem. Biophys.* (1966) **114**, 158–165.
92. Turkova, J., Mikes, O., *Biochim. Biophys. Acta* (1970) **198**, 386–388.
93. Miyata, K., Tomoda, K., Isono, M., *Agr. Biol. Chem.* (1970) **34**, 1457–1462.
94. Sanger, F., Thompson, E. O. P., Kitai, R., *Biochem. J.* (1955) **59**, 509–518.
95. Johnson, P., Smillie, L. B., *Can. J. Biochem.* (1971) **49**, 548–562.
96. Sampathnarayan, A., Anwar, R. A., *Biochem. J.* (1969) **114**, 11–17.
97. Englund, P. T., King, T. P., Craig, L. C., Walti, A., *Biochemistry* (1968) **7**, 163–175.
98. Whitaker, D. R., Roy, C., Tsai, C. S., Jurašek, L., *Can. J. Biochem.* (1965) **43**, 1961–1970.
99. Jackson, R. J., Wolfe, R. S., *J. Biol. Chem.* (1968) **243**, 879–888.
100. Nordwig, A., Jahn, W. F., *Eur. J. Biochem.* (1968) **3**, 519–529.
101. Sanger, F., Tuppy, H., *Biochem. J.* (1951) **49**, 463–481.
102. Matsubara, H., Sasaki, R., Singer, A., Jukes, T. H., *Arch. Biochem. Biophys.* (1966) **115**, 324–331.
103. Berger, A., Schechter, I., *Phil. Trans. Roy. Soc. Ser. B.* (1970) **257**, 249–264.
104. Tang, J., *J. Biol. Chem.* (1971) **246**, 4510–4517.
105. Hartsuck, J. A., Tang, J., *J. Biol. Chem.* (1972) **247**, 2575–2580.
106. Chen, K. C. S., Tang, J., *J. Biol. Chem.* (1972) **247**, 2566–2574.
107. Takahashi, K., Chang, W. J., *J. Biochem. (Tokyo)* (1973) **73**, 675–677.
108. Knowles, J. R., *Phil. Trans. Roy. Soc. Ser. B.* (1970) **257**, 135–146.
109. Herriott, R. M., Northrop, J. H., *J. Gen. Physiol.* (1934) **18**, 35–67.
110. Lokshina, L. A., Orekhovich, V. N., *Biokhimiya* (1966) **31**, 143–149.
111. Hollands, J. R., Fruton, J. S., *Biochemistry* (1968) **7**, 2045–2053.
112. Mains, G., Burchell, R. H., Hofmann, T., *Biochem. Biophys. Res. Commun.* (1973) **54**, 275–282.
113. Erlanger, B. F., Vratsanos, S. M., Wasserman, M., Cooper, A. G., *Biochem. Biophys. Res. Commun.* (1966) **23**, 243–245.
114. Sodek, J., Hofmann, T., *J. Biol. Chem.* (1968) **243**, 450–452.
115. Kitson, T. M., Knowles, J. R., *FEBS Letters* (1971) **16**, 337–338.
116. Huang, W.-Y., Tang, J., *J. Biol. Chem.* (1972) **247**, 2704–2710.
117. Clement, G. E., *Progress in Bio. Org. Chem.* (1973) **2**, 177–238.
118. Hartsuck, J. A., Tang, J., *J. Biol. Chem.* (1972) **247**, 2575–2580.

119. Silver, M. S., Stoddard, M., *Biochemistry* (1972) 11, 191–200.
120. Hunkapiller, M. W., Richards, J. H., *Biochemistry* (1972) 11, 2829–2839.
121. Neumann, H., Levine, J., Berger, A., Katchalski, E., *Biochem. J.* (1959) 73, 33–41.
122. Fruton, J. S., Fuiji, S., Knappenberger, M. H., *Proc. Nat. Acad. Sci.* (1961) 47, 759–761.
123. Greenwell, P., Knowles, J. R., Sharp, H. C., *Biochem. J.* (1969) 113, 363–368.
124. Inoye, K., Fruton, J. S., *Biochemistry* (1968) 7, 1611–1615.
125. Neumann, H., Sharon, N., *Biochim. Biophys. Acta* (1960) 41, 370–371.
126. Sharon, N., Grisaro, V., Neumann, H., *Arch. Biochem. Biophys.* (1962) 97, 219–221.
127. Kozlov, L. V., Ginodman, L. M., Orekhovich, V. N., *Dokl. Akad. Nauk, SSSR* (1967) 172, 1207–1208.
128. Shkarenkova, L. S., Ginodman, L. M., Kozlow, L. V., Orekhovich, V. N., *Biokhimiya* (1968) 33, 154–161.
129. Takahashi, M., Hofmann, T., *Biochem. J.* (1972) 127, 35P.
130. Callaghan, X. P., Shepard, J. A., Reilly, T. J., McDonald, K. J., Ellis, S., *Anal. Biochem.* (1970) 38, 330–356.
131. Bender, M. L., Kezdy, F. J., *Ann. Rev. Biochem.* (1965) 34, 49–76.
132. Bender, M. L., Chow, Y. L., Chloupek, F., *J. Amer. Chem. Soc.* (1958) 80, 5380–5384.
133. Pedersen, V. B., Foltmann, B., *FEBS Letters* (1973) 35, 250.
134. Foltmann, B., in "Atlas of Protein Sequence and Structure," Vol. 5, Dayhoff, M. O., Ed., p. D-123, Nat. Biomed. Research Found., Wash. D.C., 1972.
135. Huang, W.-Y., Tang, J., *J. Biol. Chem.* (1970) 245, 2189–2193.
136. Keil, B., *Phil. Trans. Roy. Soc. Ser. B* (1970) 257, 125–133.
137. Stepanov, V. M., Vaganova, T. I., *Biochem. Biophys. Res. Commun.* (1968) 31, 825–830.
138. Rasmussen, K. T., Foltmann, B., *Acta Chem. Scand.* (1971) 25, 3873.
139. Ottesen, M., *Ann. Rev. Biochem.* (1967) 36, 55–76.
140. Foltmann, B., *Compt. Rend. Trav. Lab. Carlsberg* (1966) 35, 143–231.
141. Kassell, B., Kay, J., *Science* (1973) 180, 1022–1027.
142. Hofmann, T., *Abstracts of Papers,* Joint Conference Chem. Inst. Canada and Amer. Chem. Soc. (1970) BCHM-006.
143. Davie, E. W., Neurath, H., *J. Biol. Chem.* (1955) 212, 515–529.
144. Olson, M. O. J., Nagabushan, N., Dzwiniel, M., Smillie, L. B., Whitaker, D. R., *Nature* (1970) 228, 438–442.
145. Smillie, L. B., in "Atlas of Protein Sequence and Structure," Vol. 5, Dayhoff, M. O., Ed., p. D-111, Nat. Biomed. Research Found., Wash. D.C., 1972.
146. Gibbs, A., McIntyre, G. A., *Eur. J. Biochem.* (1970) 16, 1–10.
147. Ong, E. B., Perlmann, G. E., *J. Biol. Chem.* (1968) 243, 6104–6109.
148. Pedersen, V. B., Foltmann, B., *FEBS Letters* (1973) 35, 255–256.
149. Harboe, M., Foltmann, B., *FEBS Letters* (1973) 34, 311.

RECEIVED September 26, 1973.

7

Serine Proteases

D. R. WHITAKER

Department of Biochemistry, University of Ottawa, Ottawa, Canada

Serine proteases are widely distributed and have many different functions. They are products of at least two evolutionary pathways, which originate in prokaryotes. Many of them resemble trypsin, chymotrypsin, elastase, or subtilisin in specificity, but serine proteases with quite different specificities have been isolated recently. A recent NMR study of a bacterial protease labelled with ^{13}C at carbon 2 of its single imidazole groups implicates a buried side chain of aspartic acid as the ultimate base for proton transfers in catalysis and eliminates a charge separation from reaction schemes for catalysis. Much of the catalytic effectiveness of serine proteases can be attributed to substrate binding, but the interactions which yield a Michaelis complex are supplemented by others which stabilize intermediates on the reaction pathway.

Serine proteases are so named because their catalytic mechanism involves the side chain of one of the enzyme's serine residues. If a neutral solution of a serine protease is exposed to a nerve gas such as diisopropyl phosphonofluoridate (DFP), this particular side chain is selectively and irreversibly esterified, and the enzyme, E, is inactivated:

$$(R - O)_2 \underset{\substack{\| \\ O}}{P} - F + H - O - CH_2 - E \xrightarrow[-HF]{}$$

$$(R - O)_2 \underset{\substack{\| \\ O}}{P} - O - CH_2 - E$$

where R is an isopropyl group. Serine proteases may be described therefore as proteases which are selectively inhibited by DFP and related organophosphorus inhibitors.

Trypsin, chymotrypsin, and elastase, the pancreatic enzymes which operate in our intestinal tract, are the most familiar examples, but serine proteases are produced by other organisms, such as invertebrates, plants, fungi, and bacteria and are not simply food-digesting enzymes. For example, thrombin and plasmin are protective agents in the blood stream. Thrombin is a key enzyme in converting fibrinogen to an insoluble blood clot while plasmin digests the clot after it has served its purpose. Cocoonase, on the other hand, is an escape agent which enables the silkworm moth to free itself from the silk fiber envelope surrounding the chrysalis. It digests sericin, the protein cement which binds the fibers together, and is secreted just before the moth is ready to emerge. All the above-mentioned enzymes are secreted as inactive zymogens which are converted to active enzymes by proteolysis. Thus, trypsin is formed when enterokinase, a serine protease of the intestinal mucosa, removes the first six residues of trypsinogen. This activation method is not confined to protease activation. The hormone, insulin, is also formed by proteolysis of its precursor, and it is likely that other peptide hormones are generated in the same way. One of the enzymes responsible for insulin release from pro-insulin is a trypsin-like enzyme which, in this operation at least, helps to control carbohydrate metabolism. In short, serine proteases do many different jobs, and the current use of the bacterial protease, subtilisin, as laundry detergent additive is simply the continuation of a very ancient trend.

Trypsin was named more than 100 years ago. It and chymotrypsin were among the first enzymes to be crystallized, have their amino acid sequences determined, and have their three-dimensional structure outlined by x-ray diffraction. Furthermore, both enzymes hydrolyze not only proteins and peptides but a variety of synthetic esters, amides, and anhydrides whose hydrolysis rates can be measured conveniently, precisely and, in some instances, extremely rapidly. As a result, few enzymes have received more attention from those concerned with enzyme kinetics and reaction mechanisms. The techniques developed by the pioneers in these various fields have enabled other serine proteases to be characterized rapidly, and the literature on this group of enzymes has become immense. It might be concluded that knowledge of serine proteases is approaching completeness and that little remains but to fill in minor details.

This paper will not necessarily refute this conclusion, but it will try to make the point that recent literature is still full of surprises. Its scope is limited to a few topics of current interest. Comprehensive accounts of serine proteases are given in two excellent monographs (*1, 2*) and critical summaries of more recent findings in an annual publication of the Chemical Society (3).

Classification of Serine Proteases

Serine proteases can be classified according to their homologies in amino acid sequence, which provide a measure of evolutionary relationships, or according to their specificity. The most readily determined significant feature of the sequence is the sequence around the active serine residue. This residue is easy to identify, for example, by treating the enzyme with ^{32}P-labelled DFP. Table I summarizes the present data at the tripeptide level.

Table I. Sequences at the Active Serine Residue of Serine Proteases

Sequence	Enzyme Source
1. Asp.Ser.Gly	Vertebrates, invertebrates, *Streptomyces griseus*, and the soil bacterium, *Myxobacter 495.*
2. Ser.Ser.Gly	*Arthrobacter sp.*, a soil bacterium.
3. Glu. Ser. Ala	Yeast (yeast carboxypeptidase).
4. Glu.Ser.Val	Baker's yeast, French beans.
5. Thr.Ser.Met	*Bacillus* species.

Sequence 1, in Table I, is the sequence of trypsin, chymotrypsin, pancreatic elastase, thrombin, and other mammalian proteases, and it occurs throughout the animal kingdom down to invertebrates as primitive as the sea anemone (4). The *Streptomyces griseus* enzymes are from Pronase, a commercial enzyme preparation. Two of its components, *Streptomyces griseus* trypsin and protease A, not only have the Asp.Ser.-Gly sequence, but they show several other homologies in sequence with the mammalian enzymes (5, 6). The same is true for the sequence of α-lytic protease, the *Myxobacter 495* enzyme (7).

Dayhoff (8), using a computer program to compare its complete sequence with those of the mammalian enzymes, concluded that the degree of homology is sufficient to establish an evolutionary relationship. A time scale for protein evolution can be established by first accepting the geological evidence that the ancestors of modern mammals diverged from the cartilaginous fishes about 400 million years ago and then by assuming that the acceptance rate of point-mutations has been constant (five point-mutations per 100 million years for this enzyme group). On this time scale, Dayhoff's homology measures show that trypsin, chymotrypsin, elastase, and thrombin became separate enzymes at about 1.5 billion years ago and that the ancestral bacterial enzymes diverged from those of the eukaryotes (organisms with nucleated cells) about 2.6 billion years ago. This latter estimate agrees very well with that based on comparisons of the cytochrome C and cytochrome C_2 sequences.

Sequence 5 is the sequence of the subtilisin enzymes. Complete amino acid sequences have been reported for two such enzymes: subtilisin BPN', from *Bacillus amyloliquefaciens* and subtilisin Carlsberg from *Bacillus subtilis* (9). Although highly homologous with one another, the sequences show no homology with those of Asp.Ser.Gly enzymes and clearly evolved independently. The sequence data is still too fragmentary to indicate the evolutionary relationships of the *Arthrobacter* enzyme (10), the yeast carboxypeptidase (11), and the enzyme with sequence 4 (12). However, the disulfide bridges in the *Arthrobacter* enzyme suggest a closer relationship with the Asp.Ser.Gly enzymes than with the bridgeless subtilisins. The active serine sequence of the yeast carboxypeptidase is noteworthy in that the same active serine sequence has been reported for horse cholinesterase and for the acetylcholinesterase of the electric eel (13).

With one recent exception, which is mentioned later, serine protease specificities tend to be related to the nature of the amino acid residue which contributes the carbonyl group to a bond which is susceptible to cleavage, *i.e.*, to residue P_1 of, for example, the polypeptide:

$$H_2N \ldots P_3 \cdot P_2 \cdot P_1 \overset{\downarrow}{\cdot} P'_1 \cdot P'_2 \ldots COOH.$$

The side chain of P_1 is usually a major participant in productive substrate binding, but carbonyl groups, amido groups, and side chains of other residues are also involved. The contributor of the other bond component, the peptide $P'_1 \cdot P'_2 \ldots COOH$ in this example, becomes a leaving group in the first step of the cleavage reaction, and it can contribute to catalysis because of its properties as a leaving group. Substrates with good binding properties for a particular enzyme are commonly referred to as specific substrates of that enzyme. The other extreme is represented by a substrate such as *p*-nitrophenyl acetate, $CH_3CO \cdot O—\langle O \rangle—NO_2$, with poor substrate binding properties but a potential leaving group. It is hydrolyzed by virtually all serine proteases and by many other enzymes.

The Commission on Biochemical Nomenclature assigns enzyme numbers to 18 serine proteases in the 1972 edition of "Enzyme Nomenclature" (14). Seven are listed as having a trypsin-like specificity, *i.e.*, their specific substrates have a positively charged lysine or arginine residue at P_1. Three are listed as having a chymotrypsin-like specificity, *i.e.*, their specific substrates have residues of tryptophan, tyrosine, phenylalanine, or leucine at P_1, *i.e.*, residues with bulky, hydrophobic side chains. Two enzymes have elastase-like specificities. They prefer a residue with a small, neutral side chain at P_1, particularly alanine. However, a substrate such as N-acetyl-L-alanine methyl ester is weakly bound and, as illustrated by the data of Thompson and Blout on elastase substrates and inhibitors (15), there is maximum binding only when residues at P_2, P_3, P_4, and P_5 can contribute to the binding. The remaining enzymes are listed as having "no clear specificity." They are less demanding about the nature of the residue at P_1 and hence have a broader specificity. The subtilisins are in this group.

Two recently isolated serine proteases have quite different specificities. One is a protease of *Staphylococcus aureus* which has a high specificity for glutamic acid residues at P_1 (16, 17). The other is the yeast carboxypeptidase listed in Table I (11). As the name indicates, it degrades polypeptides by cleaving amino acid residues from the C-terminal end of the chain—a most unexpected specificity for a serine protease. Unlike Carboxypeptidase A, it is stable, is capable of removing proline residues, and would seem to be an ideal enzyme for determining C-terminal sequences.

Substrate Binding

The first step in the reaction with substrate is the formation of an enzyme–substrate complex (Michaelis complex): E + S ⇌ E.S. Since both the enzyme and its substrate may be at extremely low molar concentrations, the equilibrium constant for the complex formation should be as high as possible, *i.e.*, the enzyme should have a high affinity for its substrate.

The structural basis of this affinity has been revealed by the x-ray crystallographers. A prominent feature of the chymotrypsin molecule is a substrate binding pocket, admirably suited for binding a substrate with a large, hydrophobic side chain at P_1 (18). Ser-189, which lies deep in the pocket, is replaced in trypsin by a negatively charged aspartic acid residue which can form an ion pair with a substrate having a positively charged arginine or lysine side chain at P_1. Elastase has two replacements: Gly-216 is replaced by a valine residue and Gly-226 by a threonine residue. These residues have bulkier side chains and effectively plug the pocket. As a consequence, elastase is more dependent on secondary binding sites beyond the pocket, which can contribute to binding at other residues. The binding site of subtilisin is much like that of chymotrypsin; substrate binding in the Michaelis complex probably extends from the carbonyl group of P_5 to the carbonyl group of P'_2 (19).

X-ray studies have also shown that substrate binding capability is a key distinction between the active enzyme and its zymogen (20, 21). Chymotrypsinogen is unable to form a substrate binding pocket. The

conversion of chymotrypsinogen to α-chymotrypsin requires four cleavages, but the critical cleavage, the one which immediately produces an
active enzyme, is a tryptic cleavage of the linkage between Arg-15 and
Ile-16. It allows a salt linkage to form between the α-amino group of
Ile-16 and the side chain of Asp-194. This salt linkage promotes the
formation of the substrate binding pocket.

Catalysis

Residues of serine, histidine, and aspartic acid will be designated
Ser-195, His-57, and Asp-102 in the following discussion. The numbers
refer to their location in the amino acid sequence of chymotrypsinogen
(22), but they have no absolute significance for other enzymes and would
occur in a different order if they had been based on the subtilisin
sequence.

The Reaction Site (Ser-195). Ser-195 is located immediately adjacent to the substrate binding pocket. It is the serine which is esterified
by DFP. It also is acylated during the hydrolysis of substrates such
as *p*-nitrophenyl acetate and, as discussed in detail by Bender and Killheffer (23) in a recent review on chymotrypsin, it was established many
years ago that hydrolysis by a serine protease is the result of an acylation
and a deacylation at Ser-195:

Acylation $R-\underset{\underset{O}{\|}}{C}-Z + HO-CH_2-E \rightarrow R-\underset{\underset{O}{\|}}{C}-O-CH_2-E + Z \cdot H$

Deacylation $R-\underset{\underset{O}{\|}}{C}-O-CH_2-E + H_2O \rightarrow HO-CH_2-E + R-\underset{\underset{O}{\|}}{C}-OH$

A considerable body of kinetic evidence (23, 24) suggests that the
acylation reaction proceeds *via* a tetrahedral intermediate which is converted to the acyl enzyme by protonation of the leaving group, Z:

$$O=\underset{\underset{Z}{|}}{\overset{\overset{R}{|}}{C}} + HO-CH_2-E \xrightarrow[-H^+]{} O^--\underset{\underset{Z}{|}}{\overset{\overset{R}{|}}{C}}-O-CH_2-E \xrightarrow{+H^+}$$

$$ZH + O=\underset{}{\overset{\overset{R}{|}}{C}}-O-CH_2-E$$

This evidence has been reinforced recently by an elegant x-ray crystallographic study of the complex formed by trypsin and a polypeptide trypsin inhibitor of the bovine pancreas (25). The complex has proven to be a tetrahedral adduct which is stabilized by hydrogen bonds between the enzyme and the leaving group and by the inability of His-57 (*see* later) to assume a conformation which would enable it to protonate the leaving group.

The acylation and deacylation reactions at the serine residue can be described mechanistically as bimolecular substitutions in which a nucleophile uses an unshared electron pair to bond to a polarized carbon atom and displace the leaving group (Z):

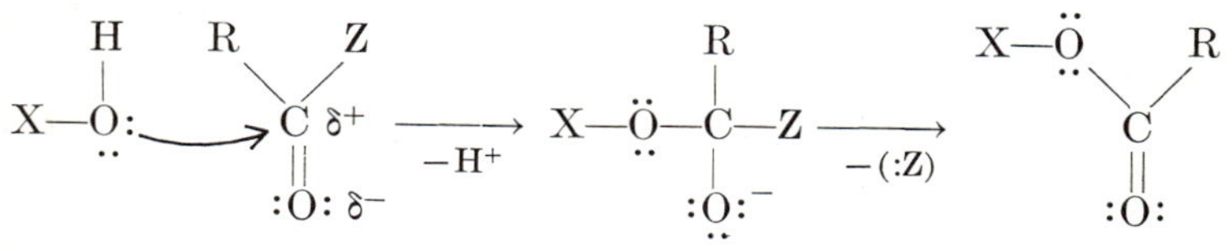

$$
X-\ddot{O}: \quad\longrightarrow\quad C^{\delta+} \xrightarrow[-H^+]{} \quad X-\ddot{O}-\overset{R}{\underset{:\overset{..}{O}:^-}{C}}-Z \xrightarrow[-(:Z)]{} \quad \overset{X-\ddot{O}}{\underset{:\overset{..}{O}:}{C}}
$$

However the nucleophiles in question—the hydroxyl group of serine for the acylation reaction and water for the deacylation reaction—are weak nucleophiles, and if the reaction is to proceed at appreciable rates, the enzyme must provide some agent to abstract a proton from the nucleophile and to donate a proton to the leaving group.

The Proton Relay (His-57 and Asp-102). Two nitrogen atoms and a carbon atom in the imidazole group of the side chain of His-57 will be designated in subsequent discussions as N-1, C-2, and N-3. These correspond to the atoms which are designated $N^{\delta 1}$, $C^{\epsilon 1}$, and $N^{\epsilon 2}$ in the system of nomenclature (26) widely used by x-ray crystallographers:

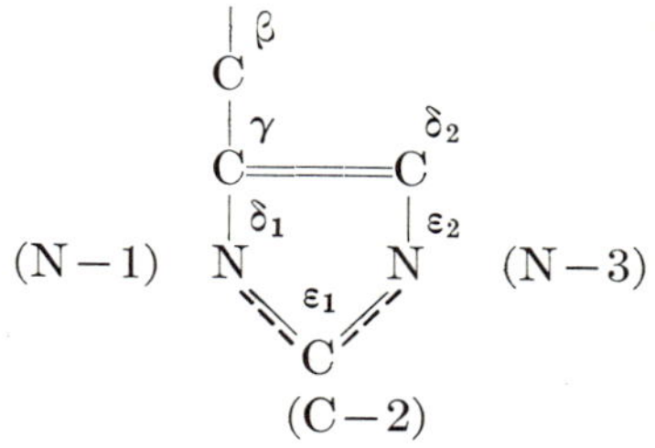

The side chain of His-57 is close enough to that of Ser-195 for a hydrogen bond to form between N-3 and the oxygen of the serine hydroxyl group. N-1 is shielded from solvent molecules. N-3 is selectively alkylated by the chloromethyl ketone and bromomethyl ketone inhibitors of serine proteases. These have the structure, R—C—CH$_2$—X, where X

$$\underset{O}{\overset{\|}{}}$$

is a chlorine or bromine atom and R—C— is an *N*-substituted aminoacyl
$\quad\quad\quad\quad\quad\quad\quad\quad\quad\quad\quad\quad\quad\quad\quad\quad$ ‖
$\quad\quad\quad\quad\quad\quad\quad\quad\quad\quad\quad\quad\quad\quad\quad\quad$ O
group chosen to direct the inhibitor to the binding site of the enzyme.
For example, an *N*-tosyl phenylalanyl group is chosen when the target
is chymotrypsin or an N^α-tosyl lysyl group when the target is trypsin (*27*).

Histidine participation in catalysis was suggested originally by the
pH dependence shown in hydrolyses of simple, neutral substrates such
as the methyl ester of an *N*-benzoyl amino acid. The hydrolysis rate
typically increases from nearly zero at pH 5 to a maximum value at pH
8. The rate of change is consistent with that of a catalytic process whose
rate-limiting step depends on an ionization with an apparent acid dissociation constant, pK_a, close to 7. This is a typical value for the dissociation of a proton from a protonated imidazole group:

$$pK_a \approx 7$$

It was recognized that protein ionizations are too dependent on the
nature of their microenvironment for a pK_a of ≈ 7 to be assigned with
certainty to histidine (*23*), but this inconclusive evidence led to a fruitful
concept, developed largely by Bender and his co-workers. They proposed that histidine functioned as a general acid–base catalyst in the
reaction—the conjugate base (neutral imidazole) being the general base
catalyst which accepts a proton from the nucleophile and the conjugate
acid (protonated histidine) being the general acid catalyst which donates
a proton to the leaving group.

One problem with such a mechanism is that the formation of the
tetrahedral intermediate entails a charge separation: the imidazole
group becomes positively charged when it accepts a proton, and the
substrate aquires a negative charge when it bonds to the nucleophile.
Such a charge separation would add greatly to the activation energy.
The enzyme could eliminate the problem if it provided an anion capable
of accepting a proton from imidazole at a neutral or weakly alkaline
pH. The enzyme acquired such a base by burying an acid, as discussed
below.

The side chain of Asp-102 is buried in the sense that it is completely
shielded from contact with water molecules. The side chain of His-57

forms part of the shield and allows a hydrogen bond to form between N-1 of its imidazole group and a carboxyl oxygen atom of Asp-102. Because N-3 of His-57 can be linked by a hydrogen bond to Ser-195, the three side chains are interconnected. Blow *et al.* (28) recognized the possible mechanistic significance of the connection between Asp-102 and Ser-195 and proposed a charge-relay mechanism for catalysis which will be discussed shortly. However, the presence of aspartic acid at the catalytic site had been demonstrated only for chymotrypsin at the time, and there was no certainty that it was a general feature of serine proteases.

The crystal structure of subtilisin BPN′ dispelled this uncertainty. As already mentioned, the subtilisins and the pancreatic enzymes are dissimilar in amino acid sequence, and they proved to be dissimilar in their gross three-dimensional structure. However, the components of their catalytic site do not differ. Both enzyme groups have the same catalytic triad with hydrogen bonds linking serine to N-3 of histidine and N-1 of histidine to a buried side chain of aspartic acid (29). Since the two enzyme groups are products of different evolutionary pathways, it follows almost inescapably that this striking homology is dictated by necessity and that the buried aspartic acid is essential for catalysis.

Blow *et al.* (28) attributed the reactivity of Ser-195 to a charge-relay mechanism operating in the following way. At a sufficiently alkaline pH, His-57 is unprotonated and no longer offsets the negative charge at Asp-102. A proton rearrangement then allows an appreciable charge transfer from Asp-102 to Ser-195, producing serine with a reactive alkoxide ion as its nucleophile:

$$
\begin{array}{ccc}
\overset{|}{C}H_2 & H\overset{|}{C}\!=\!\!=\!\overset{|}{C}H & \overset{|}{C}H_2 \\
O\!=\!C\!-\!\bar{O} \ldots HN \qquad N \ldots HO & & \longrightarrow \\
& \diagdown\!\!\diagup\!\!\diagup & \longleftarrow \\
& \underset{H}{C} &
\end{array}
$$

$$
\begin{array}{ccc}
\overset{|}{C}H_2 & H\overset{|}{C}\!=\!\!=\!\overset{|}{C}H & \overset{|}{C}H_2 \\
O\!=\!C\!-\!OH \ldots N \qquad NH \ldots \bar{O} & & \\
& \diagdown\!\!\diagup & \\
& \underset{H}{C} &
\end{array}
$$

In effect, Asp-102 is assumed to so enhance the basicity of His-57 that it can withdraw a proton from serine. However, as Bender and Killheffer (23) point out, the advantage gained in the reaction with substrate to form a tetrahedral intermediate would tend to be offset in the following step by the lessened ability of His-57 to function as the general acid catalyst which donates a proton to the leaving group.

This sort of problem underlines the need for an unequivocal determination of microscopic ionization constants. As already mentioned, the rate-limiting step in catalysis is controlled by an ionization with a pK_a of ~7. Cruickshank and Kaplan (30) used α-chymotrypsin as a test enzyme and estimated the pK_a of its two histidine residues from the pH dependence of their reaction with trace amounts of tritiated 1-fluoro-2,4-dinitrobenzene. From their data, a pK_a of 6.8 was assigned to His-57 and 6.7 to His-40, which is not involved in catalysis. This data is consistent with deprotonation of His-57 being the critical ionization for catalysis.

Hunkapiller *et al.* (31, 32) approached the problem in a radically different way. They determined the pH dependence of the parameters of the ^{13}C nuclear magnetic resonance spectrum for a bacterial protease which had been specifically enriched with ^{13}C at C-2 of its single histidine residue. The enzyme, α-lytic protease (33), is listed in Table I. The reasons for choosing it for this investigation are as follows.

1. The enzyme is produced by a bacterium, *Myxobacter* 495, which allows L-[2-^{13}C] histidine to be incorporated cleanly into the enzyme. Its original growth medium had an appreciable histidine content from the casein hydrolysate which was used as a nitrogen source. However, the organism has no absolute requirement for any individual amino acid (34), and most of the casein hydrolysate can be replaced by L-glutamate without impairing enzyme production. L-[2-^{13}C] Histidine was added at intervals during the growth cycle. Test runs with ^{14}C-labelled histidine showed no appreciable transfer of ^{14}C to other amino acid residues of the enzyme.

2. The enzyme is extremely stable. It begins to unfold reversibly when the pH is less than 4 (35), but, unlike α-chymotrypsin, it does not aggregate at low pH, autolyse at neutral pH, or undergo a major conformation change at a moderately alkaline pH. Freedom from autolysis is particularly important in ^{13}C NMR spectroscopy because of the long time periods required for signal accumulation.

3. Since the enzyme has only one histidine residue, the NMR spectrum is not confused by signals from ^{13}C-2 in histidine which is not involved in catalysis.

4. The amino acid sequence of the enzyme is homologous with those of the pancreatic enzymes and has been shown by model building to be compatible with a chymotrypsin-like three-dimensional structure and catalytic site (36). The homology in sequence is particularly well marked around His-57. The enzyme's catalytic properties are virtually indistinguishable from those of pancreatic elastase.

The reader is referred to the original papers for an account of the spectroscopic procedures. The most significant findings are in Tables II and III [taken from Hunkapiller *et al.* (*32*)]. The β-lytic protease listed in Table II is another enzyme of *Myxobacter 495*, a metallo-protease with eight histidine residues.

The data in Table II for various ^{13}C-2 imidazole compounds indicates that the chemical shift does not distinguish protonated imidazole from neutral imidazole with certainty, but the coupling constant between C-2 and its directly bonded hydrogen atom appears to distinguish them unequivocally. At pH 3.3, α-lytic protease gives three signals (Table III). The second signal listed for pH 3.3 has the coupling constant for protonated imidazole, the same chemical shift as denatured β-lytic protease in acidic solution, and although not indicated in the table, a considerably narrower line width than the other two signals, which indicates that the side chain has greater freedom of movement. This signal is assigned to the protonated imidazole which is no longer hydrogen bonded to its neighbors in the catalytic site.

The first and third signals listed for pH 3.3 are assigned respectively to neutral and protonated imidazole within the catalytic site. The enzyme at pH 5.2 and 8.2 gives a single signal with the coupling constant for neutral imidazole. Relaxation measurements at these pHs indicate that the imidazole group is held rigidly by the enzyme.

These measurements assign a pK_a of <4 to the imidazole ionization of α-lytic protease. The low pK_a indicates that the conjugate base (neutral

Table II.   Chemical Shifts and Directly Bonded Carbon–Hydrogen Coupling Constants for C-2 Carbon in Imidazole Derivatives[a]

Compound	δ (ppm (±0.04) from Me$_4$Si)		$^1J_{CH}$ (Hz (±1))	
	Cation	*Neutral*	*Cation*	*Neutral*
Imidazole	−134.05	−136.23	219	209
4-Methylimidazole	−133.17	−135.40	221	208
4-Methylimidazole (dioxane)	−133.13	−134.49	219	205
1-Methylimidazole	−135.52	−138.38	220	207
L-Histidine methyl ester	−135.08	−136.71	222	208
N-Acetyl-L-	−134.17		221	
histidine	−133.85	−139.45	220	204
(4-Imidazolyl)-	−134.25		220	
acetic acid	−133.65	−136.47	221	205
β-Lytic protease[b] (denatured)	−134.09	−136.67	218	206

[a] Measured for 1–2M aqueous solutions unless otherwise indicated.
[b] 1–2mM solution in 0.2M KCl.
[c] Reproduced from (*32*).

Table III.   Chemical Shift and Coupling Constant Values for C-2 Carbon in Histidine Residue of α-Lytic Protease

pH	Chemical Shift ($ppm \pm 0.12$ from Me_4Si)	$^1J_{CH}$ ($Hz \pm 3$)
8.2	−137.26	205
5.2	−134.79	205[a]
3.3	−134.81	208
	−134.05	222
	−132.46	218

[a] Six determinations of $^1J_{CH}$ around pH 5–6 yielded values of 203, 204, 205, 206, and 208 Hz.
[b] Reproduced from (*32*).

imidazole) is more stable and accepts a proton less readily than the conjugate base of a normal imidazole group which is fully exposed to solvent. At a pH above 4, the imidazole group is unprotonated.

More extended measurements of the chemical shift as a function of pH indicate that the environment of C-2 is perturbed by an ionization with a pK_a of 6.75. The adjoining carboxylic acid of Asp-102 is the obvious perturbant. It can be expected to ionize less readily than an unburied carboxyl group and hence to have an abnormally high pK_a value. As the unionized conjugate acid is stabilized preferentially, the conjugate base will accept a proton more readily than a normal carboxylate ion which is fully exposed to solvent. According to this assignment, the ionization with a pK_a of ≈7 which controls the catalysis rate is the ionization of Asp-102 and, because its higher pK_a indicates that Asp-102 is protonated more readily than His-57, Asp-102 is the ultimate base for proton transfer within the catalytic site.

These assignments of pK_a values allow the acyl enzyme to be formed in the following way:

(a). Predominant form at pH <6.7 (Inactive)

(b). Predominant form at pH >6.7 (Active)

(c) Tetrahedral inter-
mediate

$$CH_2 \quad HC{=\!=}CH \quad CH_2$$
$$O{=}C{-}OH \ldots N \quad NH \searrow \quad O \quad \longrightarrow$$
$$\searrow \quad \diagdown\diagup \quad Z \quad \Big| \quad R \quad -ZH$$
$$C \quad \diagdown\diagup$$
$$H \quad C$$
$$\Big|$$
$$O^-$$

(d) Acyl enzyme

$$CH_2 \quad HC{=\!=}CH \quad CH_2$$
$$O{=}C{-}O^- \ldots HN \quad N \quad O$$
$$\diagdown \quad \diagup\diagup$$
$$C \quad C{-}R$$
$$H \quad \Big|\Big|$$
$$O$$

The inactive enzyme, (a), in acidic solution is converted to active enzyme, (b), by the simultaneous transfer of a proton from aspartic acid to N-1 of His-57 and from N-3 of His-57 to the base added to the solution. Asp-102 is ionized, but His-57 remains neutral in this and in subsequent proton transfers. The reaction with substrate to give the tetrahedral intermediate, (c), does not involve a charge separation. His-57 receives a proton at N-3 and releases a proton at N-1 to neutralize the charge on Asp-102. Thus, an active enzyme with a negative charge at Asp-102 is converted to a tetrahedral intermediate with a negative charge at a new location. The reverse proton flow in the last step protonates the leaving group to give the acyl enzyme, (d).

The mechanism also accounts for Kaplan and Cruickshank's assignment of a pK_a of 6.8 to His-57 (30). As N-1 of His-57 is shielded from solvent, the reactive species for the reaction with 1-fluoro-2,4-dinitrobenzene can be expected to be (b), with an accessible lone pair of electrons at N-3. The pK_a which is measured in this procedure is the pK_a of the reaction which determines the relative proportions of (a) and (b). In this curious situation, data on the formation of DNP-histidine can measure the pK_a of an ionization of aspartic acid.

The deacylation reaction with water can proceed in the same manner:

$$CH_2 \qquad\qquad CH_2 \qquad\qquad CH_2$$
$$\Big| \qquad\qquad\qquad \Big| \qquad\qquad\qquad \Big|$$
$$O \qquad\qquad\qquad O \qquad\qquad\qquad OH$$
$$H \qquad\qquad \Big| \qquad\qquad\qquad \Big| \qquad\qquad\qquad \Big|$$
$$O \longrightarrow C{-}R \quad \longrightarrow \quad HO{-}C{-}R \quad \longrightarrow \quad HO{-}C{-}R$$
$$H \qquad\quad \Big|\Big| \quad {-}H^+ \qquad \Big|^- \quad {+}H^+ \qquad \Big|\Big|$$
$$O \qquad\qquad\qquad O \qquad\qquad\qquad O$$
$$(e) \qquad\qquad\qquad (f) \qquad\qquad\qquad (g)$$

with the proton relay taking up the proton released in the formation of the tetrahedral intermediate, (f), and releasing it to regenerate serine in the last step. The product, (g), will normally ionize to R—C—O⁻,

$$\text{R—C—O}^-$$
$$\parallel$$
$$\text{O}$$

and the negative charge at Asp-102 will help to expel it from the catalytic site.

The Stabilization of Intermediates. So far as can be determined by x-ray diffraction, Ser-195, His-57, and Asp-102 have essentially the same conformation in chymotrypsinogen as in α-chymotrypsin (*20, 21*). The zymogen cannot form a Michaelis complex but, as discussed recently by Wright (*37*), it can catalyze displacement reactions at Ser-195 by reagents such as DFP and cyanate. The rates of these chymotrypsinogen-catalyzed reactions are only $10–10^3$ times greater than the background rate from basic catalysis, whereas the rates of α-chymotrypsin-catalyzed hydrolyses are $10^8–10^9$ times the background rate when one of the enzyme's specific esters or amides is the substrate. Clearly, it is highly advantageous to supplement catalysis with a facility for capturing substrates and binding them in a conformation favorable for catalysis. As Wright points out, substrate binding must account for a large part of the $10^8–10^9$-fold rate enhancement of α-chymotrypsin. Moreover, a recent model-building study on the reaction pathway for subtilisin suggests that the interactions which yield a Michaelis complex are only the first of a series of interactions which facilitate catalysis.

Kraut and his associates had previously determined the structure of various derivatives and complexes of subtilisin BPN′. They then constructed hypothetical models of a Michaelis complex, a tetrahedral intermediate, and an acyl enzyme (*19*). The oxyanion hole illustrates the development. The oxyanion hole is an empty space (or hole) in the Michaelis complex which is occupied by a negatively charged oxygen atom (an oxyanion) in the tetrahedral intermediate. It is the focal point for hydrogen bonds from two potential hydrogen bond donors—the nitrogen atom of Ser-195 and the nitrogen atom in the side chain of an asparagine residue of subtilisin. In the Michaelis complex, the carbonyl group of the substrate's P_1 residue is close to the oxyanion hole but beyond bonding range. Two rotations—an upward swing of the side chain of Ser-195 as it attacks the carbonyl group of P_1 and an upward rotation of the carbonyl oxygen as the carbonyl group changes to a tetrahedral conformation—enable the carbonyl oxygen to enter the oxyanion hole and be stabilized by two hydrogen bonds. It was predicted that tetrahedral intermediates of the pancreatic enzymes would be stabilized in the same way, with the nitrogen atom of Gly-193 replacing the nitrogen atom of asparagine. The x-ray study of

Rühlmann *et al.* (*25*) on the complex between trypsin and pancreatic trypsin inhibitor has shown that the prediction was well founded. The complex is a tetrahedral adduct, and the oxyanion is hydrogen bonded, as predicted, to the nitrogens of Ser-195 and Gly-193.

This example is but one of several interactions which can stabilize a tetrahedral intermediate or the acyl enzyme (*19*). They lower the activation energy for the formation of intermediates and thus increase the formation of the intermediates. The net result is more efficient catalysis.

Literature Cited

1. Perlmann, G. E., Lorand, L., *Methods Enzymol.* (1970) **19.**
2. Boyer, P. D., *Enzymes*, 3rd ed. (1971) **3.**
3. Young, G. T., "Specialist Periodical Reports: Amino Acids, Peptides, and Proteins," Chemical Society, London, 1969-72.
4. Gibson, D., Dixon, G. H., *Nature (London)* (1969) **222,** 753.
5. Jurášek, L., Fackre, D., Smillie, L. B., *Biochem. Biophys. Res. Commun.* (1969) **37,** 99.
6. Johnson, P., Smillie, L. B., *Can. J. Biochem.* (1972) **50,** 589.
7. Olson, M. O. J., Nagabhusan, N., Dzwiniel, M., Smillie, L. B., Whitaker, D. R., *Nature* (1970) **228,** 438.
8. Dayhoff, M. O., "The Atlas of Protein Sequence and Structure," Vol. **5,** p. 102, National Biomedical Research Foundation, Washington, D.C., 1972.
9. Smith, E. L., DeLange, R. J., Evans, W. H., Landon, M., Markland, F. S., *J. Biol. Chem.* (1968) **243,** 2184.
10. Wählby, S., *Biochim. Biophys. Acta* (1968) **151,** 409.
11. Hayashi, R., Moore, S., Stein, W., *J. Biol. Chem.* (1973) **248,** 8366.
12. Shaw, D. C., Wells, J. R. E., *Biochem. J.* (1967) **104,** 5c.
13. Dayhoff, M. O., "The Atlas of Protein Sequence and Structure," Vol. **5,** p. 56, 1972.
14. Commission on Biochemical Nomenclature, "Enzyme Nomenclature," Elsevier, Amsterdam, 1972.
15. Thompson, R. C., Blout, E. R., *Biochemistry* (1973) **12,** 44, 47, 51, 57, 66.
16. Drapeau, G. R., Boily, Y., Houmard, J., *J. Biol. Chem.* (1972) **247,** 6720.
17. Houmard, J., Drapeau, G. R., *Proc. Nat. Acad. Sci. USA* (1972) **69,** 3506.
18. Steitz, T. A., Henderson, R., Blow, D. M., *J. Mol. Biol.* (1969) **46,** 337.
19. Robertus, J. D., Kraut, J. D., Alden, R. A., Birktoft, J. J., *Biochemistry* (1972) **11,** 4293.
20. Freer, S. T., Kraut, J., Robertus, J. D., Wright, H. T., Xuong, Ng T., *Biochemistry* (1970) **9,** 1997.
21. Wright, H. T., *J. Mol. Biol.* (1973) **79,** 1.
22. Hartley, B. S., *Phil. Trans. Roy. Soc. London* (1970) **B257,** 77.
23. Bender, M. L., Killheffer, J. V., *Crit. Rev. Biochem.* (1973) **1,** 149.
24. Fersht, A. R., Requena, Y., *J. Amer. Chem. Soc.* (1971) **93,** 7079.
25. Rühlmann, A., Kukla, D., Schwager, P., Bartels, K., Huber, R., *J. Mol. Biol.* (1973) **77,** 417.
26. Edsall, J. T., Flory, P. J., Kendrew, J. C., Liquori, A. M., Nemethy, G., Ramachandran, G. N., Scheraga, H. A., *J. Mol. Biol.* (1966) **15,** 399.
27. Shaw, E., *Enzymes* (1970) **1,** 94.
28. Blow, D. M., Birktoft, J. J., Hartley, B. S., *Nature (London)* (1969) **221,** 337.

29. Alden, R. A., Wright, C. S., Kraut, J., *Phil. Trans. Roy. Soc. London* (1970)
    B257, 119.
30. Cruickshank, W. H., Kaplan, H., *Biochem. J.* (1973) **130,** 36.
31. Hunkapiller, M. W., Smallcombe, S. H., Whitaker, D. R., Richards, J. H.,
    J. Biol. Chem. (1973) **248,** 8306.
32. Hunkapiller, M. W., Smallcombe, S. H., Whitaker, D. R., Richards, J. H.,
    Biochemistry (1973) **12,** 4732.
33. Whitaker, D. R., *Methods Enzymol.* (1970) **19,** 599.
34. Wiberger, I., Whitaker, D. R., unpublished data.
35. Leskovac, V., Whitaker, D. R., unpublished data.
36. McLachlan, A. D., Shotton, D. M., *Nature New Biol.* (1971) **229,** 202.
37. Wright, H. T., *J. Mol. Biol.* (1973) **79,** 13.

RECEIVED March 20, 1974.

8

The Sulfhydryl Proteases

IRVIN E. LIENER

Department of Biochemistry, University of Minnesota, St. Paul, Minn. 55101

The sulfhydryl proteases—papain, ficin, and bromelain—all have a sulfhydryl group at their active site and are thus readily inactivated by reagents or conditions which modify this functional group. These enzymes are used extensively in the food industry for the tenderization of meat and chill-proofing of beer, have application in the tanning and textile industries, and are used medicinally as digestive aids and debriding agents. The enzymatic and physicochemical properties and structural features of these enzymes are compared with particular emphasis on the relationship of structure to mechanism of action.

Among the proteolytic enzymes, the plant proteases are the most widely used in the food industry. Most of the plant proteases which have been studied are characterized by a free sulfhydryl group which is essential for their activity. The most important of these so-called sulfhydryl or thiol proteases are papain, ficin, and bromelain. Since the literature on these enzymes has been the subject of several recent reviews (*1, 2, 3, 4*), major emphasis is placed in this presentation on the use of these enzymes in the food industry. Some of the more recent developments relating to the structure and function of the sulfhydryl proteases are discussed.

Commercial Preparations

Papain is the main protein constituent of latex of the green fruit, leaves, and trunk of *Carica papaya*, a small softwood tree which is native to most tropical countries. Although the protein-digesting property of papain was first recorded in 1873 (*5*), the native custom of tenderizing meats by wrapping them in leaves of the papaya tree prior to cooking dates back centuries. Commercial papain is collected from full-grown but still unripe fruits by making shallow, longitudinal scratches in the fruit and allowing the collected drippings to coagulate. The coagulated latex is dried in the sun or in heated chambers to reduce the moisture content to 5–8%. Since FDA regulations do not permit the use of crude papain in food products, most commercial preparations have been further purified by precipitation with an organic solvent such as acetone. An excellent monograph on the commercial preparation of papain was published in 1963 by the Central Food Technological Research Institute, Mysore, India (6).

Proteolytic activity in the juice of the pineapple plant (*Ananas comosus*) was first reported in 1879 (7). More recently, the juice from the stem of the pineapple plant was shown to be a rich source of stem bromelain. This name is used to distinguish the enzyme from another which is derived from the fruit (8, 9). Mature pineapple stems are collected by special harvesting machines. The juice is pressed by special mills and then filtered. Most commercial preparations of bromelain have been precipitated from the stem juice by acetone.

Although ficin has been commercially available for many years, it has not attained the importance of papain and bromelain. Ficin is imported mainly from South America where it is obtained from the latex of tropical fig trees of the genus *Ficus*. There are over 2000 species of this genus, and the amount of activity varies considerably from one species to another (10). Most of the commercial preparations of ficin are derived from *F. glabrata*. The latex is collected from the trees by tapping, and the juice separated after the latex has coagulated. The juice is then spray-dried, or the protein may be precipitated with acetone.

Food, Industrial, and Medicinal Uses

Since several excellent books and monographs are available on this subject (11, 12, 13, 14, 15), only a brief survey is presented here. The principal commercial uses of the sulfhydryl proteases are summarized in Table I. Although papain is used mainly for these purposes, ficin and bromelain also have been used with essentially the same results.

Chill-Proofing Beverages. Papain is used mainly in the beverage industry for stabilizing and chill-proofing beer. When beer is made, some of the protein which is soluble at room temperature is apt to precipitate on chilling, causing a cloudy product. Because papain digests proteins in a slightly acid environment (pH 4.5), it prevents the separation of the protein in the cold beer. About 1 gram of commercial papain is added per barrel (8 ppm) prior to pasteurization at 60°C. This temperature is not severe enough to denature the enzyme to any appreciable extent.

Meat Tenderizing. This is the second largest use of papain. About one-third of all the papain sold in this country is used by the housewife

Table I.   Commercial Uses of Sulfhydryl Proteases

In Food Industry
1. Chill-proofing beverages
2. Meat tenderizing
3. Other: Milk-clotting agent, dough ingredient, protein hydrolysates

Industrial Uses
1. Leather industry
2. Textile industry
3. Cleaning agent

Medicinal Uses
1. Digestive aid
2. Prevention of post-operative adhesions
3. Debridement

for tenderizing small cuts of meat. The product used for this purpose contains in addition to papain (2%), varying amounts of salt, glucose, monosodium glutamate, flavoring agents, and spices. The powder is applied to meat before it is cooked and is forced into the tissues by stabbing with a fork. The meat is then cooked in the usual way. Muscle protein and connective tissue, primarily collagen and elastin, are digested to the point where a definite softening of the flesh is observed. Papain is relatively resistant to heat, and most of the proteolysis takes place during the early stages of cooking, the greatest breakdown of tissue occurring at 70°C. When the enzyme concentration is too high or the period of treatment too long, overtenderization and mushiness will result.

Unlike the housewife, the meat packer is faced with the problem of applying the enzyme uniformly throughout very large pieces of meat. One effective way of doing this is to inject the enzyme into the animal prior to slaughter. It will be carried by the vascular system into all parts of the body and exert its tenderizing action after the animal is dead. The recommended dosage for antemortem injection is approximately 5 ppm of commercial papain based on the live weight of the animal. An enzyme solution may also be injected into freshly slaughtered carcasses, placing the needles so that there is intramuscular distribution of the enzyme. Other techniques for tenderizing meat with papain which have been patented since 1960 are described by Wieland (*13*). The extent to which ficin and bromelain are used in place of papain is determined largely by their availability and other economic factors.

Other Food Uses. Papain has also been used to hydrolyze renderer's meat scraps to make a product which can be used for feeding farm animals. Because of their ability to coagulate casein in much the same fashion as rennet, papain and ficin have been used as milk-clotting agents in place of rennet in the production of cheese. Papain may be substituted for fungal enzymes as a dough ingredient. It serves to improve quality by lowering the viscosity of the dough. This effect results from its ability to degrade and depolymerize the gluten of the flour. In actual commercial practice, however, the thiol proteases are little used for bread making or milk clotting, at least in the United States.

Industrial Uses. Papain is used in the leather industry to prepare the sides for tanning. Its proteolytic action removes some of the undesirable proteins which adhere to the hide and thus facilitates the subsequent tanning process. In the textile industry, the treatment of wool fibers with papain has been found to reduce the shrinkage from laundering. This appears to be caused by the ability of the enzyme to destroy the elastic properties of wool protein. Because of its digestive action on protein, papain is used as a spot remover in the laundry and dry cleaning business.

Medicinal Uses. One of the earliest medicinal applications of the thiol proteases was the use of fig latex as an anthelmintic agent (*16, 17*). In more recent times papain has been used as a digestive aid in the treatment of dyspepsia and gastric distress. Its usefulness in this respect derives from the fact that papain remains active under acid conditions and is resistant to attack by pepsin. The intraperitoneal injection of sterile solutions of papain has proved to be effective in preventing post-operative adhesions. Papain is sometimes referred to as a "biological scalpel" because of its specific proteolytic action on dead tissue without affecting live tissue. For this reason it serves as a very useful debridement in the treatment of burns and the removal of scar tissue. When used in combination with antibiotics, it has proved to be effective for topical treatment of ulcerating and infected lesions.

Other Uses. The thiol proteases have proved to be extremely useful tools in studying protein structure (*18*). Their utility in this respect has been considerably enhanced by the preparation of insolubilized derivatives (*see,* for example, *19, 20, 21*).

Purification

Papain. Papain was first isolated in crystalline form from fresh latex by Balls *et al.* (*22*), but it is more conveniently isolated from commercially available dried latex by the procedure of Kimmel and Smith (*23*). Papain prepared by this procedure, however, is not fully active (*see* "Activation and Inhibition" below) and usually contains only 0.5 mole SH per mole of protein. Fully active papain containing 1 mole SH per mole of protein may be prepared by affinity chromatography on a column of an inhibitor, Gly-Gly-Tyr (Bzl)-Arg, covalently linked to agarose (*24*). Two other techniques which have been used to purify papain involve affinity chromatography on columns of agarose to which p-amino-phenylmercuriacetate (25) or glutathione-2-pyridyl disulfide (26) have been attached. The reactions involved in these techniques are depicted in Figure 1.

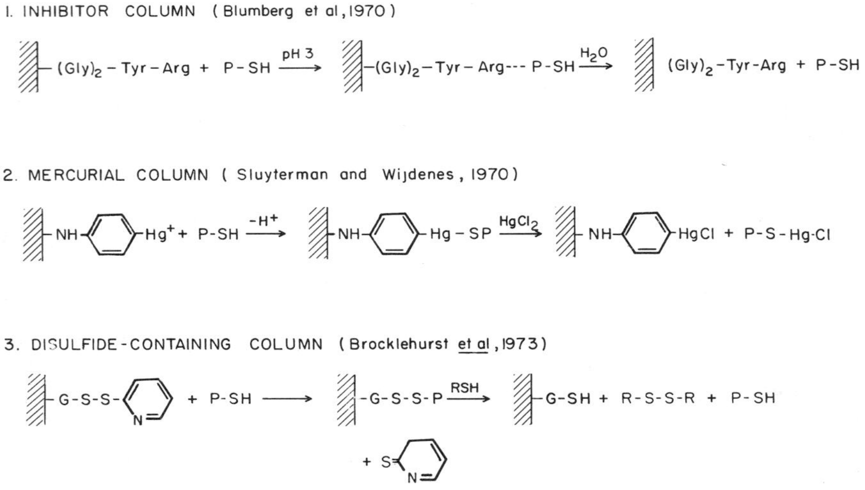

I. INHIBITOR COLUMN (Blumberg et al,1970)

2. MERCURIAL COLUMN (Sluyterman and Wijdenes , 1970)

3. DISULFIDE-CONTAINING COLUMN (Brocklehurst et al ,1973)

Figure 1.   Purification of papain by affinity chromatography

In addition to papain, at least two other proteolytic components have been shown to be present in crude extracts of commercial papaya latex, namely, chymopapain (1, 27) and papaya peptidase A (28). Since these isolated enzymes are not used commercially, they will not be considered further.

Ficin. This enzyme was first isolated in crystalline form by Walti (29) although subsequent studies have shown this preparation to be heterogeneous (30). Since several active components are generally observed when crude preparations of ficin are chromatographed (31–36), the designation of any one active component as "ficin" has been quite arbitrary. The possibility has been excluded that these multiple forms of ficin could have arisen by autodigestion from a common precursor or as artifacts of the purification procedure (32). A representative result from an attempt to purify ficin is shown in Figure 2. Despite the differences in techniques which were used in these experiments (*see* Table I of Ref. 3), it is evident that there are at least three major proteolytic components present in most preparations generally referred to as "ficin."

Bromelain. Although several investigators have reported the preparation of chromatographically pure stem bromelain (37, 38), there is some uncertainty as to whether these preparations are really homogeneous since other workers have observed several proteolytically active components to be present (*39, 40, 41, 42*). It is still not clear how much of this heterogeneity is caused by autodigestion (*43*). Bobb (*44*) recently described the purification of stem bromelain by affinity chromatography on ε-aminocaproyl-D-tryptophan methyl ester coupled to Sepharose.

An apparently homogeneous preparation of fruit bromelain can be prepared by acetone precipitation of the protein from the juice of the green or ripe fresh fruit followed by chromatography on DEAE-cellulose (*38*).

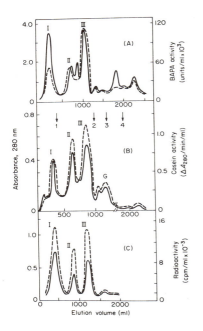

Figure 2.   Chromatography of preparations of ficin on CM-cellulose as performed in several laboratories. A: (31); B: (30); C: (55). Solid line denotes protein as measured by absorbance at 280 nm. Dotted line denotes activity on benzoyl-DL-arginine-p-nitroanilide in A, casein in B, and radioactivity of C-14 carboxymethyl group in C.

Physicochemical Properties

A comparison of the major physicochemical properties of the sulfhydryl proteases is shown in Table II. Generally speaking there do not appear to be any major differences in the size and charge distribution among the various enzymes. The data pertaining to bromelain and ficin, however, must be accepted with the reservation that the preparations used for obtaining these data may not have been homogeneous and that differences may exist with respect to the particular components which various investigators may have used. Nevertheless, despite their distinguishability by ion-exchange chromatography, it has proved difficult to differentiate among the various active components of ficin and brome-

Table II. Comparison of the Physicochemical

Property	Papain	Ficin
$S_{20, w}$	2.66	2.66[c]
$D_{20, w}$	10.23	
Molecular weight		
sedimentation	23,700	25,500
amino acid composition	23,000[e]	23,800
No. amino acid residues	212	248[b]
Carbohydrate, %	0	3.4[h]
Isoelectric point, pH	8.75	9.0
Absorbancy $A^{1\%}{}_{1cm}$ at 280 nm	25.0	21.0

[a] Unless specified otherwise, most of the data shown here are taken from Glazer and and Smith (1).
[b] (27).
[b] (45).
[d] (46)

lain on the basis of amino acid composition, amino terminal groups, substrate specificity, kinetic parameters (31–36, 39, 40), and active site sequences (48).

One of the most characteristic differences between papain and the other sulfhydryl proteases, ficin and bromelain, is the absence of carbohydrate in papain. Glycopeptides have been isolated from ficin (48) and bromelain (49, 50, 51, 52). In the case of bromelain the carbohydrate moiety composed of mannose, xylose, fucose, and N-acetylglucosamine is covalently linked to an asparagine residue of the peptide chain through a glucosamine component of the sugar moiety (51, 52).

Activation and Inhibition

Activation. Papain prepared in the presence of cysteine according to the method of Kimmel and Smith (23) is composed of three species of the enzyme: (1) active papain with a free thiol group (about 50% of the protein), (2) inactive papain in which the thiol group has formed a disulfide bond with cysteine, and (3) inactive papain in which the thiol group has been oxidized to a sulfonic acid group. Thiol proteases are activated by mild reducing agents, low molecular weight thiol compounds, and cyanide because of the formation of active papain (species 1) from disulfide-linked papain (species 2). Species 3 has undergone irreversible oxidation and cannot be activated under these conditions. Brocklehurst and Kierstan (53), however, found that when papain was isolated from dried latex in the absence of cysteine, very little active papain was obtained despite the fact that this preparation contained about 0.4 mole SH/mole protein. Papain prepared in this manner could be activated with cysteine and was indistinguishable from activated papain

Properties of the Sulfhydryl Proteases[a]

	Stem bromelain	Fruit bromelain	Chymopapain B[b]
	2.73		2.82
	7.77		
	20,000–33,200	18,000 [d]	34,500
	35,730[f]		
	179–285[g]		318
	1.46–2.1[g]	3.2[g]	
	9.55		10.4
	19.0		18.4

[e] Based on revised sequence reported by Drenth *et al.* (*47*).
[f] (*38*).
[g] (*4*).
[h] (*48*).

isolated in the presence of cysteine. On the basis of this and other evidence the authors postulate that papain exists in dried latex as a zymogen (propapain) in which the active SH group (cys-25) forms a disulfide bond with a neighboring cysteine residue, but upon treatment with a reducing agent or cyanide it undergoes intramolecular thiol–disulfide interchange to yield the active isomer (*see* Figure 3). Whether a similar mechanism of activation applies to the other thiol proteases has yet to be determined. The thiol–disulfide mechanism described here could conceivably account for the differences in amino sequence around the SH group of ficin reported by Metrione *et al.* (*54*) and Friedenson and Liener (*55*).

Inhibition. Since papain, ficin, and bromelain are all enzymes whose activity depends on a free SH group, it is to be expected that all thiol reagents act as inhibitors. Thus, α-halogen acids or amides and N-ethylmaleimide irreversibly inhibit the thiol proteases. Heavy metal ions and organic mercurial salts inhibit in a fashion that can be reversed by low molecular weight thiols, particularly in the presence of EDTA which

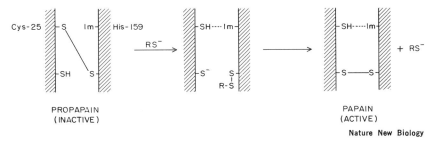

Nature New Biology

Figure 3. Activation of papain according to Brocklehurst and Kierstan (53)

chelates such metals. The chloromethyl ketone derivatives of lysine (TLCK) and phenylalanine (TPCK), which are known to react with the active site histidine residue of the serine proteases, also inhibit the thiol proteases. In this instance, the SH group is the site of reaction (*56, 57, 58, 59*). Diisopropylphosphofluoridate (DFP) has also been reported by some investigators (*60, 61, 62, 63*) to inhibit the thiol proteases, but this inactivation was subsequently found to be caused by the presence of an unidentified impurity present in some lots of DFP (*64*).

Westrik and Wolfenden (*65*) have recently reported that aldehydes, with side chains similar to those comprising the acyl portion of substrates which papain effectively hydrolyzes, were very potent inhibitors of this enzyme. Umezawa and his associates (*66*) have also recently reported that certain microorganisms (actinomyces) produced an aldehyde, acetyl-L-leucyl-L-leucyl-L-argininal (leupeptin), which inhibits papain. The structures of some of the more effective aldehyde inhibitors of papain are shown in Figure 4.

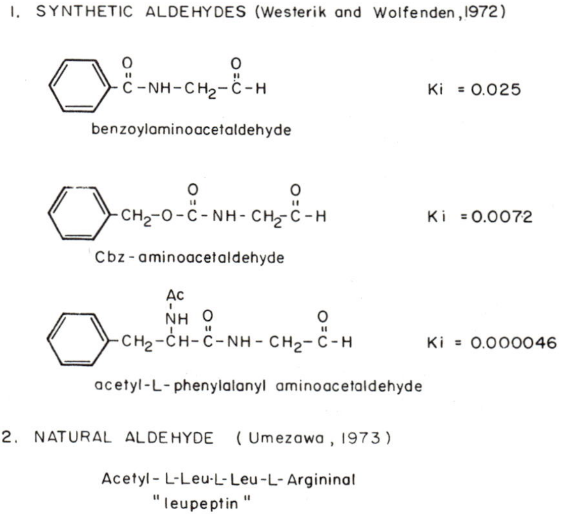

I. SYNTHETIC ALDEHYDES (Westerik and Wolfenden,1972)

benzoylaminoacetaldehyde   Ki = 0.025

Cbz-aminoacetaldehyde   Ki = 0.0072

acetyl-L-phenylalanyl aminoacetaldehyde   Ki = 0.000046

2. NATURAL ALDEHYDE   (Umezawa, 1973)

Acetyl-L-Leu-L-Leu-L-Argininal
"leupeptin"

Figure 4.   Aldehyde inhibitors of papain

Substrate Specificity

In general the thiol proteases catalyze the hydrolysis of a variety of peptide, ester, and amide bonds of synthetic substrates. Employing the general formula R′ —NH—CHR—CO—X, cleavage of the —CO—X— bond has been demonstrated when R represents the side chain of glycine, threonine, methionine, lysine, arginine, citrulline, leucine, and tyrosine.

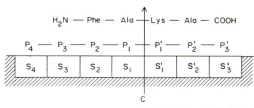

Biochemical and Biophysical
Research Communications

*Figure 5.   Specificity of papain in re-
lation to its active site (67)*

The X component of the substrate may be derived not only from amino acids but also from alcohols, ammonia, mercaptans, aniline, and *p*-nitrophenol. A benzoyl or carbobenzoxy group has generally been employed as the R′ substituent of the synthetic substrate. Examination of the peptides produced by cleavage of various proteins has confirmed the rather indiscriminate specificity of this group of enzymes.

The variety of peptide bonds cleaved by thiol proteases may be superficially interpreted as indicating a low degree of specificity. Schechter and Berger (67), however, have shown that the active site of papain can accommodate a peptide sequence of up to seven amino acid residues. One can visualize (Figure 5) the active site of papain as consisting of seven subsites where C denotes the catalytic site and the point of cleavage of the substrate. From the examination of a large number of peptides it was found that if P_2, the amino acid residue which specifically interacts with subsite S_2, is L-phenylalanine or L-tyrosine, then the cleavage of the peptide bond one amino acid residue away was considerably enhanced. These results show that the specificity of papain is determined not so much by the side chain of the amino acid containing the susceptible bond but rather by the nature of the amino acids adjacent to it.

Structure/Function Relationship

As shown in Figure 6, the similar properties that exist among the plant proteases can also be extended to the amino acid sequences which surround the active thiol group of these enzymes (68–73).

Evidence that a histidine residue is also involved at the active site of the thiol proteases is largely inferential and rests mainly on chemical (73, 74, 75), kinetic (76, 77), and crystallographic data (*see* below). For example, dibromoacetone has been used to demonstrate that a histidine residue is close enough (within 5A) to the SH group to form a covalent bridge between these two residues (*see* Figure 7). The use of active-site-directed reagents such as TLCK and TPCK, which have proved so useful in identifying a histidine residue at the active site of the serine proteases,

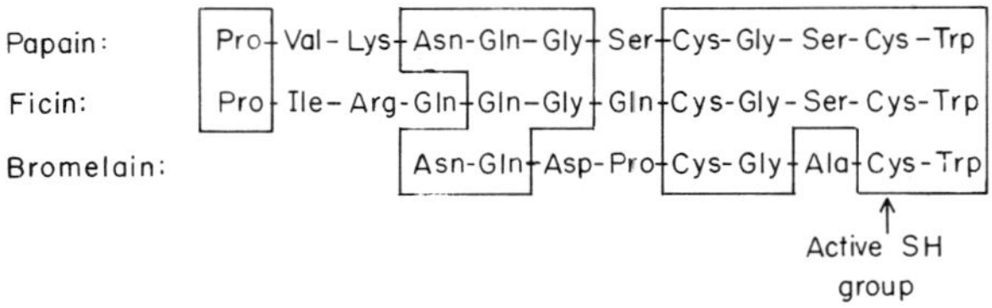

Figure 6. Amino acid sequences around the active SH group of thiol proteases

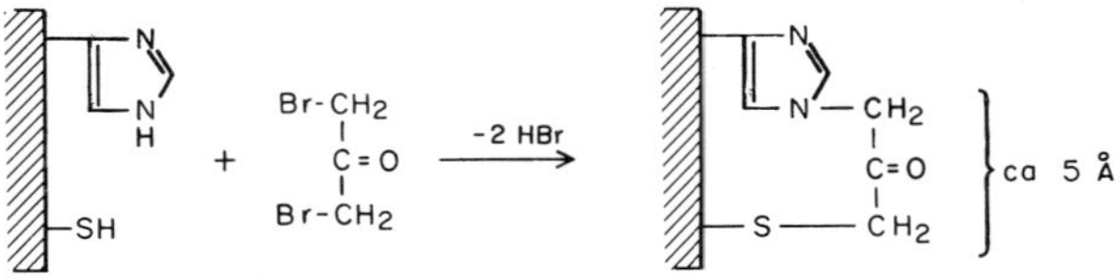

Figure 7. Reaction of thiol and histidine residues of thiol protease with dibromoacetone

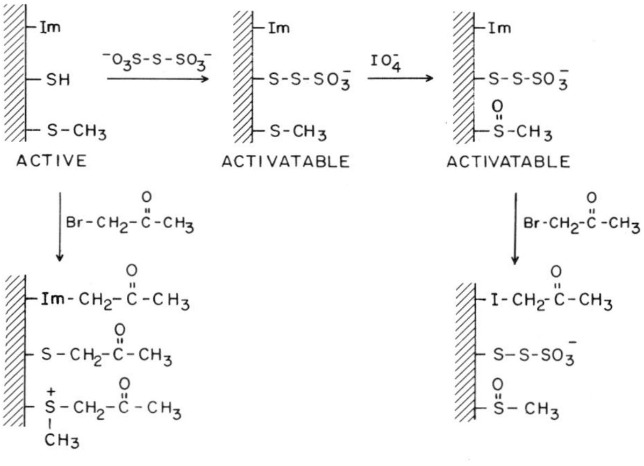

Figure 8. Reactions involved in directing the specificity of bromoacetone to the alkylation of a histidine residue in ficin

is not applicable to the thiol proteases since the thiol group seems to react preferentially with such reagents (56, 57, 58, 59). Evidence of the participation of a histidine residue at the active site of ficin was recently obtained in our laboratory by the sequence of reactions shown in Figure 8 (78). By reversible protection of the active thiol group with sodium tetrathionate and oxidation of a vulnerable methionine residue with metaperiodate, it was possible to direct the specificity of bromoacetone to the alkylation of a single histidine residue. The modification of histidine in this manner

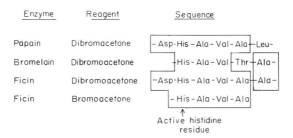

Enzyme	Reagent	Sequence
Papain	Dibromacetone	– Asp·His – Ala – Val · Ala –Leu-
Bromelain	Dibromoacetone	–His – Ala – Val – Thr –Ala –
Ficin	Dibromoacetone	–Asp·His – Ala – Val – Ala –Ala –
Ficin	Bromoacetone	– His – Ala – Val – Ala

↑
Active histidine
residue

Figure 9.   Amino acid sequences around the active histidine residue of thiol proteases

was accompanied by a complete loss of enzymatic activity. As shown in Figure 9, the amino acid surrounding the modified histidine residue was found to be virtually identical to the sequence surrounding the histidine residue which can be crosslinked to the active SH groups of papain, ficin, and bromelain with dibromoacetone (*73, 74, 75*).

The complete amino acid sequence of papain is shown in Figure 10. The structure is based on the combined efforts of several groups of investigators (*69, 72, 79*). Cysteine-25 has been identified as the active thiol

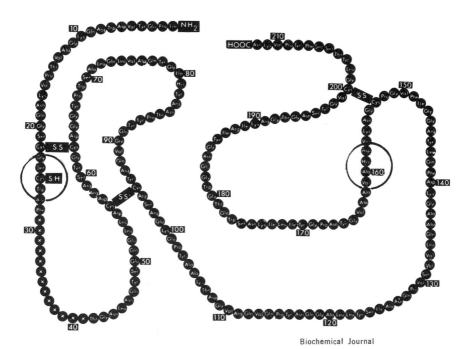

Biochemical Journal

Figure 10.   Amino acid sequence of papain (72)

group (*64*). The secondary and tertiary structure of papain based on the x-ray crystallographic studies of Drenth and associates (*47, 80*) is shown in Figure 11. The molecule has a deep cleft running diagonally down the center so that it appears to be almost two molecules. Only three segments of the chain cross the cleft, and the essential cysteine residue-25 lies to one side of the cleft. Histidine-159 lies on the other side of this cleft. Because its close proximity to cysteine-25 (about 4.5 A), it is believed to constitute the active-site histidine residue already referred to. The substrate is stereospecifically bound to groups in the cleft so that cysteine-25 and histidine-159 are properly oriented to participate in the catalytic mechanism described below.

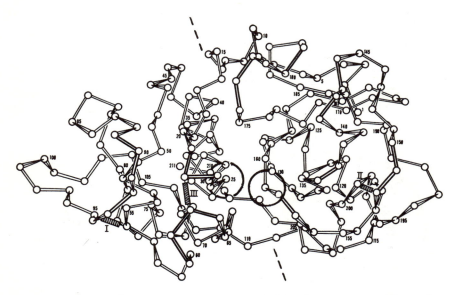

Figure 11.   Main chain conformation of papain based on x-ray crystallographic data (47, 80)

Mechanism of Action

The overall reaction pathway for the catalytic activity of the thiol proteases is best described by the scheme shown in Figure 12. This mechanism shows the formation of an enzyme–substrate complex which results in the acylation of the enzyme (to form a thiol ester) and its subsequent deacylation, the overall reaction leading to a regeneration of the enzyme, and the elimination of the products of hydrolysis.

Evidence for a thiol ester intermediate is provided by spectrophotometric data which show the formation of *N-trans*-cinnamoylpapain (*81*)

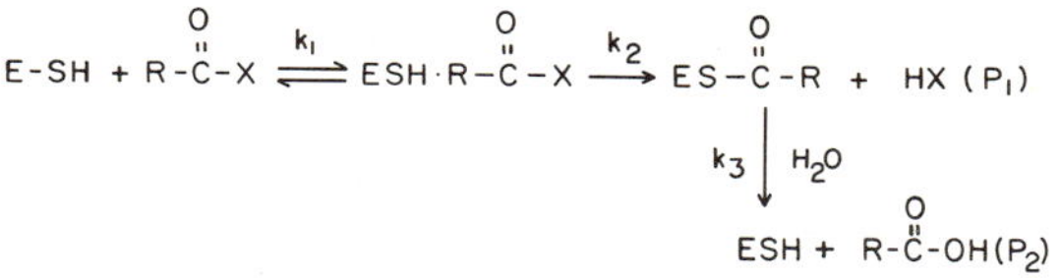

$$E-SH + R-\overset{\overset{\text{O}}{\|}}{C}-X \underset{}{\overset{k_1}{\rightleftharpoons}} ESH \cdot R-\overset{\overset{\text{O}}{\|}}{C}-X \xrightarrow{k_2} ES-\overset{\overset{\text{O}}{\|}}{C}-R + HX\ (P_1)$$

$$k_3 \downarrow H_2O$$

$$ESH + R-\overset{\overset{\text{O}}{\|}}{C}-OH(P_2)$$

Figure 12. Reaction pathway depicting the mechanism of action of papain where ESH and

$$\overset{\overset{\text{O}}{\|}}{R—C—X}$$ *are enzyme and substrate respectively and P_1 and P_2 are products formed from the substrate*

and thionohippurylpapain (82). From a study on the effect of pH on various kinetic parameters of papain (83, 84, 85), it may be concluded that a group having a pK_a of 8.5 is involved in the acylation step. This is presumably the active thiol group which reacts in its unionized form in the catalytically active enzyme. Kinetic evidence also implicates a group having a pK_a of about 4 as being involved in the acylation and deacylation steps. The precise identification of the responsible amino acid residue is a matter of controversy. Stockell and Smith (86) originally proposed that the group responsible for pK_a near 4 was a carboxyl group which from the x-ray data would most likely be aspartic-158. The chemical and x-ray evidence already referred to, however, suggest that a more likely candidate for such a group would be histidine-159. Sluyterman and Wolthers (87) believe that this residue should really have a pK_a of 9.5–10 rather than 4 because of hydrogen bonding to asparagine-175 and electro-

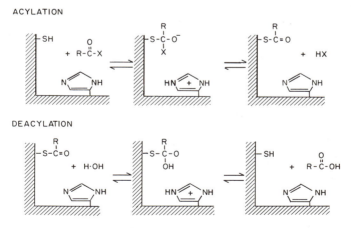

Figure 13. A possible mechanism of action for papain catalyzed hydrolysis

static interaction with aspartic acid-158. They have accordingly proposed a mechanism in which histidine-159 participates in the catalytic process as a conjugate acid. The pH-dependence of the acylation and deacylation rate constants on an apparent pK_a near 4 was considered to be caused by a carboxyl group which in its undissociated state led to an inactive conformation of the enzyme. On the other hand, Allen and Lowe (88) argue that the abnormally low pK_a of histidine-159 can be attributed to its enclosure by a hemisphere of hydrophobic residues, particularly tryptophan-177. A mechanism which portrays histidine as playing a key role in catalysis is presented in Figure 13, although other mechanisms are not necessarily precluded (87, 89).

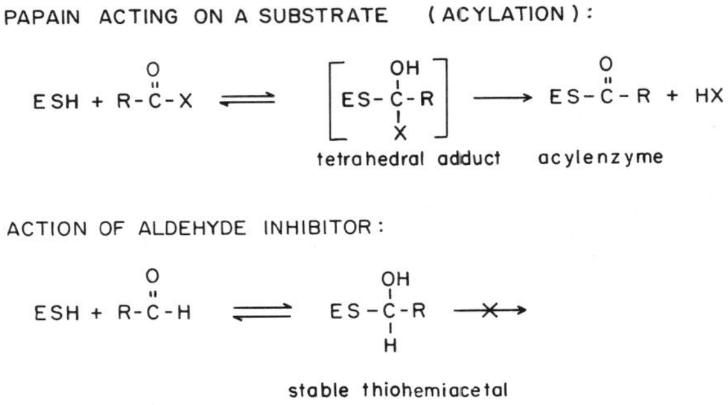

PAPAIN ACTING ON A SUBSTRATE   (ACYLATION) :

ACTION OF ALDEHYDE INHIBITOR :

Figure 14. Analogy of formation of thiol adduct with aldehyde inhibitors to formation of tetrahedral intermediate in papain catalysis

One of the key intermediates shown in this reaction scheme is the formation of a tetrahedral adduct during acylation and deacylation (84). Additional support for the formation of a tetrahedral intermedite comes from the observation already referred to—that aldehydes may act as potent inhibitors of papain. Westerik and Wolfenden (65) attribute the inhibitory effect of aldehydes to the formation of a stable thiol adduct (thiohemiacetal) analogous to the tetrahedral intermediate produced when papain acts on a substrate. This relationship is depicted in Figure 14. When the complete picture for the mechanism of catalysis by the thiol proteases finally emerges, it will no doubt be similar to the mechanism of action of the serine proteinases.

Literature Cited

1. Glazer, A. N., Smith, E. L., "Papain and Other Plant Sulfhydryl Proteolytic Enzymes" in "The Enzymes," P. D. Boyer, Ed., Vol. III, p. 501–546, Academic, New York, 1971.
2. Arnon, R., "Papain," *Methods Enzymol.* (1970) **19**, p. 226–244.

3. Liener, I. E., Friedenson, B., "Ficin," *Methods Enzymol.* (1970) **19**, p. 261–273.
4. Murachi, T., "Bromelain Enzymes," *Methods Enzymol.* (1970) **19**, p. 273–284.
5. Roy, G. C., *Calcutta Med. J.* (1873). Cited by Hwang, K., Ivy, A. C., *Ann. New York Acad. Sci.* (1951) **54**, 161.
6. Bhutiani, R. C., Shankar, J. V., Menon, P. G. K., "Papaya," Industrial Monograph no. 2, Central Food Technological Research Institute, Mysore, India, 1963.
7. Wurtz, A., Bouchet, A., *Compt. rend. hebd. Seance* (1879) **89**, 425.
8. Balls, A. K., Thompson, R. R., Kies, M. W., *Ind. Eng. Chem.* (1941) **33**, 950.
9. Heinicke, R. M., Gortner, W. A., *Economic Botany* (1957) **11**, 225.
10. Williams, D. C., Sgarbieri, V. C., Whitaker, J. R., *Plant Physiol.* (1968) **43**, 1083.
11. Reed, G., "Enzymes in Food Processing," Academic, New York, 1966.
12. DeBecze, G. I., "Food Enzymes," *Critical Reviews Food Technol.* (1970) **1**, 479.
13. Wieland, H., "Enzymes in Food Processing and Products," Noyes Data Corp., Park Ridge, N. J., 1972.
14. Whitaker, J. R., "Principles of Enzymology for the Food Sciences," Marcel Decker Inc., New York, 1972.
15. Anonymous, "Papain, a Versatile Enzyme in Industry and Medicine," S. B. Penick and Co., New York, 1956.
16. Robbins, B. H., *J. Biol. Chem.* (1930) **87**, 251.
17. Jaffe, W. G., *Trop. Diseases Bull.* (1943) **40**, 612.
18. Mihalyi, E., "Application of Proteolytic Enzymes to Protein Structure Studies," Chemical Rubber Company, Cleveland, 1972.
19. Cebra, J. J., Givol, D., Silman, I. H., Katchalski, J., *Biol. Chem.* (1961) **236**, 1720.
20. Silman, I. H., Albu-Weissenberg, M., Katchalski, E., *Biopolymers* (1966) **4**, 441.
21. Goldman, R., Kedem, O., Silman, I. H., Katchalski, E., *Biochemistry* (1968) **7**, 486.
22. Balls, A. K., Lineweaver, H., Thompson, R. R., *Science* (1937) **86**, 379.
23. Kimmel, J. R., Smith, E. L., *J. Biol. Chem.* (1954) **207**, 515.
24. Blumberg, S., Schechter, I., Berger, A., *Eur. J. Biochem.* (1970) **15**, 97.
25. Sluyterman, L. A. Ae., Wijdenes, J., *Biochem. Biophys. Acta* (1970) **200**, 593.
26. Brocklehurst, K., Carlsson, J., Kierstan, M. P. J., Crook, E. M., *Biochem. J.* (1973) **133**, 573.
27. Kunimatsu, D. K., Yasunobu, K. T., "Chymopapain B," *Methods Enzymol.* (1970) **9**, 244–252.
28. Schack, P., *Compt. rend. Trav. Lab Carlsberg* (1967) **36**, 1.
29. Walti, A., *J. Amer. Chem. Soc.* (1938) **60**, 493.
30. Sgarbieri, V. C., Gupte, S. M., Kramer, D. E., Whitaker, J. R., *J. Biol. Chem.* (1964) **239**, 2170.
31. Englund, P. T., King, T. P., Craig, L. C., Walti, A., *Biochemistry* (1968) **7**, 163.
32. Kramer, D. E., Whitaker, J. R., *J. Biol. Chem.* (1964) **239**, 2178.
33. Kramer, D. E., Whitaker, J. R., *Plant Physiol.* (1969) **44**, 1560.
34. Williams, D. C., Whitaker, J. R., *Plant Physiol.* (1969) **44**, 1566.
35. Williams, D. C., Whitaker, J. R., *Plant Physiol.* (1969) **44**, 1574.
36. Jones, I. K., Glazer, A. N., *J. Biol. Chem.* (1970) **245**, 2765.
37. Murachi, T., Yasui, M., Yasuda, Y., *Biochemistry* (1964) **3**, 48.
38. Ota, S., Moore, S., Stein, W. H., *Biochemistry* (1964) **3**, 180.

39. El-Gharbawi, M., Whitaker, J. R., *Biochemistry* (1963) **2**, 476.
40. Feinstein, G., Whitaker, J. R., *Biochemistry* (1964) **3**, 1050.
41. Ota, S., Horie, K., Hagino, F., *J. Biochem.* (Tokyo) (1969) **66**, 413.
42. Minami, Y., Doi, E., Hata, T., *Agr. Biol. Chem.* (1971) **35**, 1419.
43. Ota, S., *J. Biochem.* (Tokyo) (1968) **63**, 494.
44. Bobb, D., *Prep. Biochem.* (1972) **2**, 347.
45. Gould, N. R., Ph.D. Dissertation, University of Minnesota, 1964.
46. Ota, S., Horie, K., Hagino, F., Hasimoto, C., Date, H., *J. Biochem.* (Tokyo) (1972) **71**, 817.
47. Drenth, J., Jansonius, J. N., Koekoek, R., Swen, H. M., Wolthers, B. G., *Nature* (1968) **218**, 929.
48. Friedenson, B., Ph.D. Dissertation, University of Minnesota, 1970.
49. Murachi, T., Suzuki, A., Takahashi, N., *Biochemistry* (1967) **6**, 3730.
50. Takahashi, T., Yasuda, Y., Kazuya, M., Murachi, T., *J. Biochem.* (Tokyo) (1969) **66**, 659.
51. Scocca, J., Lee, Y. C., *J. Biol. Chem.* (1969) **244**, 4852.
52. Murachi, T., Takahashi, N., "Structure and Function of Stem Bromelain" in "Structure-Function Relationship of Proteolytic Enzymes," P. Desnuelle, H. Neurath, and M. Ottesen, Eds., p. 298, Academic, New York, 1970.
53. Brocklehurst, K., Kiersten, P. J., *Nature New Biol.* (1973) **242**, 167.
54. Metrione, R. M., Johnston, R. B., Seng, R., *Arch. Biochem. Biophys.* (1967) **122**, 137.
55. Friedenson, B., Liener, I. E., *Arch. Biochem. Biophys.* (1972) **149**, 169.
56. Bender, M. L., Brubacher, L. J., *J. Amer. Chem. Soc.* (1966) **88**, 5880.
57. Stein, M. J., Liener, I. E., *Biochem. Biophys. Res. Commun.* (1967) **26**, 376.
58. Murachi, T., Kato, K., *J. Biochem.* (Tokyo) (1967) **62**, 627.
59. Whitaker, J. R., Perez-Villasenor, J., *Arch. Biochem. Biophys.* (1968) **124**, 70.
60. Masuda, T., *J. Biochem.* (Tokyo) (1959) **46**, 1569.
61. Heinieke, R. M., Mori, R., *Science* (1959) **129**, 1678.
62. Ebata, M., Tsunoda, J. S., Yasunobu, K. T., *Biochem. Biophys. Res. Commun.* (1962) **9**, 173.
63. Gould, N. R., Wong, R. C., Liener, I. E., *Biochem. Biophys. Res. Commun.* (1963) **12**, 469.
64. Gould, N. R., Liener, I. E., *Biochemistry* (1965) **4**, 90.
65. Westerik, J. O., Wolfenden, R., *J. Biol. Chem.* (1972) **247**, 8195.
66. Umezawa, H., *Pure Applied Chem.* (1972) **33**, 129.
67. Schechter, I., Berger, A., *Biochem. Biophys. Res. Commun.* (1967) **32**, 898.
68. Wong, R. C., Liener, I. E., *Biochem. Biophys. Res. Commun.* (1964) **17**, 470.
69. Light, A., Frater, R., Kimmel, J. R., Smith, E. L., *Proc. Nat. Acad. Sci. U.S.* (1964) **52**, 1276.
70. Chao, L. P., Liener, I. E., *Biochem. Biophys. Res. Commun.* (1967) **27**, 100.
71. Husain, S. S., Lowe, G., *Biochem. J.* (1970) **117**, 333.
72. Husain, S. S., Lowe, G., *Biochem. J.* (1969) **114**, 279.
73. Husain, S. S., Lowe, G., *Biochem. J.* (1970) **117**, 341.
74. Husain, S. S., Lowe, G., *Biochem. J.* (1968) **110**, 53.
75. Husain, S. S., Lowe, G., *Biochem. J.* (1968) **108**, 861.
76. Lucas, E. C., Williams, A., *Biochemistry* (1969) **8**, 5125.
77. Lowe, G., Williams, A., *Biochem. J.* (1965) **96**, 194.
78. Gleisner, J. M., Liener, I. E., *Biochem. Biophys. Acta* (1973) **317**, 482.
79. Husain, S. S., Lowe, G., *Biochem. J.* (1970) **116**, 689.

80. Drenth, J., Jansonius, J. N., Koekoek, R., Sluyterman, L. A. Ae., Wolthers, B. G., *Philos. Trans. Roy. Soc. London B.* (1970) **257**, 231.
81. Bender, M. L., Brubacher, L. J., *J. Amer. Chem. Soc.* (1964) **86**, 5333.
82. Lowe, G., Williams, A., *Biochem. J.* (1965) **96**, 189.
83. Whitaker, J. R., Bender, M. L., *J. Amer. Chem. Soc.* (1965) **87**, 2728.
84. Lowe, G., Yuthavong, Y., *Biochem. J.* (1971) **107**, 117.
85. Williams, A., Lucas, E. C., Rimmer, A. R., Hawkins, H. C., *J. Chem. Soc. Perkin Trans.* (1972) **2**, 627.
86. Stockell, A., Smith, E. L., *J. Biol. Chem.* (1957) **227**, 1.
87. Sluyterman, A. Ae., Wolthers, B. G., *Proc. Kon. Ned. Acad. Wet. (Ser. B)* (1969) **72**, 14.
88. Allen, G., Lowe, G., *Biochem. J.* (1973) **133**, 679.
89. Polgar, L., *Eur. J. Biochem.* (1973) **33**, 104.

RECEIVED September 17, 1973. A portion of the work reported in this paper was supported by NSF grant GB-15385.

9

Metal-Containing Exopeptidases

JAMES F. RIORDAN

Biophysics Research Laboratory, Peter Bent Brigham Hospital and Department of Biological Chemistry, Harvard Medical School, Boston, Mass. 02115

The specificities of the various digestive exo- and endopeptidases suggest that they act synergistically to fulfill a major nutritional function. The concerted action of trypsin, chymotrypsin, pepsin, and carboxypeptidases A and B facilitate and ensure formation of essential amino acids. The chemical characteristics and metalloenzyme nature of two bovine exopeptidases, lens aminopeptidase and pancreatic carboxypeptidase A, indicate similarities in their mechanisms of action. However, the aminopeptidase exhibits an unusual type of metal ion activation not observed with carboxypeptidase. Chemical and physicochemical studies reveal that the latter enzyme has different structural conformations in its crystal and solution states. Moreover, various kinetic data indicate that its mode of action toward ester substrates differs from that toward peptide substrates. The active site metal atom of carboxypeptidase figures prominently in these differences.

Metal-containing exopeptidases are an important class of food-related enzymes. They can be either essential constituents regulating growth and development of various food products, or they can influence appearance, texture, or taste. They can also be effective in the nutritional utilization of such foods by aiding digestion. Little is known about the former aspect of exopeptidase biochemistry. Consequently this paper is concerned with the role of these enzymes in the degradation of dietary protein leading to the formation of amino acids.

Since the importance of dietary protein for human health and nutrition has long been recognized, it is surprising that remarkably little is known about the physiological processes involved in protein digestion. Many proteases—enzymes catalyzing the hydrolysis of peptide bonds—

degrade ingested protein to amino acids. The subsequent metabolic fate of these products has been thoroughly documented. In fact, the last decade has witnessed an almost explosive growth in the understanding of protein biosynthesis. A similar wealth of information exists concerning the interrelationships of amino acid, carbohydrate, lipid, and nucleic acid metabolism, but the means by which proteins are broken down, whether it be extracellularly in the digestive tract or intracellularly, remains a nebulous subject and constitutes a curious gap in our knowledge of the biochemistry of man.

This paper discusses the present knowledge of two classes of enzymes involved in the final stages of protein breakdown—the aminopeptidases and the carboxypeptidases. The former class of enzymes is present in a variety of sources, few of which have been well characterized. The latter includes carboxypeptidase A which, like many other enzymes present in bovine pancreas, has been the subject of extensive investigation. Both classes of peptidases catalyze the removal of amino acids from the ends of a suitable polypeptide chain. Aminopeptidases act at the end having a free α amino group while carboxypeptidases act at the end having a free α carboxyl group. Not all members of these two classes require metal ions for activity, but in each class certain enzymes do seem to require a metal ion as a cofactor. These are referred to as metalloexopeptidases and constitute the specific subject matter in this report. The representative enzymes considered are leucine aminopeptidase from bovine lens and carboxypeptidase A from bovine pancreas.

Proteolytic Enzymes and Protein Digestion

Protein digestion clearly involves more than the degradation by exopeptidases to produce free amino acids. The protein macromolecules must be first degraded to the level of oligopeptides. This is accomplished by the action of endopeptidases which catalyze the hydrolysis of peptide bonds in the middle of the polypeptide chain. The various endo- and exopeptidases that perform this function in human protein digestion are listed in Table I. Of the two proteolytic enzymes produced in the stomach, pepsin is present in the greater amount (1). It is synthesized in the form of an inactive precursor, pepsinogen, which spontaneously converts to the active form below pH 5 by an autocatalytic process involving the removal of a 41-amino acid residue peptide from the amino terminal portion of the zymogen. Several different forms of pepsin and pepsinogen have been identified. In addition another pepsin-like enzyme, gastricsin, is also present in human gastric juice (2). Both of these enzymes are endopeptidases and act best at acid pH.

Table I.   Enzymes Involved in Protein Digestion

Enzyme	*Origin*	*Site of Action*
Endopeptidases		
Pepsin	gastric mucosa	stomach
Gastricsin	gastric mucosa	stomach
Chymotrypsin	pancreas	small intestine
Trypsin	pancreas	small intestine
Elastase	pancreas	small intestine
Collagenase	pancreas	small intestine
Kallekrein	pancreas *et al.*	plasma
Exopeptidases		
Carboxypeptidase A	pancreas	small intestine
Carboxypeptidase B	pancreas	small intestine
Aminopeptidase	intestinal mucosa	small intestine

The pancreas produces a number of alkaline proteases, all in the form of zymogens (3). As with pepsin, there are at least two different forms of both chymotrypsinogen and trypsinogen (4). These enzyme precursors are excreted into the pancreatic juice which enters the duodenum *via* the pancreatic duct. They become active only after they enter the small intestine. A specific proteolytic enzyme, enterokinase, is released from the intestinal mucosa and is responsible for the activation of trypsinogen. It has little other purpose in protein digestion. It catalyzes the cleavage of a single peptide bond which releases a small peptide fragment from the amino terminus of the trypsinogen molecule. The activation of the other pancreatic endopeptidase zymogens, chymotrypsinogen and proelastase, is thought to be catalyzed by trypsin. There may be other proteolytic enzymes present in pancreatic juice, also in the form of zymogens. There is evidence for the presence of a procollagenase and kallekreinogen (5). The latter produces a proteolytic enzyme which is involved primarily in the release of kinins, such as bradykinin, from an α-globulin in plasma. This has minimal significance to dietary protein digestion. Kallekreins also originate in salivary glands and other tissues.

The two carboxypeptidases, A and B, are also excreted from the pancreas as zymogens which are then activated by trypsin. The aminopeptidases participating in protein digestion are present in the intestinal mucosa. They can be isolated in active form from mucosal extracts but it has yet to be established whether or not they are synthesized as zymogens.

In vitro studies have revealed that digestion of proteins either by pepsin or by mixtures of trypsin and chymotrypsin under otherwise physiological conditions proceeds at a much lower rate than was thought to occur *in vivo* (6). This has led to the belief that the total breakdown of proteins to amino acids in the digestive tract occurs at a fast enough rate to permit absorption (7). Pancreatic juice contains a large number of other components such as the protease inhibitors. These are thought to protect the pancreas against the possibility of *in situ* zymogen activation. However, they would likewise interfere with intestinal digestion, so some mechanism that renders them inoperative must exist. The role of such inhibitors and other substances present in the intestine in protein digestion remain to be established. Additional proteolytic enzymes which function in the overall conversion of proteins to amino acids may still be found.

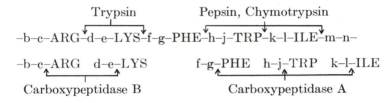

Figure 1.   Proteolytic scheme.   The arrows indicate the specificities of the various peptidases.

Despite the incomplete listing of enzymes involved in protein breakdown, it is nevertheless possible to demonstrate an interrelationship based on characteristic modes of action. As indicated in Figure 1, trypsin specifically catalyzes the hydrolysis of peptide bonds whose carbonyl portion is contributed by either a lysyl or an arginyl residue. These are the only two amino acid side chains carrying a positive charge under the pH conditions present in the duodenum. On the other hand, both chymotrypsin and pepsin catalyze the hydrolysis of peptide bonds in which the carbonyl group is contributed by a residue with an aromatic or a bulky aliphatic side chain. Because of this endopeptidase phase of protein digestion, a large number of peptides whose carboxyl terminal residues match the specificities of the two carboxypeptidases are present in the duodenum.

Carboxypeptidase B complements the specificity of trypsin and liberates arginine and lysine from the carboxyl ends of the peptides which result from the action of this endopeptidase. Carboxypeptidase A acts similarly with pepsin and chymotrypsin. The nutritional significance of this closely dove-tailed hydrolytic process is indicated in Table II.

Table II.   Essential Amino Acids

Lysine	Phenylalanine
Arginine	Tyrosine
Isoleucine	Threonine
Leucine	Tryptophan
Methionine	Valine

In the adult human, lysine is an essential amino acid. It cannot be synthesized *de novo* from non-protein sources. In the child, both arginine and lysine are essential since there is a need for more arginine than can be supplied by endogenous routes. The combination of trypsin and carboxypeptidase B seems to be a convenient evolutionary device to ensure an adequate supply of these amino acids from dietary protein. Similarly, the other essential amino acids listed in Table II are those most likely to arise from the combined action of pepsin and chymotrypsin followed by carboxypeptidase A. Hence, the overall proteolytic apparatus is apparently designed to meet the nutritional needs of man for essential amino acids.

The evolutionary pathways leading to such a scheme are, in part, reflected in the apparent structural homology which exists between these proteases. Sequence information points to a close relationship between trypsin and chymotrypsin (8), and there are marked similarities in the structures of carboxypeptidases A and B, both to each other and to pepsin. Such relationships seem to exist but are beyond the scope of this article.

Aminopeptidases

Much of the present knowledge about protein structure and function has been obtained using enzymes from the pancreas of the cow. These enzymes have been available in great quantity and in a highly purified form ever since the 1930's (9). Moreover, they have rather uncomplicated structures, and their molecular weights are not so large as to discourage sequence analysis. In contrast, the aminopeptidases, since they come from other sources, are more difficult to prepare and handle, cannot be obtained in large quantities, and generally have molecular weights around 300,000 (10). The major impetus for studying these enzymes had been their potential usefulness in sequence investigations of other proteins, but recently interest in them has revived, in part because of their metalloenzyme nature.

Aminopeptidases are present in many tissues (Table III). Leucine aminopeptidase from intestinal mucosa is very effective in catalyzing the hydrolysis of leucine from the amino terminus of peptides, polypeptides, and proteins. It also hydrolyzes leucine amide and leucine esters (10). The designation leucine aminopeptidase is somewhat of a misnomer because activity is also observed when other amino acids replace leucine. Only the L-isomers of amino acids are substrates, and the presence of a D-amino acid residue or proline in the penultimate position will retard hydrolysis (10). Enzymes having the same specificity as the intestinal aminopeptidase have been identified and/or isolated from kidney, pancreas, muscle, lens, and various bacterial sources (10). The kidney enzyme, being more readily available, has received greater attention than that from intestinal mucosa. Most likely its principal function is to act on peptide hormones circulating in the blood rather than to digest dietary protein. The function of the enzyme in lens remains to be established. Aminopeptidases A (*11*) and B (*12*) catalyze the hydrolysis of acidic and basic amino acids from peptides, respectively. Aminopeptidase M, isolated from kidney microsomes (*13*), has properties different from those of the kidney supernatant enzyme, but it catalyzes the same reaction. Aminopeptidase P only acts on substrates having a prolyl residue in the penultimate position (*14*), whereas prolyl iminopeptidase requires proline at the amino terminal position of peptides (*15*). Pyrrolidonyl peptidase removes residues specifically from an "amino terminus" lacking a free α-amino group (*16*). Cathepsin C is a dipeptidyl aminopeptidase that removes dipeptides from the ends of peptide chains (*17*) whereas true dipeptidases act only on dipeptides and tripeptidases act only on tripeptides (*10*).

Some of the more extensive studies have been with the aminopeptidase from bovine lens (*18, 19*). Though this enzyme is not involved in protein digestion, some of its characteristics are presented here since they may be closely analogous to those of the enzyme from intestinal mucosa.

The first evidence indicating that leucine aminopeptidase is a metalloenzyme was obtained only recently (*20, 21*) although metals have long been known to affect its activity (*22*). Both the lens and the kidney

Table III.   Aminopeptidases

Name	*Specificity*[a]	*Source (Location)*
Leucine Aminopeptidases	H_2N-LEU—$X-Y-$	intestinal mucosa, lens, kidney, pancreas, muscle, *Aspergillus*
Aminopeptidase A	H_2N-GLU—$X-$	kidney, serum
Aminopeptidase B	$H_2N-LYS(ARG)$—$X-$	liver
Aminopeptidase M	H_2N-LEU—$X-Y-$	kidney microsomes
Aminopeptidase P	H_2N-X—$PRO-$	*E. coli*
Cathepsin C	$H_2N-Z-PHE$—$X-$	kidney, spleen, liver
Pyrrolidonyl Peptidase	PYR—$X-$	liver, bacteria
Proline Iminopeptidase	PRO—$X-$	*E. coli*
Dipeptidases	H_2N-X—$Y-COOH$	many
Tripeptidases	H_2N-X—$Y-Z-COOH$	many

[a] The double dash (——) indicates the bond to be hydrolyzed.

Table IV. Changes in Activity on Removal and Readddition of Zinc to Lens LAP[a]

Sample	Zn/54,000 grams	Activity, %
[(LAP)Zn]	2.12	100
(LAP)	0.08	4
(LAP) + Zn	2.09	98

[a] From (18)

enzyme appear to contain two zinc atoms per subunit which, in the case of the lens enzyme, has a molecular weight of 54,000 (18). There are six subunits per molecule of enzyme, each subunit containing six sulfhydryl groups. There is evidence that a thiol may be involved in zinc binding (18).

Removal of zinc by dialysis against 1,10-phenanthroline (Table IV) abolishes activity in proportion to the zinc content (18). On readddition of zinc to two gram atoms per subunit, activity is regained completely. Other metals such as Mn^{2+}, Ca^{2+}, Mg^{2+}, and Cu^{2+} are ineffective in this regard. However, Mg^{2+} and Mn^{2+} have an activating effect on the zinc-containing enzyme.

Addition of Mg^{2+} to the zinc enzyme results in a progressive displacement of one of the two zinc atoms bound per subunit (Table V). There is an associated rise in activity which reaches 1200% of the two-zinc enzyme when the ratio of Mg^{2+} added to Zn^{2+} originally present is 667. A similar effect is given by Mn^{2+}. The kinetic parameters for leucinamide hydrolysis for these various lens metalloaminopeptidases are listed in Table VI. Each species contains one Zn^{2+} at the catalytic site and either Zn^{2+}, Mn^{2+}, or Mg^{2+} at the activating site. Changing the metal has only a minor effect on substrate binding but a marked effect on maximal activity. Mn^{2+} causes an 11-fold and Mg^{2+} a 25-fold increases in activity. However, both sites do not have to be occupied in order to have an active enzyme. The species containing only one Zn^{2+} per 54,000g is still a functional aminopeptidase.

Table V. Effect of Magnesium on Zinc Content and Activity of Lens LAP[a]

Mg Added	Mg Bound	Zn Bound	Sum	Activity
0	0.04	1.98	2.02	100
33	0.32	1.75	2.07	400
67	0.44	1.62	2.05	660
167	0.65	1.47	2.12	800
333	0.77	1.44	2.21	1100
667	0.80	1.37	2.17	1200

[a] From (18)

In spite of the large effects of Mg^{2+} or Mn^{2+} on the activity of lens aminopeptidase (and probably on the kidney enzyme as well), it is doubtful if they play any physiological role. The relative concentrations of Zn^{2+}, Mg^{2+}, and Mn^{2+} and the pH of the particular tissues in question are such as to preclude formation of any hybrid metalloenzymes *in vivo*. However, the fact that the activation does occur indicates its potential for occurring *in vivo* although it may not be brought about by Mg^{2+} or Mn^{2+}.

Table VI.   Kinetic Parameters for Lens Metalloaminopeptidases[a]

Enzyme Form	K_m, mM	V_{max} $\mu mole/min/mg$
Zn-Zn	20	82
Zn-Mn	10	905
Zn-Mg	33	2040

[a] From (*18*)

Other than the critical requirement for the metal ion, little is known about the aminopeptidase mechanism of action. There is evidence that one tyrosyl residue per subunit may be important (*23*). However, this has not been demonstrated unequivocally. Activity–pH profiles imply that two groups are required for catalysis, one of which may be the amino group of the substrate (*24*). The last aminopeptidase mechanism proposal appeared almost 10 years ago (*24*). It suggested that the metal atom polarizes the susceptible peptide carbonyl bond facilitating a general base-catalyzed attack of water at the carbonyl carbon atom. Breakdown of the resulting tetrahedral intermediate would be assisted by interaction of the peptide nitrogen with an acidic group. Neither the acid nor the general base was identified.

Similarities between the reactions catalyzed by aminopeptidases and by carboxypeptidases suggest that in addition to a metal ion, these enzymes could share other mechanistic features. A functional carboxyl group might be present at the active site of leucine aminopeptidase acting as either the general base, the acidic group, or as an aid in substrate binding. The possibility of a functional tyrosyl residue has already been mentioned. A number of avenues of approach to investigating the mechanism of aminopeptidase action are waiting to be explored.

Carboxypeptidases

Enzymes catalyzing the hydrolysis of amino acid residues from the carboxyl terminal portion of peptide chains have been known since 1929 (*25*). Since then it has been shown that pancreatic extracts contain two carboxypeptidases, one having the preference for aromatic residues and the other specific for basic residues; hence the suffixes A and B (26). Both of these enzymes are synthesized in the pancreas as zymogens, and their activation involves limited proteolysis by trypsin.

In addition to being present in human, bovine, canine, rodent, and piscine pancreatic juice, carboxypeptidase activity has also been identified in orange peel (27), amoebae (28), yeast (29), fungi (30), molds (31), the sea anemone Cribrina astemisia (32), french bean leaves (33), germinating cottonseeds (34), barley (35), spleen (36), kidney (37), connective tissue (38), and soil bacteria (39, 40). The suffixes C, Y, G, and P refer to the carboxypeptidases from citrus peel, yeast, and those having specificities for carboxy terminal glutamyl and prolyl residues, respectively. Carboxypeptidase N is present in human plasma and inactivates bradykinin by cleaving the carboxyl terminal arginyl residue (41). Dipeptidyl carboxypeptidase activity has also been observed in human plasma (42). This enzyme removes the dipeptide phenylalanylarginine from braykinin and also converts angiotensin I into the active octapeptide angiotensin II (43).

In contrast to the aminopeptidases where the most extensive studies have been carried out on enzymes involved in hormone activation, the best characterized carboxypeptidase is from bovine pancreas having A-type activity. As indicated above, this enzyme probably serves exclusively to convert dietary protein into free amino acids. Accordingly the remaining discussion will deal with recent studies on its structure and mode of action.

Procarboxypeptidase A from bovine pancreas is a trimeric protein having a molecular weight of 89,000 (44). It is thought that two of its subunits are identical and give rise to a chymotrypsin-like endopeptidase activity on treatment with trypsin. The other subunit has a molecular weight of 41,000 and requires the combined proteolytic action of trypsin and the first activated subunit (acetyltyrosine ethyl esterase) to generate ultimately carboxypeptidase A with a molecular weight of 34,600. The metabolic significance of the 6000 molecular weight fragment, which is released on activation of the immediate precursor of carboxypeptidase A, is unknown. Nor is it known whether or not human procarboxypeptidase A exists in an analogous trimeric form. Preliminary reports indicate that there is not any chymotryptic activity associated with the human zymogen (45), suggesting that the zymogen is a monomer. Monomeric procarboxypeptidase A has been found in the spiny pacific dogfish (46) and may occur in the pig (47). There may be up to four different forms of procarboxypeptidase in human pancreatic juice which give rise to two different carboxypeptidases A (48). Precise details of the activation mechanism of procarboxypeptidase A are still unknown. In both the cow and pig, it can occur in several ways. Three different bovine carboxy-peptidases have been identified, A_α, A_β, and A_γ and contain 307, 305, and 300 amino acid residues, respectively (*49*). The extra residues occur at the amino terminal region of the molecule.

Zymogens have been considered to be inactive precursors of enzymes and the activation process to involve the generation of a catalytic or substrate binding site or both (*44*). Recently, Behnke and Vallee (*50*) found that the spectral properties of cobalt-substituted procarboxypeptidase A closely resemble those of the cobalt enzyme. Since these spectra were believed to be peculiar to enzymatically active proteins (*51*), they investigated the intrinsic catalytic activity of the cobalt zymogen. Remarkably, with certain substrates, cobalt procarboxypeptidase was found to have as much activity, and in some cases even more than the native enzyme. These observations, as well as those of others (*52*), have questioned the entire concept of zymogens as inactive enzyme precursors.

Table VII. Activities of Metallocarboxypeptidases

	Peptidase[a]	Esterase[b]
	$V/V_{zinc} \times 100$	
Apo	0	0
Zinc	100	100
Cobalt	200	114
Nickel	47	43
Manganese	27	156
Cadmium	0	143
Mercury	0	86
Rhodium	0	71
Lead	0	57
Copper	0	0

[a] $0.02M$ benzyloxycarbonylglycyl-L-phenylalanine, pH 7.5, 0°C
[b] $0.01M$ benzoylglycyl-DL-phenyllactate, pH 7.5, 25°C

A metal atom is essential to the catalytic activity of carboxypeptidase A (*53*). The enzyme, as isolated, contains one gram atom of zinc per molecular weight of 34,600. Removal of the metal atom, either by dialysis at low pH or by treatment with chelating agents, gives a totally inactive apoenzyme. Activity can be restored by readdition of zinc or a number of other divalent metal ions (Table VII). The dual activity of carboxypeptidase towards peptides and esters is quite sensitive to the particular activating metal ion. Thus, the cobalt enzyme has twice the activity of the native zinc enzyme toward peptides but the same activity toward esters. Characteristic peptidase and esterase activities are also observed for the Ni^{2+} and Mn^{2+} enzymes as well while the Cd^{2+}, Hg^{2+}, Rh^{3+}, and Pb^{2+} enzymes exhibit only esterase activity. The copper enzyme has no activity at all.

Chemical modification studies with a number of site-specific reagents have identified at least three and perhaps four different amino acid residues as important to catalytic function (Table VIII). Acetylation, iodination, nitration, and azo coupling have all pointed to at least one essential tyrosyl residue which has been identified as tyrosine-248 by quantitative sequence analysis (54). Coupling with 5-diazo-1-H-tetrazole also suggests that a histidyl residue is important to peptide but not ester hydrolysis (55). Similarly, modification of a single arginyl residue specifically alters peptidase activity (56). However, substitution of one carboxyl group with a carbodiimide reagent abolishes both activities (57).

Table VIII. Changes in Peptidase and Esterase Activities[a] on Modification of Functional Residues in Carboxypeptidase A

Reagent	Percent Control Activity		Functional Residue Modified
	Peptidase	*Esterase*	
Acetyl imidazole	<2	700	Tyr
Acetic anhydride	<2	610	Tyr
Iodine	<2	500	Tyr
Tetranitromethane	14	190	Tyr
5-Diazo-1-H-tetrazole (8x)[b]	90	180	Tyr
5-Diazo-1-H-tetrazole (45x)[b]	5	180	His
2,3-Butanedione	39	98	Arg
Cyclohexyl-3-(2-morpholino-ethyl)carbodiimide	12	9	Carboxyl

[a] Conditions as in Table VII
[b] Molar excess of reagent

The complete amino acid sequence of carboxypeptidase (58) has now been established. The above chemical findings are in complete accord with those of x-ray analysis (59). Both indicate that a carboxyl, an arginyl, and a tyrosyl residue are all suitably located to interact with the substrate and to catalyze its hydrolysis in conjunction with the zinc atom. However, the conclusions drawn from studies carried out with the enzyme in solution differ significantly in one important aspect from those derived from x-ray analysis. The crystallographers note that in the absence of the substrate, the phenolic hydroxyl group of tyrosine-248 is located 17 A away from the active-site zinc atom. When the pseudosubstrate, glycyl-L-tyrosine, is added to the crystalline enzyme, an inactive complex is formed. This phenolic hydroxyl group swings around to approach the —NH— group of the potentially hydrolyzable peptide bond (59). The conformational change involves a 13-A movement of the phenolic hydroxyl group of tyrosine-248 and has been cited (60) as the best example of the induced fit theory (*61*) of catalysis. Based on these observations, the crystallographers have proposed a catalytic mechanism which incorporates a series of concerted structural changes which occur on peptide substrate binding and ultimately result in the hydrolysis of the susceptible peptide bond (*60*).

Studies in solution indicate that the conformations of tyrosine-248 in the dissolved and crystalline states may be quite different. This conclusion is based on the work of Johansen and Vallee. They have examined the absorption and circular dichroic spectral properties of a carboxypeptidase A derivative in which tyrosyl-248 is selectively labeled with the conformational probe, diazotized arsanilic acid (*62, 63*).

Zinc arsanilazocarboxypeptidase, the product of coupling crystalline carboxypeptidase with diazotized arsanilic acid, contains one arsanilazotyrosyl residue per molecule. No other residues are modified. Approximately 95% of the azotyrosine can be accounted for by the isolation of a peptide containing the label on tyrosine-248 (*54*).

In the crystalline state zinc arsanilazocarboxypeptidase is yellow, but it turns red when dissolved. These color changes are manifested in the

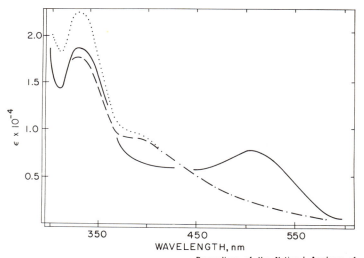

Proceedings of the National Academy of
Sciences of the United States of America

Figure 2.   Absorption spectra of zinc arsanilazocarboxypeptidase (————) and apoarsanilazocarboxypeptidase (· · ·) dissolved in 0.05M Tris·HCl–1M NaCl (pH 8.2), and of zinc arsanilazocarboxypeptidase crystals suspended in 0.05M Tris·HCl (pH 8.2) (– – –). The differences at high absorbance between the zinc arsanilazoenzyme crystals and the corresponding apoenzyme in solution are due to uncompensated light scattering and absorption flattening of the crystal suspension (62).

corresponding absorption spectra (Figure 2). Removal of the active site zinc atom from the enzyme in solution simultaneously abolishes the red color. The absorption band at 510 nm disappears, and the resultant spectrum is very similar to that of the crystalline zinc enzyme. All three forms of carboxypeptidase A—α, β, and γ—exhibit these same spectral characteristics.

The circular dichroic spectrum of the zinc azoenzyme in solution contains one positive and two negative ellipticity bands at 420, 335, and 510 nm, respectively (Figure 3). On removal of zinc, almost the entire visible CD spectrum is abolished. It is completely restored on the addition of one gram atom of zinc per mole of enzyme. This effect of zinc on both the absorption and CD spectra suggests that the red color of the enzyme in solution may be caused by the formation of a zinc-azophenol coordination complex. Closely similar color changes can be observed by adding zinc to a typical azophenol compound, tetrazolyl-N-benzyloxycar-

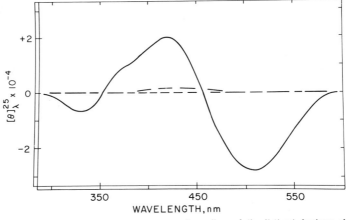

Proceedings of the National Academy of
Sciences of the United States of America

Figure 3.   Circular dichroism of zinc arsanilazocarboxypeptidase (———) and of the corresponding apoenzyme (— —) dissolved in 0.05M Tris·HCl–1M NaCl (pH 8.2). Over most of
the spectral range examined, the baseline (– – – –) is indistinguishable from the spectrum of the apoenzyme (62).

bonyltyrosine (Figure 4). The spectrum of the free azophenol resembles that of crystalline azocarboxypeptidase, and the spectrum of the metal complex is like that of the enzyme in solution.

Spectral titrations of zinc azocarboxypeptidase in solution between pH 6.2 and 8.3 generate the absorption band at 510 nm typical of the azophenol–metal complex. Above pH 8.3 this band shifts progressively to 485 nm, which is characteristic of the free azophenolate ion (Figure 5A). In contrast, the metal-free azoenzyme only forms the band at 485 nm (Figure 5B). These differences become more apparent on plotting absorbance *vs.* pH for the zinc and zinc-free enzymes. The titration of

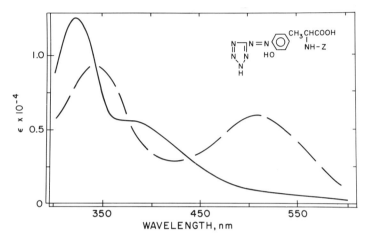

*Figure 4.   Absorption spectra of tetrazolylazo-N-carbobenzoxy-tyrosine (structure given) alone (———) and in the presence of a 10-fold molar excess of Zn²⁺ ions (— —), both at pH 7.5 (62).
Z = carbobenzoxy-*

the zinc azoenzyme at 485 or 510 nm (Figure 6A) does not fit a theoretical titration curve for a single pK while that at 560 nm readily reveals two pK values at 7.7 and 9.5, respectively (Figure 6B). Titration of the apoenzyme at 485 nm gives a sigmoid curve with a midpoint at pH 9.4 which is characteristic of the ionization of arsanilazophenol.

Circular dichroism–pH titration curves for the two azoenzymes are also quite different. It is bell shaped for the zinc enzyme at 510 nm and fits a theoretical curve for the ionization of two groups having pK values of 7.7 and 9.5, respectively. The apoenzyme at 485 nm again reveals a single pK of about 9.4.

Based on a series of studies with model azophenol-metal complexes, these absorption- and circular dichroic-pH titration data indicate the formation of an intramolecular coordination complex between arsanilazo-tyrosine-248 and the active site zinc ion where the enzyme is in solution (62, 63). This means that in solution, the phenolic hydroxyl group of tyrosine-248 would have to be much closer to the zinc ion than the 17 Å indicated by the crystallographic studies. Consequently, there must be different orientations for the arsanilazotyrosyl-248 side chain relative to the zinc ion in the crystal and in solution.

It is unlikely that the introduction of the probe itself induces these changes in protein conformation since recent studies with nitrocarboxypeptidase (64) confirm the findings of Johansen and Vallee (62, 63). Treatment of carboxypeptidase crystals with tetranitromethane exclusively nitrates tyrosine-248. In solution nitrocarboxypeptidase exhibits a visible absorption band with a maximum at 428 nm titrating with a pK of about 6.3. This abnormally low pK value (relative to model nitro-

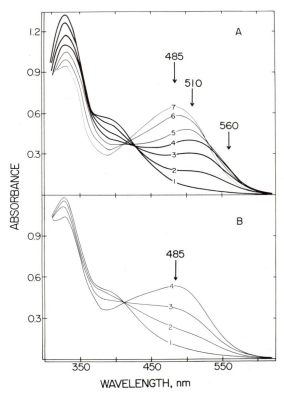

Figure 5. Effect of pH on the absorption spectrum of (A) zinc arsanilazotyr-248 carboxypeptidase$_{\alpha,\beta,\gamma}$ and (B) apoarsanilazotyr-248 carboxypeptidase$_\alpha$ both in 2mM Tris·HCl–0.5M NaCl. The numbers indicate the pH at which the spectra were recorded: (A) (1) pH 6.2; (2) pH 7.3; (3) pH 7.9; (4) pH 8.3; (5) pH 8.8; (6) pH 9.6; (7) pH 10.8. (B) (1) pH 7.5; (2) pH 8.7; (3) pH 9.3; (4) pH 9.9. At higher pH values, the apoenzyme denatures. The arrows identify maximal absorbances (or differences).

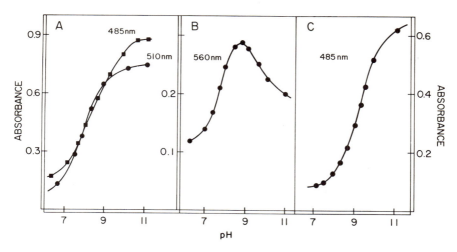

*Figure 6.   Absorbance-pH titration from the spectra in Figures 1A and B of;
(A) zinc arsanilazotyr-248 carboxypeptidase$_{\alpha,\beta,\gamma}$ at 485, 510, and; (B) 560 nm,
respectively, and of; (C) apoarsanilazotyr-248 carboxypeptidase$_\alpha$ at 485 nm.
The value at pH 11 is calculated.*

phenols) has been thought to be caused by the proximity of the nitrotyrosyl residue to a positive charge of the protein (65). Studies with a
series of nitrotyrosyl-containing copolymers support this hypothesis. The
titration curve of a copolymer of glutamic acid and nitrotyrosine has a
midpoint at pH 7.1 (Figure 7). Nitrotyrosyl ionization shifts progressively to lower pH values as the charge of the copolymer becomes more
positive. Thus, polyglutamyllysylnitrotyrosine has a nitrotyrosyl pK of

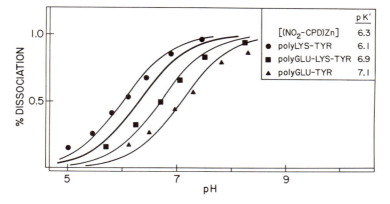

*Figure 7.   Absorbance-pH   titration   of   nitrocarboxypeptidase
(heavy line) and of nitrotyrosyl-containing copolymers. All titrations were carried out in 0.2M Tris-acetate–0.5M NaCl at 20°C.*

6.9, while the still more positively charged polylysylnitrotyrosine has a pK of 6.1 (Figure 7).

By analogy with the studies of Johansen and Vallee, it could be that the positive charge influencing the ionization behavior of nitrotyrosine-248 in nitrocarboxypeptidase is the active site zinc ion. To test this possibility, the effect of physical state on the nitrotyrosyl pK was examined. Titration of nitrocarboxypeptidase crystals from pH 6.5 to 9.5 increases absorbance at 428 nm and decreases it at 381 nm with an isosbestic point at 381 nm. However, in contrast to the enzyme in solution, the midpoint of this titration occurs at pH 8.2 rather than at pH 6.3 (Figure 8). This dramatic shift in titration behavior, brought about solely by a change in physical state of this enzyme, indicates that in nitrocarboxypeptidase as in arsanilazocarboxypeptidase, the conformations of tyrosine-248 in solution and in the crystalline state are different.

Such data suggests that a re-evaluation of current views on the structural features of carboxypeptidase catalysis is necessary. Moreover these data question whether analogous conformational differences between

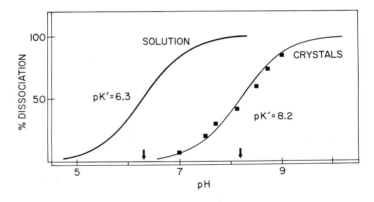

Figure 8.   Absorbance-pH titration of nitrocarboxypeptidase in solution and in the crystalline state. The arrows on the abscissa indicate the midpoints of the titration curves.

solutions and crystals escape detection for want of suitable methods. Spectrochemical probes of the type described here offer distinct opportunities in this regard.

Another difficulty which has faced the x-ray crystallographic approach to deciphering enzyme mechanisms is the inability to examine enzyme–substrate complexes. To circumvent this problem, complexes with inhibitors, products, or pseudosubstrates have been employed. In the case of carboxypeptidase A, mechanistic deductions have been based on results obtained using the very poor substrate glycyl-L-tyrosine (60).

Table IX. Differences in Mechanistic Aspects of Peptidase and Esterase Activities of Carboxypeptidase A

	Peptidase	*Esterase*
Chemical modification		
Tyrosine	decrease	increase
Arginine	decrease	no change
Metal substitution effects	k_{cat}	K_m
Phenylacetate inhibition	noncompetitive	competitive
$k_{cat}\ H_2O/k_{cat}\ D_2O$	1	2
k_{cat} *vs.* pH	sigmoid	bell-shaped

Even though this dipeptide is turned over quite slowly, the complex examined is probably a non-productive one. Furthermore an analogous ester substrate has not been found, and it is known that carboxypeptidase behaves quite differently toward ester and peptide substrates. In particular, the kinetic parameters for peptide hydrolysis for a series of metal substituted carboxypeptidases indicate that k_{cat} values can range from 6000 min^{-1} for the cobalt enzyme down to 43 min^{-1} for the cadmium enzyme (66). The K_m values on the other hand are almost totally independent of the particular metal present. The exact opposite is true for ester hydrolysis. K_m varies from 3300 M^{-1} for the cobalt enzyme to 120 M^{-1} for the cadmium enzyme while k_{cat} is essentially unchanged.

These effects of metal on the binding of the ester but not on the peptide substrate can even be visualized directly. Using stopped-flow procedures and fluorescence energy transfer as a read-out, Auld and Holmquist (66) have unequivocally demonstrated that peptides such as dansyl(glycyl)$_3$phenylalanine bind to the metal-free enzyme although they are not hydrolyzed, but their exact ester analogs, e.g., dansyl-(glycyl)$_3$-phenyllactate, do not bind to the apoenzyme. A number of other mechanistic differences have also been observed between ester and peptide substrates of carboxypeptidase A (Table IX). Such an accumulation of evidence has led to the conclusion that the enzyme uses different catalytic mechanisms for these two classes of substrates (67). One possible way to account for this difference is shown in Figure 9. The upper panel indicates that the hydrolysis of peptides involves the participation of the carboxyl group of glutamic acid-270 acting as a general base on the carbonyl carbon atom which is activated by the active site metal atom. The breakdown of the resulting tetrahedral intermediate is thought to be facilitated by the presence of tyrosine-248. Specificity arises from the orientation of the terminal carboxyl group interacting with an arginyl residue of the enzyme. At the bottom, esters, lacking the possibility of forming the hydrogen bond with tyrosine-248, seem to bind differently. Kinetic evidence (66) indicates that the esters bind through the metal atom, while chemical modification studies (56) would rule out the arginyl residue as essential for ester binding. Consequently the ester is depicted as binding through its carboxyl group to the metal atom rather than to the guanido group of arginine. The resultant decrease in effective charge on the zinc atom may account for the fact that the turnover number for ester substrates is much less than might be expected relative to the corresponding turnover number for peptide substrates (66). The role of tyrosine-248 is thus less essential to ester than to peptide hydrolysis. The analogy of the mechanistic scheme presented in Figure 9 to that given for aminopeptidase (*vide supra*) further supports the idea that many if not all the metalloexopeptidases will share the same underlying features of catalysis.

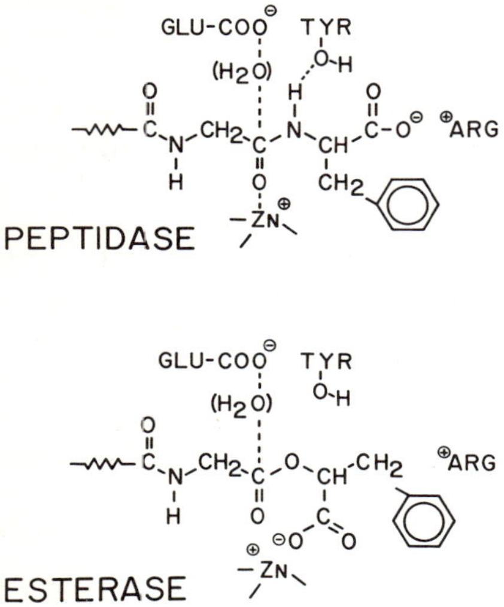

Figure 9. Mechanistic schemes for the peptidase and esterase activities of carboxypeptidase A

Summary

This discussion of the metalloexopeptidases has focused on the general role of these enzymes in the conversion of dietary proteins into amino acids. In particular, the apparent synergistic relationship which the pancreatic carboxypeptidases have with the major endopeptidases, trypsin, chymotrypsin, and pepsin, in order to facilitate formation of essential amino acids has been stressed. The chemical characteristics, metalloenzyme nature, and mechanistic details of a representative of each class of exopeptidase have been presented. Leucine aminopeptidase from bovine lens was shown to be subject to an unusual type of metal ion activation which may be representative of a more general situation. Carboxypeptidase A of bovine pancreas was discussed in terms of its three-dimensional structure, the implications of x-ray crystallography to mechanistic interpretations of enzymes, and a proposal differentiating its dual specificity toward peptides and esters. It should be emphasized that much of the information currently available on the action of metalloexopeptidases derives from studies on non-human material. In order to fully understand the role of these enzymes in human digestion, a study of our own species is necessary.

Literature Cited

1. Fruton, J. S., "The Enzymes," P. D. Boyer, Ed., Vol. 3, 3rd ed., Academic, New York, 1971.
2. Tang, J., Wolf, S., Caputto, R., Trucco, R. E., *J. Biol. Chem.* (1959) **234**, 1174.
3. Ottesen, M., *Ann. Rev. Biochem.* (1967) **36**, 55.
4. Figarella, C., Clemente, F., Guy, O., *FEBS Lett.* (1969) **3**, 351.
5. Schachter, M., *Physiol. Rev.* (1969) **49**, 509.
6. Crane, C. W., Neuberger, A., *Biochem. J.* (1960) **74**, 313.
7. Fisher, R. B., *Brit. Med. Bull.* (1967) **23**, 241.
8. Walsh, K. A., Kauffman, D. L., Sampath, K. S. V., Neurath, H., *Proc. Natl. Acad. Sci. U.S.* (1964) **51**, 301.
9. Northrop, J. H., Kunitz, M., Herriott, R., "Crystalline Enzymes," 2nd ed., Columbia University, New York, 1948.
10. Delange, R. J., Smith, E. L., "The Enzymes," Vol. 3, 3rd ed., p. 81, Academic, New York, 1971.
11. Glenner, G. G., McMillan, P. J., Folk, J. E., *Nature* (1962) **194**, 867.
12. Hopsu, V. K., Mäkinen, K. K., Glenner, G. G., *Arch. Biochem. Biophys.* (1966) **114**, 557.
13. Pfleiderer, G., Celliers, P. G., *Biochem. Z.* (1963) **339**, 186.
14. Yaron, A., Mlynar, D., *Biochem. Biophys. Res. Commun.* (1968) **32**, 658.
15. Sarid, S., Berger, A., Katchalski, E., *J. Biol. Chem.* (1959) **234**, 1740.
16. Szewczuk, A., Mulczk, M., *Eur. J. Biochem.* (1969) **8**, 63.
17. Greenbaum, L. M., "The Enzymes," P. D. Boyer, Ed., Vol. 3, 3rd ed., p. 475, Academic, New York, 1971.
18. Carpenter, F. H., Vahl, J. M., *J. Biol. Chem.* (1973) **248**, 294.
19. Lasch, J., Kudnernatsch, N., Hanson, H., *Eur. J. Biochem.* (1973) **34**, 53.
20. Himmelhoch, S. R., *Arch. Biochem. Biophys.* (1969) **134**, 597.
21. Kettmann, U., Hanson, H., *FEBS Lett.* (1970) **10**, 17.
22. Smith, E. L., *J. Biol. Chem.* (1946) **163**, 15.
23. Wachsmuth, E. D., *Biochem. Z.* (1967) **346**, 467.
24. Bryce, G. F., Rabin, B. R., *Biochem. J.* (1964) **90**, 513.
25. Waldschmidt-Leitz, E., Purr, A., *Ber.* (1929) **62B**, 2217.
26. Neurath, H., "The Enzymes," P. D. Boyer, H. Lardy, K. Myrbäck, Eds., Vol. 4, 2nd ed., p. 11, Academic, New York, 1960.
27. Zuber, H., *Nature* (1964) **201**, 613.
28. Jasumilinta, R., Maegraith, B. G., *Ann. Trop. Med. Parisit.* (1961) **55**, 518.
29. Felix, F., Labouesse-Mercouroff, J., *Biochim. Biophys. Acta* (1956) **21**, 303.
30. McConnell, W. B., Spencer, E. V., Trein, J. A., *Can. J. Chem.* (1953) **31**, 697.
31. Ito, Y., *J. Biochem. (Tokyo)* (1950) **37**, 237.
32. Takemura, S., Science Reports, Tohoku Imp. Univ. (1938) 4th Ser. **12**, 531.
33. Carey, W. F., Wells, J. R. E., *J. Biol. Chem.* (1972) **247**, 5573.
34. Ihle, J. N., Dure, L. S., III, *J. Biol. Chem.* (1972) **247**, 5034.
35. Visuri, K., Mikola, J., Enari, T.-M., *Eur. J. Biochem.* (1969) **7**, 193.

36. McDonald, J. K., Zeitman, B. B., Ellis, S., *Biochem. Biophys. Res. Commun.* (1972) **46**, 62.
37. Dehm, P., Nordwig, A., *Eur. J. Biochem.* (1970) **17**, 372.
38. Schwabe, C., *Biochemistry* (1969) **8**, 771.
39. McCollough, J. L., Chabner, B. A., Bertino, J. R., *J. Biol. Chem.* (1971) **246**, 7207.
40. Levy, C. C., Goldman, P., *J. Biol. Chem.* (1968) **243**, 3507.
41. Yang, H. Y. T., Erdos, E. G., *Nature* (1967) **215**, 1402.
42. Yang, H. Y. T., Erdos, E. G., Levin, Y., *Biochim. Biophys. Acta* (1970) **214**, 374.
43. Yang, H. Y. T., Jenssen, T. A., Erdos, E. G., *Clin. Res.* (1970) **18**, 88.
44. Neurath, H., *Federation Proc.* (1964) **23**, 1.
45. Clemente, F., DeCaro, A., Figarella, C., *Eur. J. Biochem.* (1972) **31**, 186.
46. Lacko, A. G., Neurath, H., *Biochem. Biophys. Res. Commun.* (1967) **26**, 272.
47. Folk, J. E., Schirmer, E. W., *J. Biol. Chem.* (1965) **240**, 181.
48. Kim, W. J., White, T. T., *Biochim. Biophys. Acta* (1971) **242**, 441.
49. Neurath, H., Bradshaw, R. A., Petra, P. H., Walsh, K. A., *Phil. Trans. Roy. Soc. London* (1970) **B257**, 159.
50. Behnke, W. D., Vallee, B. L., *Proc. Natl. Acad. Sci. U.S.* (1972) **69**, 2442.
51. Vallee, B. L., Williams, R. J. P., *Proc. Natl. Acad. Sci. U.S.* (1968) **59**, 498.
52. Kay, J., Kassell, B., *J. Biol. Chem.* (1971) **246**, 6661.
53. Vallee, B. L., Riordan, J. F., *Brookhaven Symp. Biol.* (1968) **21**, 91.
54. Johansen, J. T., Livingston, D. M., Vallee, B. L., *Biochemistry* (1972) **11**, 2584.
55. Sokolovsky, M., Vallee, B. L., *Biochemistry* (1967) **6**, 700.
56. Riordan, J. F., *Biochemistry* (1973) **12**, 3915.
57. Riordan, J. F., Hayashida, H., *Biochem. Biophys. Res. Commun.* (1970) **41**, 122.
58. Bradshaw, R. A., Ericcson, L. H., Walsh, K. A., Neurath, H., *Proc. Natl. Acad. Sci. U.S.* (1969) **63**, 1389.
59. Quiocho, F. A., Lipscomb, W. N., *Adv. Protein Chem.* (1971) **25**, 1.
60. Lipscomb, W. N., Hartsuck, J. A., Reeke, G. N., Jr., Quiocho, F. A., Bethge, P. H., Ludwig, M. L., Steitz, T. A., Muirhead, H., Coppola, J. C., *Brookhaven Symp. Biol.* (1968) **21**, 24.
61. Koshland, D. E., Jr., *Proc. Natl. Acad. Sci. U.S.* (1958) **44**, 98.
62. Johansen, J. T., Vallee, B. L., *Proc. Natl. Acad. Sci. U.S.* (1971) **68**, 2532.
63. Johansen, J. T., Vallee, B. L., *Proc. Natl. Acad. Sci. U.S.* (1973) **70**, 2006.
64. Riordan, J. F., Muszynska, G., Abstract 2f3, IX International Congress of Biochemistry, p. 64, Stockholm, Sweden, 1973.
65. Riordan, J. F., Sokolovsky, M., Vallee, B. L., *Biochemistry* (1967) **6**, 358.
66. Auld, D. S., Holmquist, B., *Federation Proc.* (1973) **32**, 1369.
67. Vallee, B. L., Riordan, J. F., Bethune, J. L., Coombs, T. L., Auld, D. S., Sokolovsky, M., *Biochemistry* (1968) **7**, 3547.

RECEIVED September 17, 1973.

Flavor Enzymes

THEODORE CHASE, JR.

Department of Biochemistry and Microbiology, Rutgers University,
New Brunswick, N. J. 08903

The flavorese concept—enzyme definition in terms of a flavor produced—is only a first step in biochemical understanding of flavor-producing reactions. The substrate and product must be defined chemically, and frequently the latter must be identified from among many volatile compounds which possibly contribute to flavor. Three reasonably well-understood enzymatic flavor-producing reactions are: alliinase (S-alkylcysteine sulfoxide lyase), myrosinase (thiohydroximate-O-sulfate thioglycoside hydrolase), and the reactions producing aldehydes and alcohols from unsaturated fatty acids and amino acids. Possible biosynthetic pathways to the first two substrate types are discussed. Two instances of enzymatic destruction of undesirable flavors are debittering of grapefruit juice ("naringinase," a rhamnosidase plus a glucosidase) and removal of diacetyl from beer and orange juice (diacetyl reductase).

Flavor is the interaction of taste and smell with chemical compounds in food or drink. These compounds are generally assumed to be among the many volatile components seen in gas chromatographic analysis of distillates or extracts from food or drink. Although only small amounts of most constituents are present, the full flavor usually results from the interaction of many compounds present in different amounts and with different flavor strength. Flavor chemistry is concerned with the identification and the reactions of these compounds.

Fresh food flavors are produced by biosynthetic enzymes in the living tissue and by degradative enzymes which begin acting when the tissue is cut or crushed. The flavors of cultured and fermented foods result from the enzyme action of the microorganisms involved. In some cases (*e.g.*, tea fermentation) flavor results largely from nonenzymatic reactions of compounds originally generated by enzyme action. Although cooking, especially high surface temperature processes such as frying and roasting, generates many new flavors nonenzymatically, it destroys or volatilizes many existing flavor components and inactivates the enzymes which generate them from tasteless nonvolatile precursors. Processing for preservation thus frequently robs foods, especially green vegetables, of their fresh flavor. Understanding the enzymatic processes involved in formation of the fresh flavor may make it possible to regenerate the flavor in processed food. Although nonvolatile compounds can play an important role in flavor (see the discussion of dihydrochalcones below), the greater difficulties of isolating them and the assumption that they affect only taste, not smell, has caused them to be less studied.

One pragmatic approach to the problem is the flavorese concept (1). A crude enzyme preparation from the fresh food or a closely related species is added to the processed food in the hope that the nonvolatile flavor precursors are still present and will yield the full range of fresh flavor components upon enzymatic treatment. This approach has been applied with some success to watercress (1, 2), cabbage (1, 2, 3, 4), horseradish (2, 3), onions (2, 3), carrots (2, 3, 5), peas, beans (2, 3, 6), citrus juice (2, 7), raspberries (8), tomato juice (2, 3, 9), bananas (10), and various flower fragrances (11). This would seem to be a desirable approach since the products of different enzymatic reactions are probably necessary for full flavor. However the flavor-forming activity of the enzyme preparations was variable (5, 6, 7, 9), and the flavors were not always like that of the fresh vegetable (3). The several enzymes and flavor precursors may not have been present in the normal ratios. Also, to be practical, the enzyme preparation must be inexpensive to prepare and store, the substrates must persist in the processed food, and they must be available to enzyme action. One also wonders how processed foods can be enzymatically treated under sterile conditions.

Despite the possible practical significance of such procedures, they bring flavor enzymology to about the level to which the Buchner brothers brought fermentation in 1897. The reactions involved are not necessarily understood and the enzymes are defined only in terms of the flavors produced. Development of a cell-free preparation which carries out an organoleptically desirable change may be necessary to understand the reactions involved, but it is only a first step. Better control of flavor development requires definition of the products and substrates of the enzymes involved as well as purification of the enzymes. Without this knowledge it is difficult to define the enzymes let alone discuss them at a sophisticated level.

Another, perhaps more promising, application of enzymic flavor modification is the conversion of specific undesirable flavor compounds to tasteless species. Enzymes—frequently obtained cheaply from microbial sources—do this without heat or drastic chemical treatment of the food or (especially) drink. Debittering citrus juice and reduction of diacetyl in beer are discussed here. Perhaps flavor may also be modified by using enzymic specificity to change some components of the flavor mix without affecting others.

This review concentrates on instances where both substrate and product of an individual enzyme are known and where the enzymatic reaction has been studied in a cell-free system. Enzymes discussed elsewhere in this book (mostly oxidative: lipoxygenase, polyphenol oxidase, peroxidase) are omitted.

S-Alkylcysteine Sulfoxide Lyase (alliinase), the Allium Flavor Enzyme

The Reactions. The chemical basis of the flavor of garlic (*Allium sativum*) was first investigated in 1892 when Semmler (*12*) identified dimethyl disulfide and dimethyl trisulfide in the essential oil. Cavallito *et al.* (*13, 14, 15*) identified an antibacterial component in garlic as diallyl thiosulfinate (allicin: $CH_2=CH—CH_2S(O)—SCH_2CH=CH_2$). The breakthrough to the enzymatic level was achieved by Stoll and Seebeck (*16, 17, 18, 19*) who isolated the precursor to allicin, identified it as S-allyl-L-cysteine sulfoxide (*16*), isolated the enzyme (alliin lyase, EC 4.4.1.4) converting it to allicin, pyruvate, and ammonia (*17*), and studied its substrate specificity (*18*) (*see* Figure 1).

Onions (*Allium cepa*) were shown to contain similar compounds, S-methyl and S-propyl-L-cysteine sulfoxide (*20*). The principal flavor precursor in onion is *trans*-S(+)-1-propenyl-L-cysteine sulfoxide (*21, 22, 23*). It is responsible for the lachrymatory properties and bitter taste of freshly cut onion (*22*). All these compounds were cleaved by an S-alkyl-L-cysteine sulfoxide lyase from onion (*24, 25*) which yielded pyruvate and ammonia in addition to a sulfur compound. The enzyme has also been demonstrated in *Bacillus subtilis* (*26*) and in a number of the *Cruciferae* where the only substrate known is S-methyl-L-cysteine sulfoxide (*27*). The product presumably gives rise to dimethyl disulfide which is the odor of cooked cabbage.

All possible combinations of methyl, propyl, allyl, and 1-propenyl disulfides (primarily), monosulfides, and trisulfides have been found among the volatile flavor components of onion (*28, 29, 30, 31*), garlic (*32*), caucas (*Allium victorialis*) (*33*), and other *Allium* species (*28*) although proportions vary with species. These compounds are presumably derived from the corresponding thiolsulfinates. This is accomplished either by direct decomposition by an unknown mechanism with evolution of SO_2 (*32*) or by interaction with cysteine to produce a mixed disulfide (*15*),

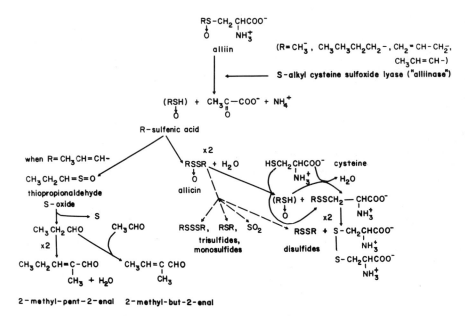

Figure 1. The alliinase reaction and subsequent nonenzymatic reactions

followed by disulfide interchange to produce volatile disulfides and cysteine. Addition of cysteine or other thiols to rehydrated onion increased
onion odor (*34*). Disproportionation to disulfides and thiosulfonates (*35*)
may be important (*36*), but thiosulfonates are absent in the distilled oil
(*32*) and are only found in small quantities in freshly cut onion (*30*).

Several questions arise—why is the first compound observed as a
product of the enzymatic reaction a dimer, except in onion (*21*), although
a monomeric product is expected? Why are the 1-propenyl disulfides
found only in small quantities although S-1-propenylcysteine sulfoxide is
the principal precursor present? Why do onions make the eyes water
when garlic does not? Why is a monomeric product found only in onions
although the enzyme from all sources acts similarly on all types of
substrate?

The answer to all these questions is found in the demonstration
(*37*) that the monomeric, unstable, lachrymatory compound produced
by onions is not 1-propenylsulfenic acid, the presumed precursor of the
dimer, but, as suggested a decade earlier (*38*), thiopropionaldehyde-S-
oxide (CH_3CH_2CHSO). The mass spectographic evidence which suggested the sulfenic acid structure (*22*) for this product is consistent with
the latter structure. The nuclear magnetic resonance and infrared spectral
evidence are consistent only with it. A hypothesis uniting all the evidence
is that the true product of the enzymatic reaction is the sulfenic acid, as

has always been supposed. If the alkyl group is anything other than 1-propenyl, the sulfenic acid immediately dimerizes to the thiolsulfinate, which yields all the other products by decomposition or reaction with cysteine and other thiols. 1-Propenylsulfenic acid can also isomerize to thiopropionaldehyde-S-oxide, the lachrymatory factor. The two reactions compete effectively. The latter compound decays to propionaldehyde with elimination of elemental sulfur, and propionaldehyde can dimerize by aldol condensation and lose water to form 2-methyl-2-pentenal or react similarly with acetaldehyde to form 2-methyl-2-butenal. All these compounds are known volatile flavor compounds of onion (*30*). The single enzymic reaction, cleavage of an S-alkyl-cysteine sulfoxide, results in many flavor components (*see* Figure 1). The compartmentalization preventing enzyme action in the intact bulb has not been elucidated.

Characteristics of the Enzyme. S-alkyl-L-cysteine sulfoxide lyases prepared from garlic (*17, 19*) and onion (*24, 25, 39*) are similar. They differ primarily in their response to pH—the garlic enzyme has a broad pH optimum from 5 to 8 and may be precipitated at pH 4.0 and redissolved without loss of activity (*17*) while the onion enzyme is sensitive to acid and is most active at pH 8.8 in pyrophosphate buffer (*24, 39*). The purified garlic enzyme showed a sharper pH optimum at pH 6.5 (*40*). The *Brassica* enzyme is most active at pH 8.5 in borate buffer (*27*) and remains soluble and active when other proteins are precipitated at pH 4.0.

Although the onion enzyme can be purified 11-fold in a particulate fraction by centrifugation at 34,800 $\times$ g (*23*), activity is also found in the soluble fraction in other species. The enzyme has otherwise been purified about fivefold from onion (*41*) and garlic (*40*), 10-fold from broccoli (*27*). The *Allium* enzymes may be fairly pure even at this degree of purification (*24, 40*).

All these enzymes have similar specificity and require the L-cysteine sulfoxide portion of the molecule. S-alkyl cysteines are not substrates (*18, 24, 27, 39*); neither are sulfoxides of N-substituted L-cysteine, β-dimethyl-L-cysteine, β-thiopropionic acid (*18*), D-cysteine (*19*), and DL-methionine (*18, 24, 39*), nor do the enzymes act on L-cysteine itself (*24, 27, 39*), cysteic acid, cysteinesulfinic acid (*24, 27*), or cycloalliin (*24, 39*). There may be some action on S-ethylcysteine sulfone (*42*).

Sulfoxides are optically active at the S atom. The natural (+) isomers are better substrates than the (−) isomers or the racemic mixture (*18, 42*). Among saturated alkyl groups, the garlic enzyme is most active on the ethyl derivative (*18*) while the onion enzyme prefers the propyl derivative (*42*). The rate differences (at substrate concentration = 0.02M) are caused solely by differences in K_m (*42*). However, the natural alkenyl sulfoxides (allyl in garlic, 1-propenyl in onion) are the best substrates for the respective enzymes (*18, 43*), and the onion enzyme has a V_{max} three times higher as well as a lower K_m (6mM) for the 1-propenyl derivative (43). At the natural levels of the methyl, propyl, and propenyl derivatives, the propenyl derivative produces 95–98% of the pyruvate (43). The broccoli enzyme is almost equally active on S-methyl (K_m = 2.7mM), ethyl, propyl, butyl, allyl, and benzyl derivatives (27). The benzyl derivative is a poorer substrate for the onion enzyme than the ethyl and propyl derivatives (24) but is not acted on by the garlic enzyme (18).

S-Alkylcysteines, while not cleaved by the onion enzymes, inhibit the cleavage of the sulfoxides (27, 42) in a partially competitive manner —1/v does not increase linearly with inhibitor concentration but approaches a maximum (42). The effect has not been explained. Other amino acids, including L-cysteine itself, did not inhibit the onion enzyme although L-cysteine did inhibit the garlic enzyme (40).

There is a general requirement for pyridoxal-5-phosphate (24, 25, 27, 44) although not all of the activity lost on dialysis is restored by adding the cofactor. This requirement explains the inhibition by hydroxylamine and hydrazine (24, 25). The reaction is a typical pyridoxal-5-phosphate catalyzed α,β-elimination with a mechanism similar to serine dehydrase and cysteine desulfhydrase (45). The coenzyme is probably bound as a Schiff base with an amino group of the enzyme since there is an absorption maximum at 415 nm in solutions of the purified garlic enzyme (40). The inhibition by L-cysteine is presumably caused by formation of a thiazolidine with the coenzyme (46). Added pyridoxal-5-phosphate also combines directly with the substrate. The dissociation constant for the complex is about $5 \times 10^{-3}M$. When this is taken into account, the dissociation constant of the holoenzyme can be shown to be about $5 \times 10^{-6}M$ (47). The higher enzyme activity in pyrophosphate buffer than in Tris or phosphate may be explained by pyrophosphate chelation of metal ions which otherwise form tighter complexes with the substrate and coenzyme (47). This decreases the availability of added coenzyme.

Inhibitors reacting with SH groups depressed the reaction only at 0.01M (iodosobenzoate, N-ethyl maleimide) or 0.05M (iodoacetate, iodoacetamide). No inhibition was obtained with $10^{-4}M$ p-chloromercuribenzoate (24).

S-Alkylcysteine C-S Lyase (EC 4.4.1.6). The endosperm of seeds of the shrub *Albizzia lophanta* contains djenkolic acid (HOOCCH(NH$_2$)-CH$_2$SCH$_2$SCH$_2$CH(NH$_2$)COOH) and an enzyme which cleaves it to pyruvate, ammonia, and highly odoriferous substances (CH$_2$(SH)$_2$?) (48, 49). The enzyme has been purified 200-fold by acid and heat treatment, (NH$_4$)$_2$SO$_4$ precipitation, and adsorption on alumina (50). It has a broad specificity and cleaves S-derivatives of L-cysteine, its sulfoxide, and sulfone. L-Cysteine itself is cleaved to pyruvate, H$_2$S, and ammonia, but D-cysteine, C-α, C-β, and N-substituted derivatives of L-cysteine and other amino acids are not substrates. Substitution at the C atom whereby the alkyl group is attached to the sulfur has a negative effect, probably by interfering with hydrogen bonding in the Schiff base intermediate between the 3 position of the pyridoxal and the imino nitrogen of the Schiff base, but such substitution does not abolish activity (even S-2,4-dinitrophenyl-L-cysteine is a substrate, which forms the basis for a convenient assay). Hydrophobic substitution at the next carbon atom is generally beneficial; the two best substrates are S-benzyl-L-cysteine and djenkolic acid. Similar enzymes in *Acacia georginae* and *Albizzia julibrizzin* are most active on the substrates naturally present in these species (*51*). The cleavage rate generally decreases with oxygen substitution on the sulfur although L-lanthionine sulfoxide is a better substrate than L-lanthionine. Pyridoxal-5-phosphate stimulated the enzyme only slightly. However, there is no reason to suspect that it is not a pyridoxal-5-phosphate dependent α,β-eliminase. The enzyme has been used to produce pyruvate and fresh flavor qualities in frozen onion, which does not have the lachrymatory qualities of fresh onions (*52*).

A similar enzyme has been purified 26-fold from the bacterium *Pseudomonas cruciviae* (*53*). It shows a constant ratio (about three) between activity on S-methyl-L-cysteine and on the sulfoxide through purification. It cleaves a wide range of S-alkyl and S(2-carboxyalkyl) cysteines but not L-cysteine itself (or other amino acids).

γ-Glutamyl Peptides. Many γ-glutamyl peptides of S-alkyl cysteines have been found in *Allium* species (*54*); however, only onion contains a sulfoxide derivative (γ-L-glutamyl-(+)-S-1-propenyl-L-cysteine sulfoxide). At first no γ-glutamyl peptidases or transpeptidases could be demonstrated in *Allium* bulbs (*54, 55*) although the combination of hog kidney γ-glutamyl transpeptidase (*56*) and *Albizzia* C-S lyase (which does not act on the peptides by itself) produced pyruvate and (presumably) flavor compounds (*52*) in dehydrated onion whose endogenous lyase had been destroyed (*55*). A γ-glutamyl peptidase was found in sprouted onion bulbs (*57, 58*) and coupled with the *Albizzia* C-S lyase to produce pyruvate from dehydrated onion (*59*). Since most of the γ-glutamyl peptides present are derivatives of cysteine rather than of its sulfoxide, this combination generates pyruvate and flavor precursors (alkyl mercaptans) from substrates inaccessible to the endogenous enzyme. It represents a real, if not necessarily economical, flavor modification by the scientific application of exogenous enzymes.

The identification of these peptides, particularly S-(2-carboxypropyl) glutathione, provides the first suggestion of a biosynthetic pathway. L-Cysteine, either free or as the γ-glutamyl peptide, might combine with methacrylic acid or its coenzyme A ester (an intermediate of valine catabolism) to yield S-(2-carboxypropyl)-cysteine or its peptide (*60, 61, 62*). Elimination of CO_2 would yield the allyl or 1-propenyl derivative, and reduction of these or H^+ displacement of CO_2 would yield the propyl derivative. No enzymes catalyzing these reactions have been isolated, but incorporation of ^{14}C-valine into S-(2-carboxypropyl) cysteine has been shown (*60*). Injection of methacrylic acid into onions stimulated formation of S-(2-carboxypropyl) cysteine. Radioactive S-(2-carboxypropyl) cysteine was a precursor of cycloalliin, the product resulting from base-catalyzed ring closure of S-(1-propenyl)-L-cysteine sulfoxide (*61, 62*). (2-Carboxypropyl) cysteine and the corresponding γ-glutamyl peptide have been found in human urine (*63*), as has S-(1-carboxy-2-methylpropyl) cysteine (isovalthine: *64*), whose predecessor in liver homogenate is the corresponding derivative of glutathione (*65*). There is a strong suggestion that S-alkylation takes place at the glutathione level. This would complicate enzymatic synthesis of the flavor precursor and suggests that peptidase plus lyase is about as complex a system as it is feasible to employ.

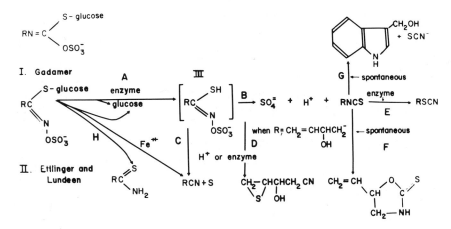

Figure 2. Enzymatic and nonenzymatic reactions of glucosinolates

Enzymatic Cleavage of Glucosinolates to Isothiocyanates

The Compounds and the Reaction. The pungent taste of mustard and other cruciferous vegetables (cabbage, horseradish, cress, etc.) is caused by isothiocyanates (*4*). These compounds also have considerable antimicrobial activity (*54*). They are formed when the plant is crushed and an enzyme (myrosinase, merosinigrate glucohydrolase, EC 3.2.3.1), localized in specific cells (idioblasts, *66*), acts (*67*) on the more generally distributed precursor compounds to produce glucose, sulfate, and isothiocyanates. These products suggested a precursor structure (Figure 2:I) to Gadamer (*68*). However it was revised by Ettlinger and Lundeen (*69*) and proved by synthesis (*70*) (Figure 2:II). The enzymatic reaction appears to be a simple β-thioglucosidase reaction (A) followed by elimination of sulfate and Lossen rearrangement of the thiohydroxamic acid to the observed isothiocyanate product (B).

At least 50 compounds with the general structure II, differing only in the R group (glucose is invariably the carbohydrate involved), have been isolated from plants (*71, 72*). The best known (and first isolated) is sinigrin (R = allyl) while others such as glucotropaeolin (R = benzyl) and sinalbin (R = *p*-hydroxybenzyl) are also important. The picturesque individual names, which were usually derived from the generic name of the plant source with the prefix gluco-, are being supplemented by the general term glucosinolate (*73*). These compounds are characteristic of and found throughout the *Cruciferae,* are widespread in several related families (*Capparidaceae, Resedaceae*), and are found in a few species of other, not necessarily related, families. (*See* Ref. 72 for a survey of the compounds known in 1968 and their occurrence.)

Enzyme(s) Involved. Early suggestions (*74, 75, 76*) that mustard seed contains two separate enzymes necessary for glucosinolate hydrolysis, a β-thioglucosidase and a sulfatase, have not been supported by recent studies (*77*). The sulfate is considered to be released nonenzymatically during Lossen rearrangement of a presumed thiohydroxamic acid O-sulfate. Gaines and Goering (*78, 79*) reported separation by DEAE-cellulose chromatography of a β-thioglucosidase and a sulfatase active on other oxime sulfates. However, these results could not be repeated (*80, 81*). Schwimmer (*82*) has suggested that sulfate release from the unstable intermediate may be catalyzed nonspecifically by any protein, but sulfate release occurs even with chromatographically purified enzyme (*81*).

During the enzyme-catalyzed reaction an intermediate with high absorbance at 245–250 nm appears (*83, 84*). Its molar extinction coefficient and half-life are calculated to be about 10,000 and 30 seconds, respectively (*83*). This is presumably the thiohydroxamate O-sulfate [*cf.*, log ϵ_{267} of phenylacetothiohydroxamate in methanol = 3.9 (*70*)]. The increase in A_{250} has a fast and a slow phase (*84*); the fast phase may represent production of one equivalent of enzyme-bound product.

A thioglucosidase which hydrolyzes glucosinolates is found in the fungus *Aspergillus sydowi* (*85*). Allyl isothiocyanate, sulfate, and glucose are products, although only the last was measured quantitatively, and a sulfatase was not sought. The mustard and *Aspergillus* enzymes show similar specificity—they cleave different glucosinolates at differing rates but not other thioglucosides, even glucosinolates lacking the sulfate group (85). Both enzymes are also active on p-nitrophenyl-β-D-glucoside (79, 85). The activity of the crude fungal preparation on other glucosides may be attributed to other glucosidases, although weak activity of the purified mustard enzyme on gentiobiose, amygdalin, phloridzin, and salicin has been reported (79). Glucose and salicin are weak competitive inhibitors [K_i = 1.5M and 0.18M, respectively (86)]; gluconolactone inhibits the fungal enzyme, with K_i about 1mM (85). The enzymes are weakly active on galacto- and xylosinolates (79) and o-nitrophenyl-β-D-galactoside (85), but not at all on mannosinolates and tetraacetyl derivatives (79).

Both are probably SH enzymes, being inhibited by I_2, Ag^+, p-chloro-mercuribenzoate and N-chlorosuccinimide, although not by iodoacetate or N-ethylmaleimide (85). Mercaptoethanol is commonly used in puri-fication (83). The enzymes are also inhibited by Co^{2+}, Mn^{2+}, and Fe^{2+}, but chelating agents are without effect (85). Neutral salts such as NaCl also inhibit the mustard enzyme (87).

The mustard enzyme is fairly stable between pH 3 and 11 while the fungal enzyme is most stable between pH 6 and 9. Both are highly active between pH 5.5 and 9 with greatest activity at pH 7.5 in phosphate buffer (85).

Mustard myrosinase is stimulated greatly by L-ascorbate (vitamin C) (77, 83, 84). Although dehydroascorbate is inactive [and ascorbate oxi-dase can thus halt the action of myrosinase (83)], the effect is not related to the reducing power of ascorbate since 2-O-methyl-L-ascorbate is also active while D-ascorbate is not (83). Enzyme saturation (K_D for L-ascorbate = 0.24mM) indicates a binding site, and various active analogs show generally higher K_Ds and slightly lower V_{max} values (83). At high concentrations ascorbate begins to inhibit (83), probably by binding to the glucose portion of the active site (88). The enzymatic hydrolysis of p-nitrophenyl-β-D-glucoside is not stimulated by ascorbate, although inhi-bition occurs at very high levels of ascorbate (88). The K_i values for glucose and salicin are also not affected in the stimulatory range (86). Ascorbate thus appears to be an effector, changing the enzyme configura-tion to improve its catalytic action on the specific substrate without itself being involved in catalysis [no change in its ultraviolet absorbance is seen (84)]. It actually increases the K_m for sinigrin, as well as the V_{max} (81), presumably by increasing the dissociation rate of products from the enzyme.

Ettlinger et al. (83) suggested that isozymes not stimulated by ascor-bate also exist. Although the two forms separated by Tsuruo et al. (81) were both stimulated by ascorbate, some of the 3–5 isozymes demon-strated by gel electrophoresis in extracts of Sinapis alba, Brassica napus,

B. juncea, and *Crambe abyssinica* (*89*) are not stimulated by ascorbate (*90*). The ascorbate-dependent (pI = 5.55) and independent (pI = 5.0) forms of *S. alba* enzyme have been separated by isoelectric focusing (*90*). It is not known whether these isoenzymes represent genetically independent species or modifications of a single genetic species.

Further Reactions. Other enzymatic and nonenzymatic reactions are associated with glucosinolate cleavage (Figure 2). When the enzymatic reaction is done at acid pH (3.0), much of the isothiocyanate is replaced by the nitrile and elemental sulfur [or H_2S (*82*)] (reaction C) (*82, 83*). This is presumably caused by an H^+-catalyzed breakdown of the thiohydroxamate-*O*-sulfate intermediate. The reaction may also be enzymatic since benzyl cyanide is produced in moistened seed powder of *Lepidium sativum* but not of *Tropaeolum majust* (*54*), and nitriles are the principal product in *C. abyssinica* seed meal up to pH 8 (*92*). Mustard myrosinase action on a boiling-water extract of *C. abyssinica* seeds does not yield nitriles. The isomers of 1-cyano-2-hydroxy-3,4-epithiobutane (*93*) (reaction D) are among the products from (S)-2-hydroxy-3-butenylglucosinolate.

Formation of the thiocyanate (*94*) is believed to be caused by enzymatic rearrangement (E) of the isothiocyanate (*54*) although direct formation from the glucosinolate has not been excluded (*95*). It occurs in homogenized *Eruca sativa* plants but not seed (*95*), and in *Lepidium sativum* seed powder but not *T. majus* (*54, 96*). The isothiocyanate is initially (10–15 sec) formed in substantial quantities but immediately decreases to a low level (*97, 98*). Addition of mustard myrosinase to heated *Lepidium* seed powder produces only isothiocyanate (*56*).

Isothiocyanates bearing a β-hydroxy substituent spontaneously cyclize (reaction F) to 5-substituted oxazolidine-2-thiones (goitrins) (*99*), which are isolated after enzyme action on β-hydroxyglucosinolates when nitrile formation does not occur (*92*). These compounds are produced in cabbage in small quantities (*100*) but are more important in turnip root (*101*), rape seed meal (*102*), and crambe meal (*93*). They have antithyroid, goitrogenic activity which prevents (non-competitively) iodine uptake in the thyroid (*103, 104*). However, the allegation that milk from cows fed largely on *Brassica* is goitrogenic (*105*) is considered overstated (*106*).

Another goitrogenic compound in cabbage is thiocyanate ion (*107*) which results from breakdown of indolylmethyl isothiocyanate produced from the corresponding glucosinolate, glucobrassicin (*97, 106, 108*) (reaction G). The 5-hydroxymethylindole thus produced can dimerize to 3,3'-diindolylmethane or form a complex (ascorbigen) with ascorbic acid (perhaps a novel type of feedback inhibition of ascorbate-dependent myrosinase). The effect of thiocyanate, unlike that of goitrins, may be reversed by increasing iodine in the diet.

These reactions are of considerable commercial importance. Rape seed (*Brassica campestris* and *B. napus*) is a major oil seed [five million tons were produced in 1966 (*109*)]. The cake left after oil extraction is limited in its use as livestock feed by the toxicity of the glucosinolate degradation products. The same considerations apply with crambe meal. The acute toxicity of the nitriles ($LD_{50} \simeq 175$ mg/kg in rats) is about eight times that of the goitrins, and the chronic toxicity in feeding experiments (meal as 10% of the ration) was also much more severe (*109*). Using these meals in animal feed therefore requires removal of the nitriles and goitrins or prevention of their formation.

Another economic problem of glucosinolates is the burnt flavor of milk from cattle who have eaten the cruciferous weed *Coronopus didymus*. This is caused by benzyl methyl sulfide or benzyl mercaptan formed in the cow from the glucotropaeolin of the weed (*110*), presumably *via* the thiocyanate. Finally, ferrous ion catalyzes the nonenzymatic breakdown (H) of glucosinolates to nitriles and thioamides (*111*).

Biosynthesis. The biosynthetic pathway from amino acids to glucosinolates has been largely worked out, principally by the Prairie Regional Laboratory group by incorporating labeled precursor compounds into glucosinolates. Figure 3 shows the biosynthesis of sinigrin, although many steps have been worked out with glucotropaeolin (benzyl glucosinolate) to illustrate chain lengthening by addition of acetate units [a feature limited to *Cruciferae* and *Resedaceae* (*72*)] and R group modification at the glucosinolate level. Chain lengthening appears to occur by transamination to the keto acid (II), condensation with acetyl CoA, and reactions analogous to those lengthening ketovaline to ketoleucine (*112*). Chain lengthening of methionine may introduce up to eight CH_2 units while phenylalanine is lengthened by only one unit (*72*). Only one intermediate, 3-benzylmalate (analog of V), has been demonstrated as a glucosinolate precursor (*113*).

Incorporation of natural amino acids (*114, 115, 116, 117, 118*) and homologs (*119, 120, 121*) without further chain lengthening (VII to XVI) proceeds with retention of the α-amino nitrogen (*119, 120, 122*). An enzyme catalyzing the oxidation and decarboxylation of the N-hydroxyamino acid VIII to the aldoxime XI (*123, 124*) has been purified 1400-fold (*125*). It is stimulated by FMN, O_2 uptake is observed, and the α-keto acid oxime V is not used as a substrate (*124, 125*). Decarboxylation may occur *via* the α-nitroso acid IX. Incorporation of the nitro compound XIII (*126*) presumably occurs *via* the *aci*-nitro compound XII which was suggested by Ettlinger and Kjaer (*72*) as an intermediate. The addition of thiols to *aci*-nitro compounds is known (*127*). RSH is unidentified, but cysteine is a good precursor (*128*). RSH is not 1-thioglucose (*128, 129*). The enzyme which transfers glucose from UDPG to XIV has been purified 20-fold (*130*). Only thiohydroxamates are substrates (but not acetothiohydroxamate), and thymidine diphosphate glucose is the only analog of UDPG used (1/10 as well). It is a sulfhydryl enzyme.

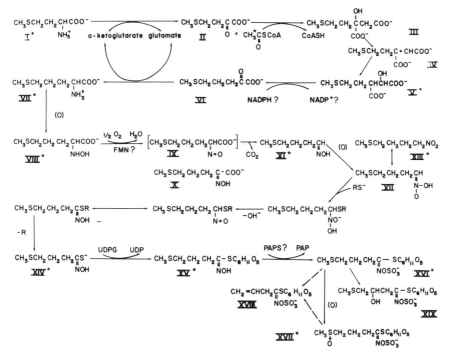

Figure 3. Biosynthetic pathway to glucosinolates. Starred compounds have been demonstrated as glucosinolate precursors.

The source of the sulfate transferred to XV in the final step of glucosinolate biosynthesis proper has been suggested, but not proved, to be 3'-phosphoadenosine-5'-phosphosulfate (PAPS) (*132*). The methylthio compound XVI and methylsulfinyl compound XVII are interconvertible, so it is not known which loses the methylthio group to form sinigrin (*131*). Allylglycine and 2-amino-5-hydroxyvaleric acid are not precursors of sinigrin (*120, 132*). β-Hydroxylation also occurs at the glucosinolate level [as, XVI to XIX (*133, 134*)] although formation of the methyl ester in 3-methoxycarbonylpropyl glucosinolate occurs at the amino acid level (2-aminoadipate) (*135*). Prevention of glucosinolate formation in oil seeds is probably best sought through plant breeding (*136*).

Formation of Aldehydes and Conversion to Alcohols

The Reactions. The most prominent volatile flavor compounds, especially in fruits, are the carbonyls and alcohols. Long chain aldehydes
such as *n*-hexanal, *trans*-2-hexenal, and *cis*-3-hexenal in many fruits (especially bananas) (*137, 138, 139, 140*) are formed by an O_2-dependent (*143*)
cleavage of unsaturated fatty acid [linoleic, linolenic (*144*)], probably by
hydroperoxidation by an enzyme resembling lipoxygenase and attack by
an aldehyde–lyase on these intermediates (*145*) (Figure 4). *Cis*-3-hexenal
is considered responsible for the "green" flavor of string beans (*146*).

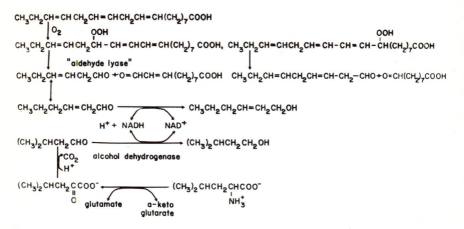

Figure 4.   Generation of flavor aldehydes and alcohols

Aldehydes related to common amino acids (3-methylbutanal from
leucine, 2-methylpropanal from valine, phenylacetaldehyde from phenylalanine) are formed by enzymatic decarboxylation of the corresponding
keto acids, which in turn are reversibly related to the amino acids by
transamination [*i.e.*, the keto acids are both degradation products of
amino acids (*147, 148*) and intermediates in their synthesis (*149*). A
third possibility—non-enzymatic oxidation of amino acids to aldehydes by
enzymatically produced *o*-quinones—is established (*150, 151*) but is not
discussed here.

The aldehydes are then either reduced to alcohols or oxidized to
carboxylic acids, both of which may be incorporated into esters (*145*).
Even without considering other flavor compounds [at least 46 volatile
flavor compounds are known in tomatoes (*138, 152*)], identification of the
reactions involved is only a step toward understanding the complex controls responsible for the production of a few major volatiles and many
minor ones, with consequent variations in flavor.

Enzymes of Aldehyde Production. Lipoxygenases are discussed elsewhere in this volume. The aldehyde–lyase has only recently been obtained even in crude extracts (*145*). However, use of 13- and 9-hydroperoxyoctadecanoic acids to form C_6 and C_9 aldehydes and corresponding ω-oxoacids is evidence that the pathway is correct (*145*). As with alliinase and myrosinase these reactions occur only when leaves or fruit are cut or crushed (*138, 143, 144, 145*). If they are heated before blending or blended under N_2 (*138*), little carbonyl compound is formed (*142*). H_2O_2 inhibits (*138*) this reaction as expected for lipoxygenase (*153*). *trans*-2-Hexenal is formed nonenzymatically from *cis*-3-hexenal (*138*).

The production of aldehydes and alcohols from the hydrophobic amino acids has been studied primarily in yeast (*147, 148, 149, 154*) since the fusel oils produced are significant secondary products of alcoholic fermentation (*149*). Also important are the studies of Morgan and coworkers on the malty flavor defect in milk which is produced by *Streptococcus lactis* var. *maltigenes* (*155–161*). This flavor is principally caused by 3-methylbutanal (*155*) which is produced by transamination of leucine to α-ketoisocaproate and decarboxylation (*157*). Labeling experiments showed that this compound is produced from leucine by tomato extract (*9, 162, 163*). The corresponding 3-methylbutanol and 3-methylbutyl acetate are produced similarly in banana (*145, 163*). Similar transformations of valine and phenylalanine are also carried out by banana slices (*145*).

In all cases the keto acids seem to be formed by typical α-ketoglutarate-linked, pyridoxal phosphate-dependent transaminases (EC 2.6.1.6, etc.) (*9, 154, 156, 157*). There has been little study of isolated, presumably specific enzymes in connection with flavors, although the leucine and alanine aminotransferases of tomato have been precipitated with $(NH_4)_2SO_4$ (*164, 165*). Transaminase activity in *Saccharomyces cerevisiae* has a pH optimum of 7.2 (*154*), and α-ketoglutarate is the only amino group recipient (*154, 166*). Only aspartate and amino acids with hydrophobic side chains are acted on (*154*).

The decarboxylation of branched chain keto acids is apparently carried out by the same decarboxylase (EC 4.1.1.1, Mg^{2+} and thiamine pyrophosphate dependent) which is active on pyruvate in yeast (*154, 167*) and orange juice (*168, 169*). The latter enzyme acts on the keto acids derived from valine and isoleucine at about 15% of the pyruvate rate (*169*). Its activity declines rapidly (80% in 1 hour) in juice (pH 3.4). The malty flavor defect produced by *S. lactis* var. *maltigenes*, however, appears to be caused by a specific decarboxylase activity not found in other strains (*158, 160*) since simultaneous decarboxylation of pyruvate and α-ketoisocaproate was almost additive (*163*) while α-ketoisocaproate and α-ketoisovalerate competed with each other. Decarboxylation of pyruvate and α-ketobutyrate (but not of α-ketoisocaproate and α-keto-isovalerate) was accompanied by O_2 uptake and partially inhibited by arsenite. This indicates that most of the pyruvate decarboxylation is probably done by a dihydrolipoate-dependent pyruvate dehydrogenase. Since both activities were irretrievably lost on dialysis, it was not possible to demonstrate a requirement for thiamine pyrophosphate. The α-keto acid decarboxylases of tomato and banana do not appear to have been studied.

Alcohol Dehydrogenases. Oranges contain an NAD^+-linked alcohol dehydrogenase (EC 1.1.1.1) (173) and an $NADP^+$-linked geraniol dehydrogenase which is separable from the former by $(NH_4)_2SO_4$ precipitation (171). Similarly, potatoes contain an aliphatic alcohol dehydrogenase, 20 times more active with NAD^+ than with $NADP^+$ (172), an $NADP^+$-dependent terpene alcohol dehydrogenase, and an $NADP^+$-dependent aromatic alcohol dehydrogenase most active on benzaldehyde (173). Rhodes (174) found that many plant tissues (apple, avocado, banana, carrot, cauliflower, grapefruit, lemon, melon, orange, parsnip, potato, spinach, swede, and tomato) used both NADH and NADPH in CH_3CHO reduction. He was unable, even with polyacrylamide gel electrophoresis at two pH levels, to resolve NADH and NADPH dependent activities from melon. The two cofactors showed mutual competition when both were present.

These enzymes are probably similar to other alcohol dehydrogenases in possessing an SH group necessary for activity. Orange geraniol dehydrogenase is inhibited by iodoacetamide (171), and melon alcohol dehydrogenase is inhibited by p-chloromercuribenzoate (174). Orange geraniol dehydrogenase is inhibited by o-phenanthroline and by added Zn^{2+}.

The most extensively investigated vegetable alcohol dehydrogenases are from oranges (170) and peas (175, 176). In both cases activity is higher on α,β-unsaturated alcohols than on saturated alcohols; the highest activity occurs with trans-2-hexenal. This is a V_{max} effect (170) and is related to the equilibrium constant of the reaction, which is 100-fold further toward the aldehyde for α,β-unsaturated compounds. The α,β-unsaturated aldehydes are poorer substrates than the corresponding saturated aldehydes. Alcohol substrate affinity for the orange enzyme is lowest with n-butyl alcohol and increases with chain length. However aldehyde affinity continues to decrease with chain length (170). Methyl branching of the substrate decreases the rate of pea enzyme action unless the methyl group is beyond an α,β double bond (176). The pea enzyme is inhibited competitively by imidazole and by fatty acids.

The initial rates of action on the various substrates are probably less important to flavor than the equilibrium concentrations which depend on the equilibrium constant of the particular alcohol–aldehyde pair, the

NAD$^+$-NADH ratio, and the pH (since H$^+$ is a product of alcohol reduction). The ethanol/acetaldehyde ratio in orange juice is about 100 (*169, 170*). This is what is expected from the NAD$^+$/NADH ratio (*177*), but for what pH? The pH of the subcellular environment of the dehydrogenases is not that of the blended fruit [*e.g.*, geraniol dehydrogenase is inactivated below pH 5 (*171*)]. The pH optimum for aldehyde reduction rate is typically near neutrality; that for alcohol oxidation much higher (9.0). On the basis of the incomplete evidence available it is likely that the alcohol dehydrogenases—at least in pea, orange, and potato—have broad enough specificity to bring all alcohol–aldehyde pairs present to the equilibrium (greatly favoring the alcohol, except for α,β-unsaturated species) expected from the NAD$^+$/NADH and NADP$^+$/NADPH ratios and the pH, once intracellular compartmentalization is broken down. It is not likely that one flavor alcohol can be oxidized specifically without affecting another related alcohol, at least by the plant dehydrogenases naturally present. The relative amounts of different alcohol–aldehyde pairs probably depend on the levels of the precursors (unsaturated fatty acids, amino and keto acids). However, generalized oxidation of a small part of the alcohols to aldehydes might be possible.

Enzymatic Debittering of Citrus Juice: "Naringinase"

Enzymatic Debittering. Citrus fruits contain many flavanone glycosides (Figure 5:I). Hesperidin, in sweet oranges (*Citrus sinensis*) and lemons, is tasteless; naringin, in grapefruit (*C. paradisi*) and natsudaidai oranges (*C. natsudaidai*), and neohesperidin, in Seville oranges (*C. aurantium*), are intensely bitter. Hesperidin and neohesperidin have the same R groups of the B ring (R$_1$ = OCH$_3$, R$_2$ = OH; in naringin R$_1$ = OH, R$_2$ = H). The structural element responsible for bitterness is the rhamnose attachment to glucose in the disaccharide, α-1,2 in naringin and neohesperidin (neohesperidose), and α-1,6 in hesperidin (rutinose) (*178, 179*). The 6-O-methylglucosylflavanone is also bitter, indicating that the 6-rhamnosyl substituent in hesperidin acts in part by blocking access to the C-3 and C-4 hydroxyl groups which are essential for bitterness (*179*). The unsubstituted glucosides II, naringenin-7-β-D-glucoside (prunin) and hesperetin-7-β-D-glucoside, are weakly bitter.

Most of the bitter glucosides are in the albedo and segment membranes so that commercial production of grapefruit juice without removal of naringin is possible. Production of juice from natsudaidai oranges (*180*) and restoration of the pink color to pink grapefruit juice by addition of pulp (*181*) require debittering.

An enzyme hydrolyzing naringin to the aglycone naringenin (III) and the disaccharide was isolated from celery seeds by Hall (*182*) and found also in leaves of the shaddock, *C. grandis*. Ting (*183*) observed that the commercial pectic enzyme preparation Pectinol 100-D hydrolyzed naringin to naringenin, glucose, and rhamnose. Thomas *et al.* (*184*) selected a fungal culture for this purpose. The enzyme preparation (naringinase C) was optimally active at pH 4.0–4.5, 40°–50°C. Glucose (2%; or δ-gluconolactone, 0.25%) strongly inhibited naringenin production but not debittering. This suggested the presence of two enzymes acting successively: a rhamnosidase, not inhibited by glucose, and a glucose-inhibited β-glucosidase (EC 2.3.1.21). Only rhamnosidase action is necessary for satisfactory debittering because its product (prunin) is not bitter enough to render the juice unacceptable (*181*). Dunlap *et al.* (*185*) separated the two enzymes by continuous flow paper electrophoresis.

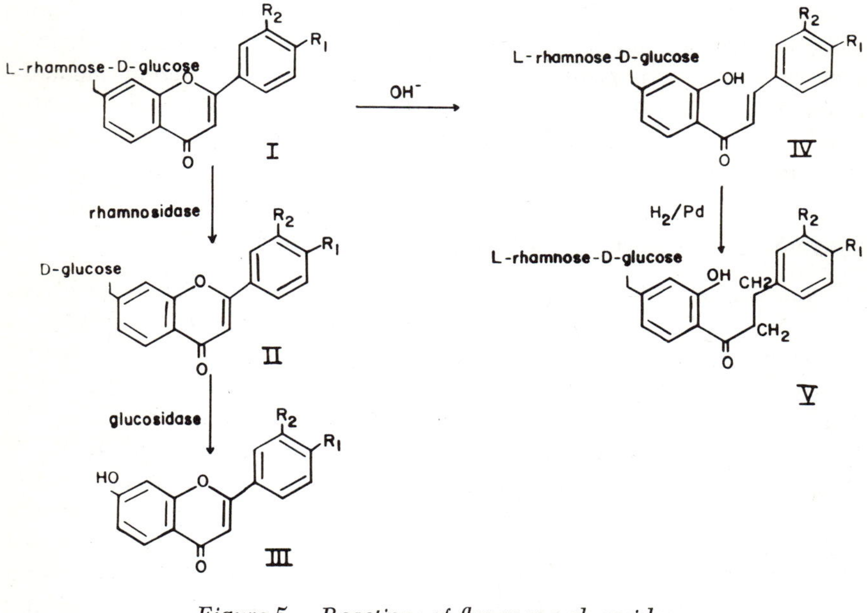

Figure 5.   Reactions of flavanone glycosides

Okada *et al.* similarly purified extracellular flavanone glycosidases produced by *Aspergillus niger*. They reported separate enzymes which produced rhamnose from naringin and hesperidin, respectively (*186, 187*). These would be α-1,2 and α-1,6 rhamnosidases (though the authors have other ideas of the rhamnosyl–glucose linkage). The same authors purified a flavanone β-glucosidase which was also active on salicin and phenyl β-glucoside (*188*). This enzyme, like that of the Rohm and Haas preparation (*184*), was inhibited by glucose and by galactose, maltose, and mannose (*189*). Rhamnosidase action on naringin was inhibited by rhamnose but not by other sugars. The naringin rhamnosidase and the glucosidase were again most active at pH 4.5, 50°C (*186, 188*) while the hesperidin rhamnosidase was most active at pH 3.5, 60°C (*186*).

A similar enzyme, characterized definitely as an α-rhamnosidase, was purified by Kamiya *et al.* (*190*), and a rhamnosidase could be induced in *Klebsiella aerogenes* (*191*). A *p*-nitrophenyl rhamnosidase from *Corticium rolfsii* was only weakly active on naringin (*192*).

Enzymatic debittering of citrus juice by rhamnosidase action is a practical process although not now economically important in the United States. It could be a particularly attractive use for an immobilized enzyme.

Dihydrochalcones. Basic cleavage of the heterocyclic ring of flavanone glycosides and catalytic hydrogenation of the resulting chalcone (IV) produces dihydrochalcone V (*193*). The dihydrochalcone from naringin is intensely sweet, comparable to saccharin; 10 times sweeter than cyclamates and 300 times sweeter than sucrose (*179*). The sweetness, however, is slow in onset and disappearance. With the removal of cyclamates from general use and current suspicion of saccharin as a potential carcinogen, there is much interest in dihydrochalcones as noncarcinogenic (*179*) artificial sweeteners. Neohesperidin dihydrochalcone is even sweeter (20 times as sweet as naringin dihydrochalcone), but neohesperidin itself is not commercially available. Hesperidin dihydrochalcone, like hesperidin, is tasteless, but hesperetin-7-glucoside dihydrochalcone is about as sweet as naringin dihydrochalcone. It may be produced by rhamnosidase action on hesperidin (*194*) and chemical conversion to the dihydrochalcone.

Although the search for enzymes to convert flavanone glycosides to dihydrochalcones has not been successful (*195*) and the chemical conversion is facile (*193, 196*), several possibilities for enzymatic modification exist besides conversion of hesperidin to hesperetin-7-glucoside. Conversion of naringin (or its dihydrochalcone) to neohesperidin (or its dihydrochalcone) by enzymatic hydroxylation and chemical methylation and enzymatic rhamnosylation of hesperetin-7-glucoside or its dihydrochalcone at the 2 position (*197*) would yield neohesperidin dihydrochalcone.

Diacetyl

Diacetyl (2,3-butanedione) is the principal desirable flavor component of butter (*198*) but gives an offensive flavor to beer (*199, 200*), frozen orange juice (*201, 202*), and wine (*203*). It is produced by yeast (*202, 204*) and bacteria (*203, 204, 205, 206*) and degraded by irreversible reduction to acetoin (3-hydroxy-2-butanone, acetyl methylcarbinol) with oxidation of NADH (*207, 208*). This may be reduced reversibly to 2,3-butanediol (*207, 208*), possibly by the same enzyme (*208*). Understanding the enzymes is therefore desirable, whether one seeks to maintain diacetyl in butter and cheese cultures or destroy it in beverages.

Unfortunately diacetyl formation is still not well understood. Acetoin formation occurs either by nonspecific interaction of acetaldehyde with the α-hydroxyethyl thiamine pyrophosphate intermediate in pyruvate decarboxylation (*209*) or by decarboxylation of α-acetolactate (*210*), which in turn arises either from interaction of pyruvate with α-hydroxyethyl thiamine pyrophosphate (*211*) or as a specific intermediate in valine biosynthesis (*212, 213*). Diacetyl does not appear to be formed directly from acetoin (*208, 214*). It is formed from α-acetolactate, in absence of cells, by O_2 oxidation (*215*), and even under N_2 (*216*), although an oxidation must occur. It is also formed from acetyl CoA (*217, 218*), probably by interaction with α-hydroxyethyl thiamine pyrophosphate [*cf.* stimulation by acetyl CoA addition to a solution of pyruvate and pyruvate decarboxylase (*215*)]. It is not known whether this involves a specific enzyme or is a mere side reaction.

Diacetyl reductase (acetoin dehydrogenase, 1.1.1.5) is widespread in bacteria (*207, 208, 219, 220, 221*), including the species (*Streptococcus diacetilactis, Lactobacillus casei*) used to prepare cultured dairy products. Mutants lacking diacetyl reductase also fail to synthesize diacetyl (*222*). The enzyme has been purified 30-fold from *L. casei* (*223*). The activity was not fully separable from an NADH oxidase activity, and the enzyme appeared to be a flavoprotein. Maximum activity was at pH 4.5. The NADH oxidase activity is associated with diacetyl reductase in other sources.

Tolls *et al.* (*224*) sought to use crude diacetyl reductase from *Aerobacter aerogenes* or yeast to remove diacetyl from beer. Diacetyl was removed from aqueous solution, but the enzyme was inhibited 42% by 3.3% ethanol and was unstable at low pH. These problems, and the more serious one of NADH regeneration in substrate quantities, were solved by immobilizing the enzyme and yeast cells as a source of NADH in gelatin sheets (*225*). Complete removal of diacetyl from beer and 66% removal from orange juice was achieved in 96 hours at 5°C. The gelatin flakes were readily removed from the beer and could be reused once or twice with decreased effectiveness. The results suggest that the *A. aerogenes* enzyme played only a minor part with immobilized yeast cells effecting adequate removal without it. NADH passing from the yeast cells to the enzyme is unlikely to return adequately to the yeast cells for re-reduction. The system could be improved by decreasing permeability of the yeast cells to NADH while retaining permeability to diacetyl and ethanol (substrate for reduction of NAD^+). Alternatively, a system could be devised whereby the coenzyme is attached to a solid support but able to swing between bound diacetyl reductase and alcohol dehydrogenase. Immobilizing yeast cells is simpler if less elegant.

Conclusions

If there is any general insight to be gained by discussing this diverse group of enzymes together, it is that positive modification of flavor (*i.e.*, formation of desirable flavors) is possible, but complicated by economic difficulties. Negative modification (removal of undesirable flavors) is much more attractive, especially if no cofactor is required (*cf.* the problem of NADH regeneration for diacetyl reductase). Food scientists should call the attention of enzymologists to other undesirable compounds whose enzymatic removal would be economically important.

Acknowledgments

I am indebted to Joseph H. Bruemmer for communicating results before publication and to Stephanie Kut for assistance in searching the literature.

Literature Cited

1. Hewitt, E. J., Mackay, D. A. M., Konigsbacher, K. S., Hasselstrom, T., *Food Technol.* (1956) **10**, 487.
2. Konigsbacher, K. S., Hewitt, E. J., Evans, R. L., *Food Technol.* (1959) **13**, 128.
3. Schwimmer, S., *J. Food Sci.* (1963) **28**, 460.
4. Bailey, S. D., Bazinet, M. L., Driscoll, J. L., McCarthy, A. I., *J. Food Sci.* (1961) **26**, 163.
5. Heatherbell, D. A., Wrolstad, R. E., *J. Agr. Food Chem.* (1971) **19**, 281.
6. Miller, T. D., "Contract Research Report No. 1," Contract *DA19-129-QM-1094*, Quartermaster Food and Container Institute, Chicago, 1958.
7. Attaway, J. A., Metcalf, F., *Fla. State Hort. Soc.* (1966) **79**, 307.
8. Weurman, C., *Food Technol.* (1961) **15**, 531.
9. Yu, M.-H., Olsen, L. E., Salunkhe, D. K., *Phytochemistry* (1968) **7**, 561.
10. Hultin, H. O., Proctor, B. E., *Food Technol.* (1962) **16**, 108.
11. Konigsbacher, K. S., Hewitt, E. J., *Ann. N.Y. Acad. Sci.* (1964) **116**, 705.
12. Semmler, F. W., *Arch. Pharmacol.* (1829) **230**, 443.
13. Cavallito, C. J., Buck, J. S., Suter, C. M., *J. Amer. Chem. Soc.* (1944) **66**, 1950.
14. Small, LaV. D., Bailey, J. H., Cavallito, C. J., *J. Amer. Chem. Soc.* (1947) **69**, 1710.
15. *Ibid.*, (1949) **71**, 3565.
16. Stoll, A., Seebeck, E., *Helv. Chim. Acta* (1948) **31**, 189.
17. *Ibid.*, (1949) **32**, 197.
18. *Ibid.*, (1949) **32**, 866.
19. Stoll, A., Seebeck, E., *Advan. Enzymol.* (1951) **11**, 377.
20. Virtanen, A. I., Matikkala, E. J., *Acta Chem. Scand.* (1959) **13**, 1898.
21. Carson, J. F., Lundin, R. E., Lukes, T. E., *J. Org. Chem.* (1966) **31**, 1634.

22. Spare, C. G., Virtanen, A. I., *Acta Chem. Scand.* (1963) **17**, 641.
23. Schwimmer, S., *Phytochemistry* (1968) **7**, 401.
24. Kupiecki, F. P., Virtanen, A. I., *Acta Chem. Scand.* (1960) **14**, 1913.
25. Schwimmer, S., Carson, J. F., Makower, R. U., Mazelis, M., Wong, F. F., *Experientia* (1960) **16**, 449.
26. Murakami, F., *Vitamins (Kyoto)* (1960) **20**, 126.
27. Mazelis, M., *Phytochemistry* (1963) **2**, 15.
28. Saghir, A. R., Mann, L. K., Bernhard, R. A., Jacobson, J. V., *Proc. Amer. Soc. Hort. Sci.* (1964) **84**, 386.
29. Bernhard, R. A., *J. Food Sci.* (1968) **33**, 298.
30. Boelens, M., de Valois, P. J., Wobben, H. J., van der Gen, A., *J. Agr. Food Chem.* (1971) **19**, 984.
31. Brodnitz, M. H., Pollock, C. L., Vallon, P. P., *J. Agr. Food Chem.* (1969) **17**, 760.
32. Brodnitz, M. H., Pascale, J. V., Van Derslice, L., *J. Agr. Food Chem.* (1971) **19**, 273.
33. Nishimura, H., Fujiwara, K., Mizutani, J., Obata, Y., *J. Agr. Food Chem.* (1971) **19**, 992.
34. Schwimmer, S., Guadagni, D. G., *J. Food Sci.* (1967) **32**, 405.
35. Barnard, D., *J. Chem. Soc.* (1957) 4675.
36. Carson, J. F., in "Chemistry and Physiology of Flavors," H. W. Schultz, E. A. Day, L. M. Libbey, Eds., Avi, Westport, p. 390, 1967.
37. Brodnitz, M. H., Pascale, J. V., *J. Agr. Food Chem.* (1971) **19**, 269.
38. Wilkens, W. F., Ph.D. Thesis, Cornell University, 1961.
39. Schwimmer, S., Mazelis, M., *Arch. Biochem. Biophys.* (1963) **100**, 66.
40. Mazelis, M., Crews, L., *Biochem. J.* (1968) **108**, 725.
41. Schwimmer, S., *Methods Enzymol.* (1969) **17**, B:475.
42. Schwimmer, S., Ryan, C. A., Wong, F. F., *J. Biol. Chem.* (1964) **239**, 777.
43. Schwimmer, S., *Arch. Biochem. Biophys.* (1969) **130**, 312.
44. Goryachenkova, E. V., *Doklady Akad. Nauk SSSR* (1952) **87**, 457.
45. Snell, E. E., *Vitam. Horm.* (1958) **16**, 77.
46. Schirch, L., Mason, M., *J. Biol. Chem.* (1962) **237**, 2578.
47. Schwimmer, S., *Biochim. Biophys. Acta* (1964) **81**, 377.
48. Gmelin, R., Hasenmeier, G., Strauss, G., *Z. Naturforsch.* (1957) **12**, 687.
49. Hansen, S. E., Kjaer, A., Schwimmer, S., *C.R. Trav. Lab. Carlsberg Ser. Chim.* (1959) **31**, 193.
50. Schwimmer, S., Kjaer, A., *Biochim. Biophys. Acta* (1960) **42**, 316.
51. Mazelis, M., Fowden, L., *Phytochemistry* (1973) **12**, 1287.
52. Schwimmer, S., Guadagni, D. G., *J. Food Sci.* (1968) **33**, 193.
53. Nomura, J., Nishizuka, Y., Hayaishi, O., *J. Biol. Chem.* (1963) **238**, 1441.
54. Virtanen, A. I., *Phytochemistry* (1965) **4**, 207.
55. Schwimmer, S., *J. Agr. Food Chem.* (1971) **19**, 980.
56. Orlowski, M., Meister, A., *J. Biol. Chem.* (1965) **240**, 338.
57. Austin, S. J., Schwimmer, S., *Enzymologia* (1971) **40**, 273.
58. Schwimmer, S., Austin, S. J., *J. Food Sci.* (1971) **36**, 807.
59. *Ibid.*, (1971) **36**, 1081.
60. Suzuki, T., Sugii, M., Kakimoto, T., *Chem. Pharmacol. Bull.* (1962) **10**, 328.
61. Granroth, B., Virtanen, A. I., *Acta Chem. Scand.* (1967) **21**, 1654.
62. Granroth, B., *Ann. Acad. Sci. Fenn. Ser. AII* (1970) **154**, 1.
63. Mizuhara, S., Oomori, S., *Arch. Biochem. Biophys.* (1961) **92**, 53.
64. Oomori, S., Mizuhara, S., *Biochem. Biophys. Res. Commun.* (1960) **3**, 343.
65. Kowaki, T., Oomori, S., Mizuhara, S., *Biochim. Biophys. Acta* (1963) **73**, 553.
66. Guignard, L., *J. Bot.* (1890) **4**, 385.
67. Boutron, F., Fremy, E., *Ann. Chem.* (1840) **34**, 230.
68. Gadamer, J., *Arch. Pharm.* (1897) **235**, 44.

69. Ettlinger, M., Lundeen, A. J., *J. Amer. Chem. Soc.* (1956) **78**, 4172.
70. *Ibid.*, (1957) **79**, 1764.
71. Kjaer, A., *Organ. Sulfur Compounds* (1961) **1**, 409.
72. Ettlinger, M., Kjaer, A., in "Recent Advances in Phytochemistry," T. J. Mabry, R. E. Alston, V. C. Runeckles, Eds., pp. 89-144, Appleton-Century-Crofts, New York, 1968.
73. Dateo, G. P., Jr., Ph.D. Thesis, Rice University, 1961.
74. Euler, H. v., Erikson, S. E., *Fermentforsch.* (1926) **8**, 518.
75. Sandberg, M., Holly, O. M., *J. Biol. Chem.* (1932) **96**, 433.
76. Neuberg, C., Schoenebeck, O. V., *Biochem. Z.* (1933) **265**, 223.
77. Nagashima, Z., Uchiyama, M., *Nippon Nogei Kagaku Kaishi* (1959) **33**, 44.
78. Gaines, R. D., Goering, K. J., *Biochem. Biophys. Res. Commun.* (1960) **2**, 207.
79. Gaines, R. D., Goering, K. J., *Arch. Biochem. Biophys.* (1962) **96**, 13.
80. Calderon, P., Pedersen, C. S., Mattick, L. R., *J. Agr. Food Chem.* (1966) **14**, 665.
81. Tsuruo, I., Yoshida, M., Hata, T., *Agr. Biol. Chem.* (1967) **31**, 18.
82. Schwimmer, S., *Acta Chem. Scand.* (1960) **14**, 1439.
83. Ettlinger, M., Dateo, G. P., Jr., Harrison, B. W., Mabry, T. J., Thompson, C. P., *Proc. Nat. Acad. Sci. (Wash.)* (1961) **47**, 1875.
84. Schwimmer, S., *Acta Chem. Scand.* (1961) **15**, 535.
85. Reese, E. T., Clapp, R. C., Mandels, M., *Arch. Biochem. Biophys.* (1958) **75**, 228.
86. Tsuruo, I., Hata, T., *Agr. Biol. Chem.* (1968) **32**, 1420.
87. *Ibid.*, p. 479.
88. *Ibid.*, p. 1425.
89. MacGibbon, D. B., Allison, R. M., *Phytochemistry* (1970) **9**, 541.
90. Henderson, H. M., McEwen, T. J., *Phytochemistry* (1972) **11**, 3127.
91. Vose, J. R., *Phytochemistry* (1972) **11**, 1649.
92. VanEtten, C. H., Daxenbichler, M. E., Peters, J. E., Tookey, H. L., *J. Agr. Food Chem.* (1966) **14**, 426.
93. Daxenbichler, M. E., VanEtten, C. H., Wolff, I. A., *Biochemistry* (1965) **4**, 318.
94. Gmelin, R., Virtanen, A. I., *Acta Chem. Scand.* (1959) **13**, 1474.
95. Schluter, M., Gmelin, R., *Phytochemistry* (1972) **11**, 3427.
96. Virtanen, A. I., *Arch. Biochem. Biophys. Suppl.* (1962) **1**, 200.
97. Virtanen, A. I., Saarivirta, M., *Suomen Kemistilehti* (1960) **B35**, 102.
98. Saarivirta, M., Virtanen, A. I., *Acta Chem. Scand.* (1963) **17**, Suppl. 1, S74.
99. Greer, M. A., *J. Amer. Chem. Soc.* (1956) **78**, 1260.
100. Altamura, M. R., Long, L., Jr., Hasselstrom, T., *J. Biol. Chem.* (1959) **234**, 1847.
101. Astwood, E. B., Greer, M. A., Ettlinger, M. G., *J. Biol. Chem.* (1949) **181**, 121.
102. Raciszewski, Z. M., Spencer, E. Y., Trevoy, L. W., *Can. J. Technol.* (1955) **33**, 129.
103. Greer, M. A., Astwood, E. B., *Endocrinology* (1948) **43**, 105.
104. Greer, M. A., *Recent Progr. Horm. Res.* (1962) **18**, 187.
105. Clements, F. W., Wishart, J. W., *Metab. Clin. Exp.* (1963) **5**, 623.
106. Virtanen, A. I., "Final Report on Investigations on the Alleged Goitrogenic Properties of Milk," Biochemical Institute, Helsinki, 1963.
107. Langer, P., Michajlovski, N., *Z. Physiol. Chem.* (1958) **312**, 31.
108. Gmelin, R., Virtanen, A. I., *Ann. Acad. Sci. Fenn. Ser. AII* (1961), No. **107**.
109. VanEtten, C. H., Daxenbichler, M. E., Wolff, I. A., *J. Agr. Food Chem.* (1969) **17**, 483.

110. Walker, N. J., Gray, I. K., *J. Agr. Food Chem.* (1970) **18**, 346.
111. Austin, F. L., Gent, C., Wolff, I. A., *Can. J. Chem.* (1968) **46**, 1507.
112. Strassman, M., Ceci, L. N., *J. Biol. Chem.* (1963) **238**, 2445.
113. Underhill, E. W., *Can. J. Biochem. Physiol.* (1968) **46**, 401.
114. Kutacek, M., Prochazka, Z., Vares, K., *Nature* (1967) **194**, 393.
115. Schraudolf, H., Bergmann, F., *Planta* (1965) **67**, 76.
116. Underhill, E. W., Chisholm, M. D., Wetter, L. R., *Can. J. Biochem. Physiol.* (1962) **40**, 505.
117. Benn, M. H., *Chem. Ind.* (1962) 107.
118. Benn, M. H., Meakin, D., *Can. J. Chem.* (1965) **43**, 1874.
119. Underhill, E. W., *Can. J. Biochem.* (1965) **43**, 179.
120. Matsuo, M., Yamazaki, M., *Biochem. Biophys. Res. Commun.* (1966) **24**, 786.
121. Chisholm, M. D., *Phytochemistry* (1972) **11**, 197.
122. Underhill, E. W., Chisholm, M. D., *Biochem. Biophys. Res. Commun.* (1964) **14**, 425.
123. Tapper, E. A., Butler, G. W., *Arch. Biochem. Biophys.* (1967) **120**, 719.
124. Underhill, E. W., *Eur. J. Biochem.* (1967) **2**, 61.
125. Kindl, H., Underhill, E. W., *Phytochemistry* (1971) **7**, 745.
126. Matsuo, M., Kirkland, D. F., Underhill, E. W., *Phytochemistry* (1972) **11**, 697.
127. Copenhaver, J. W., U.S. Patent **2,786,865**; *Chem. Abstr.* (1957) **51**, 13920.
128. Wetter, L. R., Chisholm, M. D., *Can. J. Biochem.* (1968) **46**, 931.
129. Underhill, E. W., Wetter, L. R., *Plant Physiol.* (1969) **44**, 584.
130. Matsuo, M., Underhill, E. W., *Phytochemistry* (1971) **10**, 2279.
131. Chisholm, M. D., Matsuo, M., *Phytochemistry* (1972) **11**, 203.
132. Chisholm, M. D., Wetter, M., *Can. J. Biochem. Physiol.* (1966) **44**, 1625.
133. Lee, C., Serif, G., *Biochim. Biophys. Acta* (1971) **230**, 462.
134. Underhill, E. W., Kirkland, D. F., *Phytochemistry* (1972) **11**, 1973.
135. Chisholm, M. D., *Phytochemistry* (1973) **12**, 605.
136. Youngs, C. G., *Oleagineux* (1965) **20**, 679.
137. Drawert, F., Heimann, W., Embarger, R., Tressl, R., *Ann. Chem. Pharm.* (1966) **694**, 200.
138. Kazienac, S. J., Hall, R. M., *J. Food Sci.* (1970) **35**, 519.
139. Stone, E. J., Hall, R. M., Kazienac, S. J., "Abstracts of Papers," 162nd National Meeting, ACS, 1971.
140. Grosch, W., *Z. Lebensm. Unters. Forsch.* (1967) **135**, 75.
141. Forss, D. A., Dunstone, E. A., Ramshaw, E. H., Stark, W., *J. Food Sci.* (1962) **27**, 90.
142. Fleming, H. P., Cobb, W. Y., Etchells, J. L., Bell, T. A., *J. Food Sci.* (1968) **33**, 572.
143. Nye, W., Spoehr, M. A., *Arch. Biochem.* (1943) **2**, 23.
144. Major, R. I., Thomas, M., *Phytochemistry* (1972) **11**, 611.
145. Tressl, R., Drawert, F., *J. Agr. Food Chem.* (1973) **21**, 560.
146. Hewitt, E. J., *J. Agr. Food Chem.* (1963) **11**, 14.
147. Ehrlich, F., *Chem. Ber.* (1907) **40**, 1027.
148. Neubauer, O., Fromherz, K., *Z. Physiol. Chem.* (1911) **70**, 326.
149. Guymon, J. F., *Develop. Ind. Microbiol.* (1966) **7**, 88.
150. Saijo, R., Takeo, T., *Agr. Biol. Chem.* (1970) **34**, 227.
151. Co, H., Sanderson, G. W., *J. Food Sci.* (1970) **35**, 160.
152. Viani, R., Bricout, J., Marion, J. P., Müggler-Chavan, F., Raymond, D., Egli, R. H., *Helv. Chim. Acta* (1969) **52**, 887.
153. Mitsuda, H., Yasumoto, K., Yamamoto, A., *Arch. Biochem. Biophys.* (1967) **118**, 664.
154. SentheShanmuganathan, S., *Biochem. J.* (1960) **74**, 568.
155. Jackson, H. W., Morgan, M. E., *J. Dairy Sci.* (1954) **37**, 1316.

156. MacLeod, P., Morgan, M. E., *J. Dairy Sci.* (1955) **38**, 1208.
157. *Ibid.* (1956) **39**, 1125.
158. *Ibid.* (1958) **41**, 908.
159. Morgan, M. E., Lindsay, R. C., Libbey, L. M., Pereira, R. L., *J. Dairy Sci.* (1966) **49**, 15.
160. Tucker, J. S., Morgan, M. E., *Appl. Microbiol.* (1967) **15**, 694.
161. Sheldon, R. M., Lindsay, R. C., Libbey, L. M., Morgan, M. E., *Appl. Microbiol.* (1971) **22**, 263.
162. Yu, M. H., Salunkhe, D. K., Olson, L. E., *Plant Cell Physiol.* (1968) **9**, 633.
163. Myers, M. J., Issenberg, P., Wick, E. L., *Phytochemistry* (1970) **9**, 1963.
164. Yu, M. H., Spencer, M., *Phytochemistry* (1969) **8**, 1173.
165. *Ibid.* (1969) **9**, 341.
166. Bigger-Gehring, L., *J. Gen. Microbiol.* (1955) **13**, 45.
167. Green, D. E., Herbert, D., Subrahmanyan, V., *J. Biol. Chem.* (1941) **138**, 227.
168. Davis, P. L., Roe, B., Bruemmer, J. H., *J. Food Sci.* (1973) **38**, 228.
169. Roe, B., Bruemmer, J. H., unpublished data.
170. Bruemmer, J. H., Roe, B., *J. Agr. Food Chem.* (1971) **19**, 266.
171. Potty, V. H., Bruemmer, J. H., *Phytochemistry* (1970) **9**, 1003.
172. Davies, D. D., Patil, K. D., Ugochukwu, E. N., Towers, G. H. N., *Phytochemistry* (1973) **12**, 523.
173. Davies, D. D., Ugochukwu, E. N., Patil, K. D., Towers, G. H. N., *Phytochemistry* (1973) **12**, 531.
174. Rhodes, M. J. C., *Phytochemistry* (1973) **12**, 307.
175. Eriksson, C. E., *Acta Chem. Scand.* (1967) **21**, 304.
176. Eriksson, C. E., *J. Food Sci.* (1968) **33**, 525.
177. Bruemmer, J. H., *J. Agr. Food Chem.* (1969) **17**, 1312.
178. Horowitz, R. M., Gentili, B., *Arch. Biochem. Biophys.* (1961) **72**, 191.
179. Horowitz, R. M., Gentili, B., *Symp. Foods: Carbohyd. Their Roles,* [Pap.] (1968) **5**, 253.
180. Fukumoto, J., Okada, S., *Kagaku (Kyoto)* (1963) **18**, 614.
181. Griffiths, F. P., Lime, B., *J. Food Technol.* (1959) **13**, 430.
182. Hall, D. M., *Chem. Ind.* (1938) 473.
183. Ting, S. V., *J. Agr. Food Chem.* (1958) **6**, 546.
184. Thomas, D. W., Smyth, C. V., Labbee, M. D., *Food Res.* (1958) **23**, 591.
185. Dunlap, W. J., Hagen, R. E., Wender, S. H., *J. Food Sci.* (1962) **27**, 597.
186. Okada, S., Kishi, K., Higashihara, M., Fukumoto, J., *Nippon Kogei Kagaku Kaishi* (1963) **37**, 84; *Chem. Abstr.* (1965) **62**, 12081.
187. *Ibid.,* p. 142.
188. *Ibid.,* p. 146.
189. Okada, S., Yano, M., Fukumoto, J., *Nippon Kogei Kagaku Kaishi* (1964) **38**, 246; *Chem. Abstr.* (1965) **62**, 15004.
190. Kamiya, S., Esaki, S., Hamm, M., *Agr. Biol. Chem.* (1967) **31**, 133.
191. Barker, S. A., Somers, P. J., Stacey, M., *Carbohyd. Res.* (1965) **1**, 106.
192. Kaji, A., Ichimi, T., *Agr. Biol. Chem.* (1973) **37**, 431.
193. Horowitz, R. M., Gentili, B., U.S. Patent 3,087,821, 1963.
194. Horowitz, R. M., Gentili, B., U.S. Patent 3,583,894, 1971.
195. Ciegler, A., Lindenfelser, L. A., Nelson, G. E. N., *Appl. Microbiol.* (1971) **22**, 974.
196. Linke, H. A. B., unpublished data.
197. Eveleigh, D. E., personal communication.
198. Kluyver, A. J., Van Niel, C. B., Derx, H. G., *Biochem. Z.* (1929) **210**, 231.
199. West, D. B., Lautenbach, A. L., Becker, K., *Proc. Amer. Soc. Brew. Chem.* (1952) 81.
200. Latimer, R. A., Glenister, P. R., Koepple, K. G., Dallos, T. C., *Tech. Quart. Master Brew. Assn. Amer.* (1969) **6**, 24.

201. Baisel, C. G., Dean, R. W., Kitchel, R. L., Rowell, K. M., Nagel, C. W., Vaughn, R. H., *Food Res.* (1954) **19**, 633.
202. Murdock, D. I., *J. Food Sci.* (1964) **29**, 354.
203. Fornachon, J. C. M., Lloyd, B., *J. Sci. Food Agric.* (1965) **16**, 710.
204. Burger, M., Glenister, P. R., Becker, K., *Proc. Amer. Soc. Brew. Chem.* (1957) 110.
205. Clausen, M. H., *C. R. Trav. Lab. Carlsberg* (1903) **6**, 64.
206. Shimwell, J. L., Kirkpatrick, W. F., *J. Inst. Brew.* (1939) **45**, 137.
207. Seitz, E. W., Sandine, W. E., Elliker, P. R., Day, E. A., *Can. J. Microbiol.* (1963) **9**, 431.
208. Strecker, H. J., Harary, I., *J. Biol. Chem.* (1954) **211**, 263.
209. Holzer, H., Beaucamp, K., *Biochim. Biophys. Acta* (1961) **46**, 225.
210. Juni, E., *J. Biol. Chem.* (1952) **195**, 715.
211. Holzer, H., *Angew. Chem.* (1961) **73**, 729.
212. Owades, J. L., Maresca, L., Rubin, G., *Proc. Amer. Soc. Brew. Chem.* (1959) 22.
213. Magee, P. T., de Robichon-Szulmajster, H., *Eur. J. Biochem.* (1968) **3**, 507.
214. Portno, A. D., *J. Inst. Brew.* (1966) **72**, 193.
215. Pette, J. W., *12th Internat. Dairy Congr.* (1949) **2**, 572.
216. Suomalainen, H., Ronkainen, P., *Nature* (1968) **220**, 792.
217. Speckman, R. A., Collins, E. B., *J. Bacteriol.* (1968) **95**, 174.
218. Chang, L. F., Collins, E. B., *J. Bacteriol.* (1968) **95**, 2083.
219. Elliker, P. R., *J. Dairy Sci.* (1945) **28**, 93.
220. Parker, R. B., Elliker, P. R., *J. Dairy Sci.* (1953) **26**, 843.
221. Seitz, E. W., Sandine, W. E., Elliker, P. R., Day, E. A., *J. Dairy Sci.* (1963) **46**, 186.
222. Burrows, C. D., Sandine, W. E., Elliker, P. R., Speckman, C., *J. Dairy Sci.* (1970) **53**, 131.
223. Branen, A. L., Keenan, T. W., *Can. J. Microbiol.* (1970) **16**, 947.
224. Tolls, T. N., Shovers, J., Sandine, W. E., Elliker, P. R., *Appl. Microbiol.* (1970) **19**, 649.
225. Thompson, J. W., Shovers, J., Sandine, W. E., Elliker, P. R., *Appl. Microbiol.* (1970) **19**, 883.

RECEIVED November 5, 1973.

Structure and Mechanism of Copper Oxidases

JOSEPH E. COLEMAN

Department of Molecular Biophysics and Biochemistry, Yale University, New Haven, Conn. 06510

Copper is important in plant metabolism because it functions as a prosthetic group necessary for the action of a number of oxidative enzymes and electron transfer proteins. Both the physicochemical properties of these molecules (MW 11,000–120,000) and the environment of the copper ion show great variation among this group of proteins. Polyphenol oxidase is a colorless enzyme and contains Cu(I) or magnetic dipole-dipole coupled Cu(II) pairs. Fungal laccase is an intensely blue protein containing four Cu(II) ions in different environments. Among the electron transfer proteins stellacyanin and azurin contain a single "blue" Cu(II) ion with highly unusual ESR and optical spectra—characteristics shared by one of the Cu(II) sites in laccase. The optical, magnetic, redox, and oxygen-binding properties of the copper centers are discussed as they relate to the possible mechanism of action of these important plant proteins.

Among the enzymes which catalyze oxidations involving molecular oxygen as a cosubstrate, the copper-containing oxidases have engaged the interests of biochemists for many years. In part this has been caused by the interesting coordination chemistry of the enzyme-copper itself; a chemistry that has been shown by various spectroscopic techniques to be very different from that of most simple copper coordination compounds. The most readily obtainable examples of these copper oxidases have come from plants, fungi, and bacteria although their mammalian counterparts also catalyze important reactions in the animal cell. The plant copper oxidases (including electron transfer proteins) and some of their properties are listed in Table I along with the related proteins from animal cells.

The first group are the plant oxidases which use molecular oxygen as the final electron acceptor and yield H_2O or H_2O_2 as a product. Even

Table I. Physicochemical Properties of Copper Proteins

Enzyme	Source	MW	Cu or Cofactor Mole/Mole	Reference
Plant copper oxidases				
Laccase	*Rhus vernicifera*	120,000	4 Cu	*(1,2)*
	Rhus succedanea	120,000	4 Cu	*(3)*
	Polyporus versicolor	62,000	4 Cu	*(4,5)*
	lac trees	141,000 130,000	4.5-6 Cu	*(6)*
Ascorbate oxidase	cucumber	146,000	6 Cu	*(7)*
Tyrosinase	mushroom	120,000 4 subunits	4 Cu	*(8)*
Tyrosinase	*Neurospora crassa*	30,000 to 120,000	1 Cu/ subunit	*(9)*
ᴅ-Galactose oxidase	*Polyporus circinatus*	75,000	1 Cu	*(10)*
Diamine oxidase	pea seedlings	96,000	1 Cu 1 Pyri-doxal-P	*(11)*
Copper-containing plant and bacterial electron transfer proteins				
Azurins	1. *Pseudomonas aeruginosa*	16,400	1 Cu	*(12–17)*
	2. *Pseudomonas denitrificans*	16,500	1 Cu	
	3. *Pseudomonas fluorescens*	14,010	1 Cu	
Stellacyanin	*Rhus vernicifera* (40% carbohydrate)	20,000	1 Cu	*(18–20)*
Plastocyanin	spinach	11,500	2 Cu	*(21,22)*
Oxygen-carrying copper proteins of arthropoda and molusca				
Hemocyanin	*Loligo pealei*	3,750,000	100 Cu/ mole	*(23–26)*
Hemocyanin	*Cancer magister*	950,000	24–26 Cu/ mole	*(27)*
Hemocyanin	snail	6,700,000	200 Cu/ mole	*(28)*
Mammalian copper proteins				
Tyrosinase	mammalian melanoma	64,000	2 Cu	*(29–34)*
Pre-tyrosinase	*Bombyx mori* (hemolymph)	80,000 40,000 subunit	2 Cu	*(35)*
Cytochrome oxidase	bovine heart	200,000– 1,000,000	1 Cu/ heme	*(36–38)*

Table I. Continued

Enzyme	Source	MW	Cu or Cofactor Mole/Mole	Reference
Ceruloplasmin	human plasma	151,000	8 Cu	(39–42)
		143,000	7 Cu	(43)
		132,000[a]	6 Cu	(44)
Monoamine		132,000[b]	6 Cu	(45, 46)
oxidase	bovine plasma	255,000	4 Cu	(47)
	bovine mito-	200,000–	n Cu	(48)
	chondria	300,000		
Urate oxidase	mammalian liver	120,000	1 Cu	(49)
Superoxide	mammalian			
dismutase	erythrocytes	32,000	2 Cu	
Superoxide			2 Zn	(50–52)
dismutase from				(53)
higher plants	green peas	31,500	2 Cu	
			2 Zn	

[a] X-ray data
[b] SDS gel

among this group there is remarkable variation including the intensely blue laccase which appears to contain several types of copper; tyrosinase, a colorless protein which appears to contain copper pairs; and diamine oxidase, which uses pyridoxal phosphate as a cofactor in addition to the copper.

The group of small plant proteins, azurin, stellacyanin, and plastocyanin, appear to be electron transfer proteins. They are listed because they share a type of copper site with the intensely blue representatives of the first class like laccase and ascorbate oxidase. The evidence that they participate in plant electron transfer chains remains circumstantial. Azurins, for example, purify along with well-known respiratory chain proteins like cytochrome C. A good deal of evidence exists, however, that plastocyanin is important in the photoreduction of NADP (*see* below).

Hemocyanin, the rather distantly related oxygen-carrying protein from arthropods and moluscs, is included since recent data show that the copper pairs in this protein are probably closely related to those of the fungal enzyme, tyrosinase. This suggests that the two molecules are evolutionarily related (54).

Lastly the copper oxidases that appear in mammals are listed. Many of these seem similar to their counterparts in the plant kingdom with the exception of some forms of superoxide dismutase. In mammals it is a zinc–copper enzyme and catalyzes the dismutation of the O_2^- radical according to the reaction $O_2^- + O_2^- + 2H^+ \rightarrow H_2O_2 + O_2$. This reaction is believed necessary for the maintenance of aerobic metabolism (55). In mamalian erythrocytes and liver cytosol the enzyme has a molecular weight of 32,000 and is made up of two identical subunits each containing a Zn(II) and a Cu(II) ion (55). Peptide maps derived from tryptic digests of the human superoxide dismutase show more spots than expected if the subunits were identical (56). In addition, labelling of the two free sulfhydryl groups per dimer in the presence of 8M urea or 7M guanidine HCl followed by limited proteolysis appears to give two different carboxymethylcysteine-containing peptides, one of which is present in insoluble material (57). A peptide obtained from the insoluble material by further digestion with chymotrypsin gave a different amino acid composition than the soluble peptide. Such data indicate that the subunits may not be identical, but more complete information on primary structure is required before a definitive statement can be made since both peptide mapping and alkylation experiments can be misleading.

In bacteria, on the other hand, the similar dimeric superoxide dismutase contains two Mn(II) ions per mole and no copper or zinc (55). In contrast, the enzyme from higher plants (eukaryotes) appears to be the Zn–Cu dimeric protein. The enzyme from green peas has been isolated as a Zn–Cu protein with molecular weight 31,500 (53) (Table I). Of great interest is the evidence that the manganese superoxide dismutase is present in chicken liver mitochondria (58).

Although the enzyme proteins in this list are classed together because of their copper content, some of their enzymatic and physicochemical features vary greatly from one member of the group to another. This suggests that molecular details of the mechanism of action may be very different among these copper oxidases. The suggestion is supported by the great differences among these proteins in the physicochemistry of the copper ion.

This paper summarizes briefly the physicochemistry and enzymology of plant copper oxidases with particular emphasis on polyphenol oxidase and laccase. A brief comparative discussion of other naturally occurring copper proteins and artificial copper proteins is appropriate when discussing the physicochemistry of the copper site itself. In the case of the copper proteins listed in Table I, we know a great deal more about the copper site than about the physicochemistry of the rest of the protein molecule. This is primarily a result of the availability of sophisticated spectroscopic techniques such as optical spectroscopy (both absorption and circular dichroism) and electron spin resonance which are applicable to the electronic transitions of the copper ion. On the other hand, protein chemistry has progressed more slowly. Many of the proteins are large and complex multisubunit enzymes, difficult to purify, and often unstable. There are several excellent reviews on this group of proteins (59, 60, 61, 62).

The Copper Sites

The detailed physicochemistry of the copper sites in the resting enzymes and during the interaction of these sites with various ligands or substrates has been determined by many spectroscopic techniques. Although the copper ions and their first coordination sphere occupy a relatively small fraction of the total molecular volume, in most of these proteins the copper site is either directly involved in binding molecular oxygen or transferring electrons from substrate to acceptor. Therefore, the physicochemistry of the copper sites is reviewed first since this reveals chemistry that ultimately must be related to the mechanism of action.

The nature of the copper sites varies so much among these proteins that it has become customary to class them according to the spectroscopic properties of the copper, especially the electron spin resonance parameters. This is done in Table II by listing $g_{\parallel}$, $g_{\perp}$, and the splitting of the copper nuclear hyperfine lines in the parallel region of the spectrum, $|A_{\parallel}|$. Both parameters are sensitive to features of the copper complex such as the symmetry of the ligand field and the degree of covalency of the copper–ligand bonds.

"Blue" Copper Proteins. A number of these proteins are intensely blue in color when highly purified. Their molar extinction coefficients are all out of proportion to their copper content. This is based on visible extinction coefficients observed in normal small chelate complexes of $Cu(II)$ which have molar extinction coefficients around $100M^{-1}cm^{-1}$ in the visible region near 600 nm. Indeed, the copper complexes in some of these proteins have molar extinction coefficients of several thousand per copper ion near 600 nm; $\sim3500M^{-1}cm^{-1}$ for azurin (*60, 76*) and $4100M^{-1}cm^{-1}$ per mole for fungal laccase (*60, 76*). The visible absorption and circular dichroism (CD) spectra of azurin from *Pseudomonas aersuginosa* and laccase from *Polyporus versicolor* are shown in Figure 1 (*76*). Thus, "blue" copper appears in some of the large multimeric oxidases as well as in the small monomeric plant "electron transfer proteins" containing a single $Cu(II)$ ion per monomer. This similarity disproved the hypothesis that this spectral species represented an electronic transition associated with a $Cu(I)$–$Cu(II)$ pair since pairs could not be present in the monomeric proteins (*17*). Another possible reason for increased transition probabilities is the distortion from the axial symmetry present in the usual square-planar $Cu(II)$ complexes. At least two laboratories have shown that it is possible to calculate both the optical and ESR parameters (*see* below) of these unusual copper sites by assuming considerable distortion of four ligands out of the plane toward a tetrahedron (*77, 78*).

Table II. Optical and ESR Spectral

Protein	Absorption Spectrum	
	λ_{max} nm	ε M^{-1}cm^{-1}
"Blue" copper oxidases		
Laccase	720	2000
(*Polyporus versicolor*)	610	4600
	440	800
	330	3000
Ascorbate oxidase (cucumber)	760	3600
	607	9700
	425	—
	330	3600
Ceruloplasmin (human plasma)	794	2200
	610	11,300
	459	1200
	332	4100
Cytochrome oxidase (bovine heart)	830—"blue", "non-blue" and	
"Blue" plant electron transfer proteins		
Azurin	820	390
(*Pseudomonas aeruginosa*)	625	3500
	467	270
Stellacyanin	845	700
(*Rhus vernicifera*)	604	3820
	448	554
Plastocyanin (spinach)	770	3300
	597	9800
	460	1180
"Non-blue" copper in enzymes		
Laccase (*Polyporus versicolor*)	obscured by blue copper	
Ceruloplasmin	obscured by blue copper	
Urate oxidase		
D-Galactose oxidase		
Monoamine oxidase	480	
Diamine oxidase	500	
Superoxide dismutase (green pea)	680	290
Superoxide dismutase (bovine)	680	—250
	615	—170
Cu(II)—substituted enzymes		
Cu(II) Carboxypeptidase A	790	<100
Cu(II) Carbonic anhydrase B	900	75
	750	100
	590	50
Cu(II) Carbonic anhydrase B + CN$^-$	900	80
	700	130
Cu(II) Alkaline phosphatase (*E. coli*)	750	—100

Parameters of Copper Proteins

ESR Spectrum			E'_o *volts*	*Reference*
$|A_\|| \| cm^{-1}$	$g_\||$	$g_\perp$		
0.0090	2.190	2.052 2.033	0.76	*(60)*
0.005	2.22	2.06		*(7,63)*
0.0083	2.21	2.05	0.5–0.6	*(39–41)*

ESR non-detectable

0.006	2.26	2.052	0.328	*(17,60)*
0.0031	2.288	2.08 2.025	0.30	*(18–20)*
0.0063	2.226	2.053	0.37–0.39	*(21,22)*
0.0194	2.24	2.05		*(4,5,60)*
0.0136	2.28	2.04		*(39–42,60)*
0.0185	2.28	2.04		*(64)*
0.016	2.22	2.05		*(65)*
0.016	2.29	2.06		*(66)*
0.014	2.23	2.04		*(53)*
0.014	2.268	2.087		*(67)*
0.019	2.24	2.06		*(68)*
0.0152	2.295	2.071		*(69)*
0.0188	2.199	2.053		*(69)*
0.0168	2.270	2.050		*(70)*

Table II.

Protein	Absorption Spectrum	
	λ_{max} nm	ε $M^{-1}cm^{-1}$
Proteins containing ESR non-detectable copper in the resting state		
Tyrosinase (mushroom)	colorless or faint green	
Tyrosinase + H_2O_2	600	700
	345	10,000
Tyrosinase + NO	yellow green	
Hemocyanin (*Loligo pealei*)	580	370
	345	8900
Hemocyanin + NO (*Helix pomatia*)	yellow green	

It has been suggested that a site with a tetrahedral distortion would be desirable for a Cu(II) ion undergoing reversible reduction-oxidation since the Cu(I) ion prefers tetrahedral coordination. Thus the tendency toward conformational change on reduction would be lessened (*60*).

An additional feature of the "blue" Cu(II) site is revealed by the CD (Figure 1) which shows that there may be as many as five or six

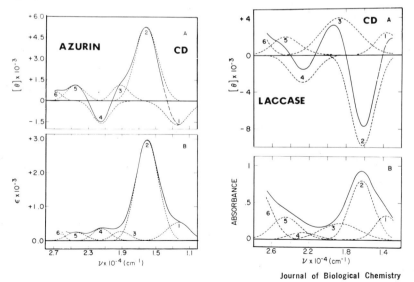

Journal of Biological Chemistry

Figure 1. Visible CD (A) and absorption spectra (B) of azurin from Pseudomonas aeruginosa *(left) and laccase from* Polyporus versicolor *(right) (76).*

Continued

	ESR Spectrum		E'$_o$ volts	Reference
$|A|_{||}$ cm^{-1}	$g_{||}$	$g_\perp$		
		$g = 2$ less than 2% of the copper		(71)
		no ESR		(72)
		$g = 2$ $(\Delta m = 1)$		(73,74)
		$g = 4$ $(\Delta m = 2)$		
		no ESR		(23–26)
		$g = 2$ $(\Delta m = 1)$		(75)
		$g = 4$ $(\Delta m = 2)$		

additional small amount of
$A_{||} = 0.0132\ g_\perp = 2.08\ g_{||} = 2.30$

visible and near infrared absorption bands for these "blue" copper proteins (76). This is three more than expected from the simplest treatment of Cu(II) in a tetragonal field. The high energy of some of these bands (Figure 1) suggest charge transfer from ligand to metal. This in turn suggests that the chromophore is a combination of *d-d* and charge-transfer transitions which makes the designation of a particular band as one or another artificial. This could greatly affect the intensity of all the bands.

The highly unusual ligand field around the blue Cu(II) ions must relate in some way to the enzymatic function of this copper, but as yet the mechanism of electron transfer in these proteins or the mode of oxygen binding in the oxidases is not well understood. Disappearance of the blue absorption maximum and the ESR signal during reduction of laccase by substrate shows that the blue copper is reduced (60, 79). Recent kinetic data on fungal laccase show that this type I or blue copper is rapidly reduced by substrate (rapidly enough to be on the direct reaction pathway) and that all electrons appear to enter the enzyme through this type I copper (60, 80). The precise nature of the copper interactions in this enzyme is complicated by the fact that this protein contains at least two other types of copper: Cu(II), which shows more normal spectral and ESR parameters (*see* below), and ESR non-detectable copper which may consist of coupled Cu(II)–Cu(II) pairs (*see* below) (60).

Even less is known about the detailed chemistry of the electron transfer process and the structure of the small blue proteins. Unfortunately an x-ray structure is not yet available for any of these small proteins. Aside from their isolation from plant organelles containing electron transfer systems, there is some enzymological evidence relating plasto-cyanin to the operation of the photoreduction system in spinach chloro-plasts (81). Sonication rapidly destroys the ability of the chloroplasts to photoreduce NADP. This loss appears to be coincident with the loss of plastocyanin. Readdition of the latter protein can restore the photore-duction of NADP to sonicated chloroplasts (81). Also the photoxidation of reduced cytochrome C in detergent-treated chloroplasts requires a soluble protein with the characteristics of plastocyanin in addition to the chlorophyll-containing particle (82). *Scenedesmus* cells grown in a cop-per deficient medium show a much reduced ability to photoreduce CO_2 with H_2 (83). There is a light-induced change in absorption of certain green plants around 600 nm which is attributed to the oxidation-reduction of plastocyanin (84).

Electron Spin Resonance of "Blue" and "Non-Blue" Cu(II). Most simple mononuclear Cu(II) complexes, when sufficiently magnetically dilute to prevent spin-spin interaction, have characteristic ESR spectra which reflect the axial symmetry of the square-planar or tetragonally dis-torted coordination geometry usually preferred by Cu(II). Although the ligands in the plane may be different nuclei, such complexes appear in most cases to have approximate axial symmetry and $g_x = g_y \neq g_z$ (where g_z is designated $g_{||}$, the g factor parallel to the symmetry axis, and $g_x = g_y = g_{\perp}$ are the g factors perpendicular to the symmetry axis). These g factors describe the ESR absorption when the applied field is parallel and perpendicular, respectively, to the symmetry axis of the coordination complex. Two areas of absorption are observed, one cor-responding to $g_{||}$, containing one-third of the total absorption and usually split into four separate lines by the nuclear spitting induced by the copper nucleus of spin 3/2. Two-thirds of the spin occurs in the region corre-sponding to $g_{\perp}$. The copper nuclear hyperfine splitting is generally much smaller in this region and is often not resolved. The copper ions in a number of copper proteins show this typical ESR spectrum. This is illustrated by the ESR spectrum of bovine superoxide dismutase in Fig-ure 2. The copper nuclear hyperfine splitting in the parallel region, $|A_{||}|$, is ~150 gauss or 0.016 cm^{-1}—typical of model Cu(II) complexes. Close inspection of this signal reveals a slight inequality of g_x and g_y which reflects some rhombic distortion around the copper ion. Otherwise the physicochemistry of the copper site does not appear unusual. This inequality of g_x and g_y is much clearer in ESR spectra taken at 35 GHz (85). Copper of this type is referred to as non-blue since it has much lower extinction coefficients near 600 nm (~100M^{-1}cm^{-1}/Cu for super-oxide dismutase, Table II) than the blue copper described above (Figure 1).

As with the visible absorption spectra, the ESR spectra differentiate between the more normal type of copper and the blue copper sites. The ESR spectrum of stellacyanin from the lac tree is shown in Figure 2. Three g factors are required to calculate this spectrum from the spin

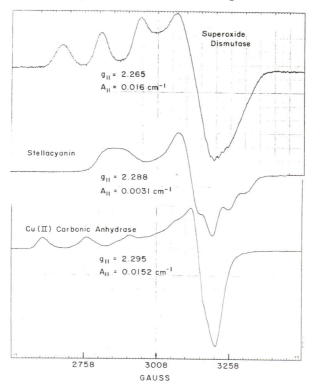

Figure 2.   ESR spectra of bovine superoxide dismu-tase, stellacyanin, and Cu(II) human carbonic anhy-drase B.   Conditions: 5 × 10⁻⁴M protein, 0.01M tris, pH 8, 111°K, 9.15 GHz.

Hamiltonian, showing that the site contains considerable distortion from the usual axial symmetry (*86*). The other important finding is an extremely small copper nuclear hyperfine splitting in the $g_{\|}$ region. The four copper nuclear hyperfine lines in the $g_{\|}$ region are lumped in one peak of this first derivative spectrum at 2850 gauss. Computer simulation of this spectrum shows that the $|A_{\|}|$ value is 0.003 cm⁻¹ (*77, 78, 86*), one-fourth to one-fifth as large as that of more normal copper complexes. This may also be related to distortions in the complex (*86*), but the reasons for this extremely small nuclear hyperfine splitting constant are not entirely clear.

Another interesting comparison of a property of Cu(II) in a non-blue coordination site (bovine superoxide dismutase) and a blue site (stellacyanin) is the natural circular dichroism and the magnetically induced circular dichroism of these Cu(II) chromophores (Figures 3A and B). It has been suggested that the less symmetrical sites occurring in the blue proteins may lead to large intrinsic circular dichroism. It is true that the CD associated with the 600-nm bands of the blue proteins (Figures 1 and 3B) is considerably larger than most dissymmetric Cu(II) complexes, such as peptide complexes. The CD of the blue proteins is also larger than that of the Cu(II) chromophore in superoxide dismutase. It is also believed, on the basis of the ESR, to have some distortion from strictly axial symmetry (85). The latter in turn is considerably larger than that observed for a Cu(II)-substituted Zn(II) protein like carbonic anhydrase, which has an axially symmetric ESR.

It is tempting to relate the natural CD in some manner to the intrinsic symmetry of the complex. However, there are indications, especially in Co(II)-substituted proteins, that the dissymmetry of the surrounding protein potential field may be more important than the intrinsic symmetry of the coordination complex in determining the magnitude and sign of the natural CD (87). Thus symmetry conclusions based on natural CD must be treated with caution.

Magneto circular dichroism (MCD) (or the molecular Zeeman effect), on the other hand, is not conformationally sensitive and depends only on the nature of the ground and excited states in a particular electronic transition. In transition metal chromophores the nature of the ground and excited states depends on the symmetry of the ligand field. While the application of MCD is somewhat limited because it only responds secondarily to conformational change, it can be valuable in determining symmetry if two different symmetries give ground or excited states with different magnetic properties. This is the case in Co(II) complexes, where tetrahedral Co(II) has large MCD, $[\theta] \cong 10^4$ deg cm^2/decimole, while for octahedral Co(II) $[\theta] \cong 10^2$ deg cm^2/decimole (87). Thus MCD has been valuable in distinguishing the two geometries (87).

The difference in magnitude of the MCD for octahedral and tetrahedral Co(II) results from a second-order effect. Most of the MCD magnitude in tetrahedral Co(II) arises from C terms derived from the splitting of the ground state, $^4A_2(F)$. This occurs not because it is magnetically degenerate to begin with, but because it becomes so by mixing with other higher energy states caused by the strong spin-orbital coupling which occurs in tetrahedral Co(II) (88).

We have examined a number of Cu(II) complexes and proteins and have found that the Cu(II) absorption bands in most environments undergo little induction by the magnetic field. Usually a relatively small asymmetric negative MCD corresponding to the visible absorption bands is shown. Bovine superoxide dismutase is a typical example (Figure 3A). While the ESR parameters of superoxide dismutase and stellacyanin suggest considerable differences in the complex, the MCD associated with the visible bands of the two proteins is similar (Figure 3), although of somewhat larger magnitude in stellacyanin. The prominent absorption band at 450 nm in stellacyanin, which has large natural CD, is not induced by the magnetic field (Figure 3B). The same lack of induction is observed for the band at 350 nm in superoxide dismutase (Figure 3A). The latter would appear to be a charge-transfer band (89). The 450-nm band in stellacyanin may also be a charge-transfer transition. There appear to be at least three additional bands above 500 nm. The CD suggests that two occur under the main absorption envelope at 604 nm (Figure 3B). The distinction between the lower energy bands near 600 nm (in the region of the more normal *d-d* transitions) which are induced by the magnetic field and the high energy non-induced bands requires further analysis. Perhaps the distortions in the blue copper sites are not particularly large, and some of the alterations of g and $|A_{\parallel}|$ values may come from delocalization of the electrons (86). The latter is a desirable property of a copper complex participating in electron transfer (90).

Cu(II) has been used to explore the metal-binding sites of a number of Zn(II) metalloenzymes by substituting Cu(II) for the native Zn(II) ion. In all cases examined thus far, the ESR signals of these derivatives show that these sites create a coordination geometry around the Cu(II) similar to normal Cu(II) complexes. This is illustrated by the ESR spectrum of Cu(II) carbonic anhydrase B in Figure 2C. While the metal-binding sites of these Zn(II) proteins have high affinities for some first transition period metals (91), none of the sites have the peculiar constraints that characterize the blue copper sites in a number of the naturally occurring copper proteins.

Ligand Superhyperfine Interaction. Another property of the ESR spectrum of Cu(II) which is often valuable in studying Cu(II) proteins is the splitting induced by the ligand nuclear spins. In many Cu(II) complexes the ligand–metal bonding involves considerable sigma bonding, so the free electron has a finite probability of being at the ligand nuclei. Often the ESR signal is then additionally split according to the quantized values that can be assumed by the ligand nuclear spin, in the same way as the copper nucleus of spin 3/2 splits the signal. The common ligand nuclei encountered in biological systems have such characteristic spin values, magnetic moments, and super hyperfine coupling

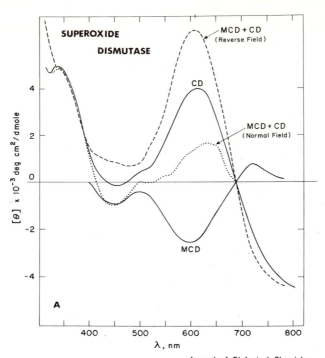

Journal of Biological Chemistry

Figure 3A.   MCD and CD of Zn(II)-Cu(II) bovine superoxide dismutase. Natural CD (—); MCD + CD, normal field (····); MCD + CD, reverse field (- - -); MCD = CD − (MCD + CD, reverse field). Conditions: pH 6, 25°C, field = 43.5 kgauss (89).

constants that the species of ligand atom coordinating the metal ion can be identified from the splitting pattern.

Two examples in Figure 4 involve a Cu(II)-substituted carbonic anhydrase (69, 92). In the cyanide derivative of this enzyme, five super hyperfine lines are seen on each copper hyperfine line (Figure 4A). This splitting is caused by two magnetically equivalent nitrogen nuclei as indicated by the splitting of ~20 gauss and the five lines. Each nitrogen nucleus should give a three-line spectrum since ^{14}N has a spin of 1. Thus two nitrogen nuclei might be expected to give rise to nine lines. However, if they are magnetically equivalent, the line spacings coincide, resulting in five observable lines. Since $^{12}C^{15}N$ does not change the spectrum (Figure 4A), the nitrogen nucleus of the cyanide does not contribute to the splitting. ^{15}N has a spin of 1/2, and the substitution of ^{15}N for ^{14}N should reduce the number of lines. Thus the nitrogen ligand nuclei are protein nuclei, probably two of the histidyl residues identified by x-ray crystallography as coordinating the Zn(II) ion (93).

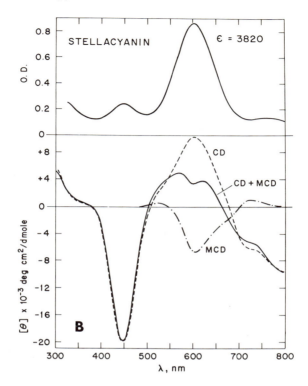

Figure 3B.   Absorption (——), CD (- - -), and MCD (—·—) of stellacyanin.   MCD = (CD + MCD, normal field) − CD.   Conditions: pH 8, 0.01M tris, 25°C, field = 43.5 kgauss.

It can be shown that cyanide does coordinate the Cu(II) through the carbon atom by using $^{13}C^{14}N^-$ to form the complex. ^{12}C has no nuclear spin and thus does not interact with the free electron. ^{13}C, on the other hand, has a spin of 1/2, which further spits the copper hyperfine lines into doublets (Figure 4B). This type of ligand identification has been used for some Cu(II) complexes and proteins including the non-blue Cu(II) of laccase (60) (*see* below).

Multiple Types of Copper in One Enzyme. When the blue copper oxidases were first investigated, attention naturally centered on the copper responsible for the intense blue color and the highly unusual ESR signal. Since both the unique optical and ESR parameters of the Cu(II) disappeared when the proteins were denatured, any more normal Cu(II) detected by ESR in these proteins was most often attributed to denatured enzyme (60). However, as more extensive studies on homogeneous samples of the blue copper oxidases were carried out, it became clear that

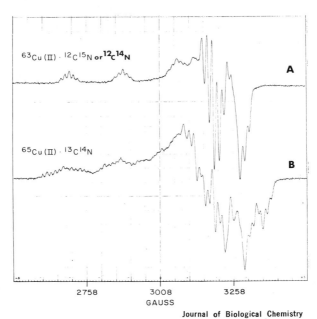

Figure 4. ESR spectra of Cu(II) human carbonic anhydrase B plus $^{12}C^{14}N^-$ or $^{12}C^{15}N^-$ (A) and $^{13}C^{14}N^-$ (B). Conditions: pH 8, 0.01M tris, 111°K, 9.15 GHz (92).

as much as 50% of the naturally occurring copper in oxidases like laccase and ceruloplasmin was non-blue. For example, in purified fungal laccase only part of the 4 gram atoms of copper present per mole gives rise to the ESR spectrum with the typical small $|A_{||}|$ (Figure 5). As shown in Figure 5, additional copper is present with more normal copper nuclear splitting in the $g_{||}$ region of the ESR spectrum (Table II). A computer simulation of the ESR spectrum (Figure 5B) shows that the blue and non-blue copper are present in almost equal amounts (94). Similar findings have been reported for ceruloplasmin (60, 61).

In the case of *Polyporus* laccase, Malkin *et al.* (95) have differentially removed the non-blue Cu(II) from the protein. This inactivates the enzyme but leaves the intense blue color intact. The activity and original copper content can be restored by adding Cu(II) and ascorbate (95). Anions such as F⁻ and CN⁻ appear to inhibit by reacting with the non-blue copper (66). Fluoride, for example, appears to react exclusively with the non-blue Cu(II) since the super hyperfine lines from the fluoride nucleus appear exclusively on the non-blue Cu(II) hyperfine lines in the ESR spectrum, and the blue Cu(II) hyperfine lines remain unaltered (Figure 6) (96). Figure 6 is an ESR spectrum taken at a microwave frequency of 35 GHz which gives better resolution of the two types of copper hyperfine lines in the $g_\parallel$ region. Cyanide appears to reduce the blue copper since its ESR signal disappears, and the remaining ESR signal [40% of the original Cu(II)] with more normal $|A_\parallel|$ values appears to reflect the non-blue copper. This signal, however, is considerably modified and shows extensive nitrogen ligand nuclear super hyperfine lines (60).

The experiments outlined above and the reversible reduction of the blue copper by substrate established that both types of ESR-detectable copper appear to participate in the mechanism of action of laccase. However when the integrated signal intensity was used to quantitate the ESR-detectable copper, it accounted for only about one-half of the total amount of copper present in the enzyme as determined by chemical methods or atomic absorption (60). Only two out of the four copper atoms present in a molecule of fungal laccase give a detectable ESR signal. The reasons for the presence of ESR non-detectable copper are outlined below in the discussion of tyrosinase and hemocyanin. Magnetic susceptibility studies have established that the ESR non-detectable copper in *Polyporus* laccase is diamagnetic and not caused by line broadening

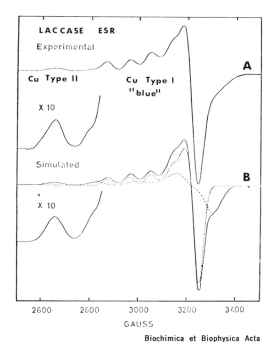

Biochimica et Biophysica Acta

*Figure 5.   Experimental and simulated ESR
spectra at 9.2 GHz of* Polyporus *laccase* (94)

(97). The presence of this ESR non-detectable copper has been established in other blue copper oxidases including *Rhus* laccase (60), ceruloplasmin (98), and cytochrome oxidase (99).

The function of the diamagnetic copper in the blue copper oxidases is not clear. Anaerobic titrations of *Polyporus* laccase with a number of

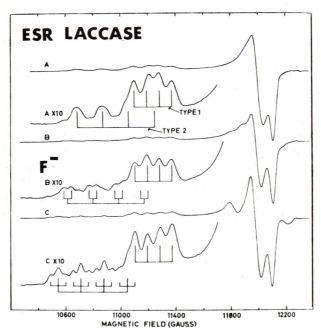

Figure 6. The reaction of Polyporus laccase with F^- as studied by ESR at 35 GHz. (A) Native laccase; (B) laccase plus 1 equivalent of NaF; (C) laccase plus 15 equivalents of NaF. The hyperfine splitting due to two forms of Cu(II) and the superhyperfine splitting due to F^- are indicated (96).

reducing agents have demonstrated that the blue Cu(II) is linearly reduced with the addition of ~3.5 electron equivalents of reductant (95). This implies that there are nearly two additional reducible sites in the enzyme besides the two ESR-detectable Cu(II) ions. On this basis Malkin *et al.* (95) have argued that the remaining two sites may be a spin-paired Cu(II)–Cu(II) couple (60, 95). This pair may be associated with the 330-nm absorption band in the protein since in the F^- treated enzyme, the 610-nm band associated with the blue copper can be reduced separately from the 330-nm band with the first electron equivalent (60, 95). The second and third equivalents then reduce the 330 nm. Rapid flow measurements show, however, that the loss of the 330-nm band during anaerobic reduction is too slow to be directly on the catalytic pathway (*60*) although the rate of reduction could be artificially slow in the absence of O_2 (*see* Radical Mechanisms, page 293 for substrate–oxygen interactions).

ESR-Non-detectable Copper. Tyrosinase or polyphenol oxidase from mushrooms is the classic example of a copper protein which produces no ESR signal. In the pure state this enzyme also shows no visible absorption spectrum except for a broad, poorly resolved band between 320 and 360 nm on the long wavelength side of the protein absorption (*71, 100*). It is interesting in terms of recent findings that purification has almost always been associated with the removal of a yellow-green material (*101, 102*). It may have been a form of the enzyme in which the Cu sites were altered as in the modified tyrosinase (described below). The lack of color and ESR signal suggested that copper in the resting enzyme was in the Cu(I) state. In addition, no ESR or spectral evidence has ever been found for reversible oxidation of the copper during the enzymatic reaction. Since the mushroom enzymes are multimeric—probably a tetramer containing four copper ions in the active form (*8*)—it has long been speculated that more than one copper was involved in each active site. The binding of ^{14}C-benzoic acid to the mushroom enzyme has suggested that there may be only two inhibitor (substrate) binding sites per tetramer (*102*). The tyrosinase isolated from *Neurospora crassa*, however, readily dissociates into monomers of MW 35,000 (*9*). This supported the idea that the lack of an ESR signal was due to the presence of Cu(I) rather than pairs of spin-paired Cu(II) ions. It did not rule out the possibility, however, that the active form of the *Neurospora* enzyme is a dimer.

Before discussing more recent findings on the existence of Cu(II) pairs in tyrosinase, it is appropriate to discuss the classic example of a protein in which copper pairs are thought to exist—namely hemocyanin. It is the oxygen-carrying protein of the hemolymph of certain molusca and arthropoda. In oxygenated form, this is a striking blue protein with the intensity of the visible absorption near 600 nm intermediate between that of the blue copper discussed above and non-blue copper (Table II). The spectrum of hemocyanin is also distinguished by the presence of an intense absorption band, $\epsilon = {\sim}10{,}000$, at 345 nm. Both bands are shown in Figure 7 for hemocyanin from *Loligo pealei* (*23, 24, 25, 26*). Both bands are presumably associated with the chromophore at the O_2 binding site since they and their associated Cotton effects disappear in the deoxygenated form of the protein (*23, 24, 25, 26*). Despite its blue color, oxyhemocyanin shows no ESR signal indicating that monomeric Cu(II) does not exist. The actual electronic structure of the copper–oxygen chromophore has been a matter of extensive discussion (*103*), particularly

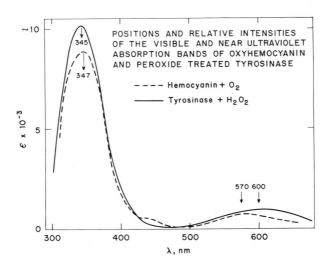

*Figure 7.   Absorption spectra of oxyhemocyanin (- - -)
from* Loligo pealei *and the H_2O_2 derivative of mushroom tyrosinase (—). Redrawn from (23, 24, 25, 26)
and (72).*

concerning the formal valence state of the copper. Binding studies with
CO and O_2 have established that the Cu/O_2 or Cu/CO ratio is 2, sug-
gesting that two copper ions are present at each O_2 binding site.
Structures I or II are usually drawn for the oxygen complex.

$$Cu-O=O-Cu \quad \text{or} \quad Cu-\overset{\displaystyle O}{\underset{\displaystyle O}{\|}}-Cu$$

$$\text{I} \qquad\qquad\qquad \text{II}$$

Structure III is more likely for the CO complex rather than a bridge

$$\begin{array}{c} O \\ \| \\ C \\ \diagup \diagdown \\ Cu-Cu \end{array}$$

$$\text{III}$$

(*105*). Either the close copper ions have unpaired electrons which
are spin paired, or the expected ESR lines are broadened by contact
interaction. The colorless deoxy form of the protein may contain Cu(I)
although the formal valence state in either form of hemocyanin is not
absolutely settled (*see* below for NO derivative).

Recent data on derivatives of both hemocyanin and tyrosinase have established great similarities between the copper in both proteins. In certain chemically modified forms of the enzyme, magnetic dipole-dipole coupled $Cu(II)$ pairs exist in both proteins.

As reported by Jolley *et al.* (*72*), the reaction of tyrosinase with equivalent amounts of H_2O_2 produces a yellow-green enzyme with a chromophore very similar to that observed in hemocyanin (Figure 7). Both the visible and near ultraviolet bands are present. Titration of the enzyme with H_2O_2 shows that $\sim$one equivalent of H_2O_2 per two copper ions is required to produce the colored derivative. The H_2O_2 derivative becomes colorless when it is deoxygenated, and the chromophore is regenerated on reoxygenation, a pattern analogous to oxy- and deoxy-hemocyanin.

The second type of derivative is produced by the anaerobic reaction of both hemocyanin and tyrosinase with NO. Both proteins give yellow-green products with unusual ESR signals (*54*). The $Cu(II)$ nitric oxide complex, $CuNO^+$, is blue. It has been pointed out that nitric oxide will react with $Cu(I)$ and oxidize the copper to produce $Cu(II)$ which then forms the blue $CuNO^+$ (*105*). ESR absorption for the NO derivatives occurs both in the region of $g = 2$ and $g = 4$ (*54*). In the $g = 2$ region (shown in Figure 8 for the NO-treated mushroom tyrosinase), there is a moderate signal from a typical $Cu(II)$ monomer with four copper hyperfine lines in the $g_{||}$ region, but underlying it is a broad smooth signal.

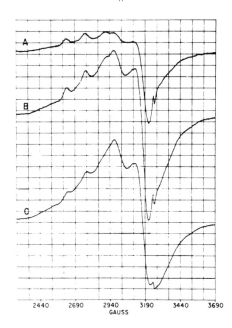

Proceedings of the National Academy of Sciences of the United States of America

Figure 8.   ESR spectra near $g = 2$ *of the nitric oxide derivative of mushroom tyrosinase.* $T = 14°K$, *(A) 0.1 mW; (B) 1 mW; (C) 10 mW (54).*

This is brought out by increasing the microwave power (Figure 8C) which saturates the Cu(II) monomer signal and makes the broad featureless signal more prominent. At $g = 4$ there is another well-resolved signal in the ESR spectra of the NO derivatives of both proteins which shows seven hyperfine lines (Figure 9). The $g = 2$ signals are allowed

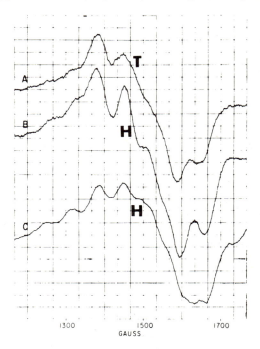

Proceedings of the National Academy of
Sciences of the United States of America

Figure 9.   ESR spectra near $g = 4$ *of the nitric oxide derivatives of mushroom tyrosinase (A) and hemocyanins from* Helix pomatio *(B) and* Cancer magister *(C).* T $= 14°K$, power $= 200$ mW (54).

ESR transitions in which the change in magnetic quantum number is 1; $\Delta M = 1$. The $g = 4$ signal is a forbidden transition in which $\Delta M = 2$. It is expected to occur when magnetic dipole-dipole coupled Cu(II) ion pairs are present (61). The broad featureless signal at $g = 2$ is also compatible with magnetic dipole-dipole coupled Cu(II) pairs. The small amount of overlying Cu(II) monomer signal may arise from disrupted copper sites. Schoot Uiterkamp and Mason (54) have determined that the signals show normal curie behavior between 230° and 14°K, suggesting that significant spin exchange does not occur between the two ions. On the basis of magnetic dipole-dipole interaction alone, the distance between the Cu(II) ions of the pair is estimated to be 6 A (54).

These findings suggest that the copper sites in these seemingly functionally unrelated proteins are quite similar and that the two proteins may be evolutionarily related. A tyrosinase or pretyrosinase has been isolated from the hemolymph of the larvae of the silk moth which does show reversible association into large aggregates (35) (Table I). This protein is blue, with $\varepsilon_{650} = 290$, and MW = 80,000. At least part of the copper is Cu(II) since it gives an ESR signal at $g = 2$. The published signal, however, is complex and may be composed of overlapping species.

Enzymology of Tyrosinase and Laccase

In recent years kinetic and mechanistic studies have been done on the copper oxidases. Recent work has concentrated on laccase and tyrosinase, and since these two enzymes are good examples of the complexities involved, the rest of this paper concentrates on the enzymology of these two proteins. They also give rise to interesting comparisons since they have some substrates in common (*e.g.,* catechol) but differ in certain aspects of their physicochemistry and mechanism.

Both enzymes are prominent in the early history of oxidative biochemistry. The early work of Bertrand and Keilin and Mann on laccase and of Bourquelot, Bertrand, and Kubowitz on tyrosinase has been adequately covered in excellent reviews of both enzymes [tyrosinase (106); laccase (107)]. Much of the early work was on the nature of the reactions catalyzed by these enzymes, and only brief mention is made here as to the nature of the reactions.

The distinguishing feature of tyrosinase is that it catalyzes the oxidation of monohydric phenols, like tyrosine, to the dihydric form and dihydric phenols, like DOPA and catechol, to the corresponding quinones. The striking biological effects of this enzyme arise from quinones which polymerize to produce the darkening of various plants on injury and melanin in mammals. The relative oxidation rates of several dihydric phenols by tyrosinase are given in Table III.

Laccase shares with tyrosinase the ability to oxidize dihydric phenols, like catechol, to the corresponding quinones. Whether the enzyme is active in the oxidation of monohydric phenols like cresol is a matter of controversy since the purity of preparations said to catalyze such oxidations has been questioned (107). Much more important is the ability of laccase to oxidize various aminophenols, like p-phenylenediamine, which is in fact the best substrate for the enzyme (Table III). The enzyme is important commercially because it oxidizes some complex

Table III. Substrates for Tyrosinase

Tyrosinase (102)

Substrate	V_{max} mole sec^{-1}/mole tetramer (120,000) (mushroom)
L-Dopa	884
Pyrocatechol	4600
4-SCN-catechol	240
4-COCH$_3$-catechol	26
4-CHO-catechol	8
4-CN-catechol	0.108
4-NO$_2$-catechol	0.096

Natural Substrate

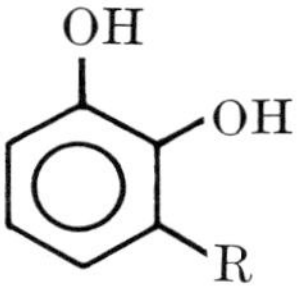

substituted catechols in the latex of the lac tree (Table III), producing lacquer. That process was elegantly described by Yoshida in 1883 (108). The discussion here is not exhaustive but summarizes briefly kinetic and mechanistic studies which relate to the electron transfer mechanism in these enzymes.

The dominant feature of tyrosinase is that it has both cresolase and catecholase activity. Laccase has a very clear catecholase activity, but its cresolase activity is not so clear. Mason, Fowlks, and Peterson (109) used $^{18}O_2$ to label 3,4-dimethylphenol during the tyrosinase-catalyzed oxidation of this compound and showed that the source of the oxygen introduced into the phenol in the phenolase reaction was molecular oxygen according to Reaction 1.

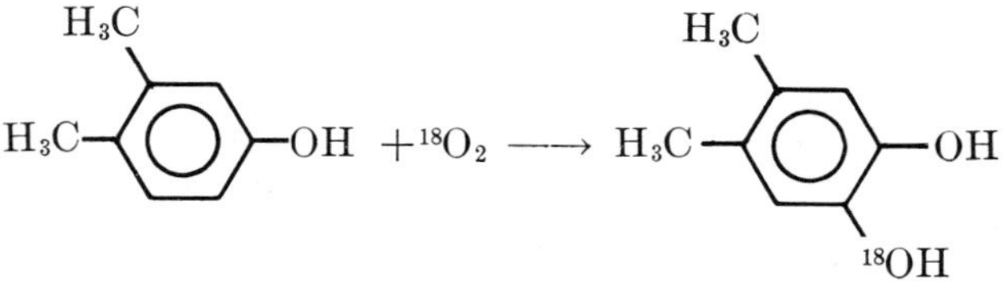

$$(1)$$

The subsequent catecholase activity of tyrosinase requires easy removal of electrons from the phenolic oxygens at the 1 and 2 positions

and Laccase Velocities of Oxidation

Laccase (107)

Substrate	V_{max} $\mu liters\ O_2/hr/\mu mole$ (Rhus vernicifera)	V_{max} *mole sec^{-1}/mole of 2 dimers (120,000)* (Polyporus versicolor)
p-phenylene diamine	21.7	
N,N-dimethyl-p-phenylene diamine	17.6	
Hydroquinone	7.6	
N-phenyl-phenylenediamine	3.5	
O-phenylenediamine	3.2	
p-aminophenol	1.04	
Pyrocatechol	0.95	
4-NO$_2$-catechol		1.38

$$R = - (CH_2)_{14}CH_3$$
$$- (CH_2)_7CH=CH-(CH_2)_5-CH_3$$
$$- (CH_2)_7CH=CH-CH_2-CH=CH-(CH_2)_2-CH_3$$
$$- (CH_2)_7CH=CH-CH=CH-CH_2-CH=CH-CH_3$$

since the oxidation becomes progressively harder as electron-withdrawing substituents are placed in the 4 position (Table III). Both k_{cat} and K_m for the substrates given in Table III follow reasonable Hammett relations (*102*). 4-Nitrocatechol is the most slowly oxidized substrate. Initially it was reported that it was not oxidized by tyrosinase. It does, however, provide a convenient spectrophotometric assay, especially since the quinone product is stable over a long time in contrast to other quinones.

The spectral changes which occur during complete oxidation of 4-nitrocatechol by tyrosinase are shown in Figure 10A. Oxidation of the same substrate by fungal laccase is shown in Figure 10B. Although the same product is formed, laccase is seven times more active than tyrosinase in oxidizing this substrate on an equivalent molecular weight basis ('120,000) (Table III). Of course these enzymes are quite different in many respects; one is a blue copper oxidase while the other is the classic Cu(I) enzyme. Both, however, catalyze oxidations using molecular oxygen as the final electron acceptor reducing it to water. Thus much of the work on these enzymes has concentrated on the mechanism of coupling the one- or possibly two-electron transfers in the oxidation of substrate to the four-electron transfer required to reduce oxygen to water. The pathway by which this is accomplished is the essence of the mecha-

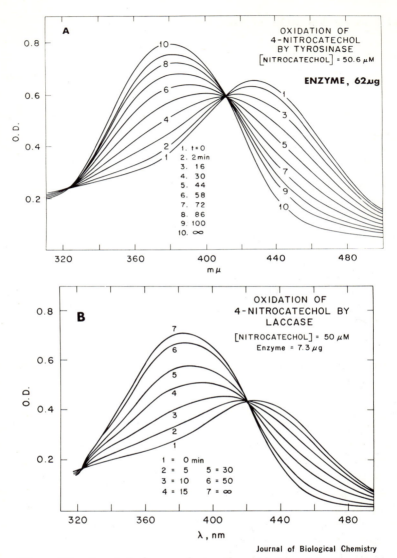

Figure 10.   Spectral changes during the oxidation of 4-nitrocate-chol by mushroom tyrosinase (A) and fungal laccase (B) (102)

nism of action. Various kinetic and physicochemical data relating to this question are available for both enzymes, and an attempt is made below to summarize these comparatively.

The Copper Sites of Tyrosinase and Laccase. The information suggesting that copper pairs exist at the active site of tyrosinase was reviewed on page 285. Tyrosinase has been the classic Cu(I) enzyme, and there was no evidence that the copper underwent reversible oxidation-reduction during the reaction. There is no. some doubt because of the work on the H_2O_2- and NO-treated enzyme which suggests that Cu(II) pairs may actually exist under certain circumstances although oxidation may have occurred in such derivatives.

Laccase on the other hand has been quoted as a clear case of reversible oxidation of the copper by the substrate during reaction since the blue copper and its attendant ESR signal are abolished by substrate. The sequence of electron transfer is more complex, however, as shown by the extensive work of Malmstrom and co-workers on fungal laccase (*60*). This enzyme actually contains four copper ions per mole of protein (MW = 60,000), only one of which is blue. This accounts for the ESR with the small $|A_{||}|$. One other is clearly Cu(II), giving rise to a normal Cu(II) ESR (Figure 5). The other two copper ions are not detectable by ESR. However, when the blue copper is reduced anaerobically by substrate, four electron equivalents are required to abolish the 610-nm absorption which decreases linearly (*95*). Thus two additional electron accepting sites exist in the protein besides the obvious two Cu(II) ions. All four are clearly involved in the oxidation of substrate. The evidence suggests that the two additional electron centers are the two ESR-non-detectable coppers. Since they can be reduced, Malkin *et al.* (*95*) suggest that they are paired Cu(II) ions rather than Cu(I).

Radical Mechanisms. In the oxidation of the substrates for both laccase and tyrosinase it might be expected that single electron transfers would result in the production of free radicals as intermediates. The presence of free radicals during substrate oxidation has been sought in the case of both enzymes, but only for laccase has evidence suggested that a substrate free-radical is formed by removal of one electron from the substrate by the enzyme (*110*). The ESR spectrum of *o*-benzosemi-quinone, which is observed when catechol is oxidized by tyrosinase (*111*), is attributed to the non-enzymatic equilibration between the fully oxidized quinone and the substrate, producing the semiquinone (*111*). It has been suggested that the electron transfers take place in pairs during the tyrosinase oxidations (*112*). Kinetic studies on the mechanism of tyrosinase action are outlined below.

Rapid-flow ESR experiments by Nakamura first demonstrated a signal from a catechol semiquinone during the oxidation of this substrate by *Rhus* laccase (*6*). In extensive rapid-flow ESR studies of the reduction of the copper ESR signal of fungal laccase by *p*-phenylenediamine, Broman *et al.* (*110*) found a free radical signal which arose simultaneously with the reduction of the blue copper signal. In their experiments the radical decayed rapidly with oxygen exhaustion because of the rapid turnover of laccase. On the other hand with ceruloplasmin (containing similar blue copper), which oxidizes the same substrate more slowly, the radical signal was identical to that reported for the positive radical of p-phenylenediamine, $[NH_2C_6H_4NH_2]^+$. The simultaneous loss of the Cu(II) ESR signal and the appearance of the substrate free-radical suggests that the blue Cu(II) was the initial electron acceptor from the substrate in a one-electron transfer. The loss of the radical signal on oxygen exhaustion appeared to be a non-enzymatic process. While blue copper may be the initial acceptor, subsequent work has shown that the reduction of the enzyme distributes four electron equivalents into the protein before the blue Cu(II) is completely reduced (60, 95). This may relate to the final necessity of transferring four electrons to molecular oxygen.

    Tyrosinase Kinetics. K_m values for the oxidation of a series of catechols are relatively small: from $26 \times 10^{-5}M$ for the most rapidly oxidized substrate to $4 \times 10^{-6}M$ for the most poorly oxidized catechol, 4-NO$_2$-catechol (Table IV). In the absence of precise knowledge of the mecha-

Table IV. K_m and K_i Values for Mushroom Tyrosinase (102)

Substrate Varied	Inhibitor	K_m M $\times$ 10^5	K_i M $\times$ 10^6
L-DOPA	—	26.3	—
Pyrocatechol	—	19.4	—
4-SCN-catechol	—	8.1	—
4-COCH$_3$-catechol	—	2.7	—
4-CHO-catechol	—	1.5	—
4-NO$_2$-catechol	—	0.43	—
Pyrocatechol	benzoic acid	—	1.0 (competitive)
Pyrocatechol	cyanide	—	14.7 (noncompetitive)
O$_2$	benzoic acid	—	3.0 (noncompetitive)
O$_2$	cyanide	—	13.0 (competitive)

nism or of the rate-limiting step, the meaning of K_m is unclear, but the magnitudes are only slightly larger than the dissociation constant for the binding of the competitive inhibitor (*i.e.* competitive with catechol), ^{14}C-benzoic acid, as measured by equilibrium dialysis (102). Kinetic studies with pyrocatechol are complicated by an irreversible inactivation of the enzyme during the reaction, believed to result from a nucleophilic addition of o-benzoquinone to a group on the enzyme (113). Nucleophilic addition reactions are characteristic of o-benzoquinone (113). It has been shown that ^{14}C-labeled monophenols do become covalently attached to the enzyme during the oxidation (114). This inactivation is a first-order process, so initial reaction rates can be determined from back extrapolation of first-order plots (102). The 4- and 4,5-substituted quinones do not undergo facile nucleophilic additions, and thus the oxidations of 4- and 4,5-substituted catechols are not accompanied by the troublesome enzyme inactivation (*102, 113*). In many cases the quinones are so stable that the reactions may be followed for long times spectrophotometrically [(*102*), *see* nitrocatechol oxidation above].

Our laboratory has done kinetic studies on the mushroom enzyme with the catechols listed in Table III, and a brief summary of results is given here. Inhibition studies with cyanide and benzoic acid show that benzoic acid is competitive with catechols but noncompetitive with the cosubstrate oxygen. In a reciprocal fashion, cyanide is competitive with O_2 but non-competitive with the catechol substrate (*102*). The kinetic constants are given in Table IV. Since a likely hypothesis is that CN^- combines with the copper site, it would appear that O_2 binds near the copper and that there is an adjacent site for the catechol or phenol substrate.

It is necessary to know the topological relationship between the bound substrate, copper, and enzyme-bound oxygen to understand the mechanism completely. Unfortunately few direct methods lend themselves to the study of the O_2 interaction, and the kinetic data must be interpreted with caution. The latter does reveal, however, some features of the O_2 interaction. The K_m for oxygen determined with rapidly oxidized catechol substrates is around $10^{-5}M$ (*102, 115*). Since the K_m values for O_2 did not appear to vary for several substrates, it was suggested that O_2 binding was independent of substrate, and oxygen probably bound first (*115*). However, if oxidation rates as a function of O_2 concentration are examined with catechols showing a much wider range of oxidation rates, different oxygen saturation behavior is observed (*102*). The double reciprocal plots for the oxidation of pyrocatechol, 4-acetylcatechol, and 4-formylcatechol at several partial pressures of O_2 are shown in Figure 11. It is clear that oxygen saturation behavior varies widely between them. For the more slowly oxidized substrates the slopes of the double reciprocal lines increase rapidly with lowered O_2 tension while the rapidly oxidized pyrocatechol shows a series of almost parallel lines. This latter behavior might suggest a ping-pong mechanism but could also result from a sequential mechanism in which the turnover rate is much faster than the rate of dissociation of the first enzyme–substrate complex formed (*102*). The data on the slowly oxidized catechols which show convergent plots (Figure 11) suggest a sequential mechanism. An oxygen concentration as low as 2.8% in the gas phase barely desaturates the enzyme in the presence of 4-thiocyanatocatechol (*102*). Calculations from this data show that the K_m for oxygen seems to vary with the structure of the substrate (Table IV). The range for K_{O_2} is $1.4 \times 10^{-4}M$ to $1.4 \times 10^{-5}M$. This suggests that substrate binding influences oxygen binding. It also implies that if the reaction is ordered, O_2 does not bind first. While substrate may be necessary for the formation of the proper final

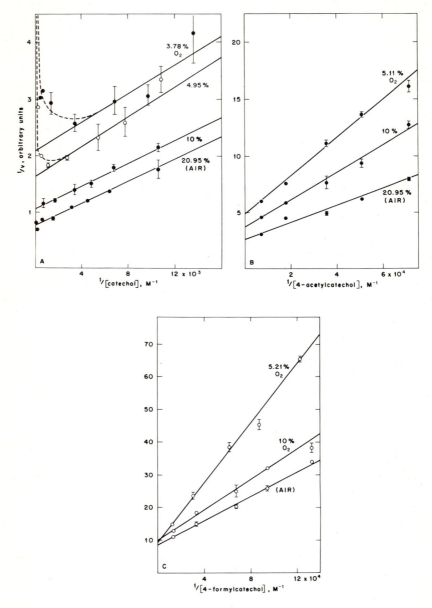

Journal of Biological Chemistry

*Figure 11.   Lineweaver-Burk plots of the oxidation of pyrocatechol (A),
4-acetylcatechol (B), and 4-formylcatechol (C) by mushroom tyrosinase
at several oxygen tensions (102)*

reactive complex between enzyme, catechol, oxygen, and copper (perhaps also an activator molecule, *see* below), it does not necessarily mean that substrate induces O_2 binding to the enzyme. Study of the interaction of O_2 and CO with this enzyme has been greatly hindered by the fact that no spectral difference (optical or ESR) of any kind has ever been detected between tyrosinase in the absence of O_2, in the presence of O_2, or in the presence of CO (71). CO is an inhibitor of the enzyme.

On the other hand, the discovery of the colored H_2O_2 derivative of the enzyme does provide some information on oxygen complex formation with the enzyme (72). If the H_2O_2/Cu ratio does not exceed 1.0, the enzyme remains active against its substrates, catechol, butyl catechol, and *p*-cresol (72). Addition of these substrates does abolish the 345-nm band (shown in Figure 7 for the H_2O_2 complex), suggesting that the colored complex is at least in equilibrium with a catalytic intermediate (72). In addition the colored complex requires the presence of O_2. Reduction of the oxygen tension abolishes both the 345-nm and 600-nm bands of the H_2O_2-treated enzyme, bands which return when oxygen is readmitted (72). This cycle is analogous to oxy- and deoxyhemocyanin. Thus the H_2O_2 derivative of tyrosinase can be oxygenated in the absence of substrate. Whether this complex with a modified Cu(II) center is anything like the normal complex is unclear, but it can apparently lead to an active complex with the addition of substrate.

Kinetics of Copper Binding. The spectroscopic silence of the copper in this enzyme has been frustrating for the direct study of the copper binding site. It has been possible, however, to study the binding process by ESR since the ESR signal of Cu(II) disappears when the ion is bound to apotyrosinase (112, 116). The latter can be prepared by removal of the native copper with cyanide (116). Rapid-freezing ESR studies by Kertesz *et al.* (116) show that immediately after mixing equimolar concentrations of Cu(II) and apoenzyme, an ESR signal of Cu(II) is present which is heterogenous and shows that the Cu(II) is bound to the enzyme (116). Hence a rapid initial binding step must occur. A minimum estimate for the second-order velocity constant describing this initial binding step is $k_{on} > 5 \times 10^6 M^{-1} sec^{-1}$ (116).

After this step the Cu(II) signal, which shows the presence of Cu(II) in two different magnetic environments, decays slowly to a residual small signal typical of the native enzyme. The best analysis of the spectrum shows that one type of Cu(II) signal, Cu_1, decays within minutes, $t_{1/2} = 5$ min. The second type, Cu_2, decays more slowly with $t_{1/2} = 2\frac{1}{2}$ to 3 hr. Activity is restored to the apoenzyme by a first-order process with a half-time corresponding to that of the slowly disappearing Cu(II) signal (116). Thus reconstitution of the active protein from Cu(II) and the apoenzyme is a complex, slow process which involves copper in non-equivalent environments, at least initially. This may be caused by intermediates on the path to the reconstitution of a site containing a copper pair. The process could involve either reduction of both copper ions to Cu(I) or spin-pairing or induction of contact interactions between two Cu(II) ions. The conditions would account for the lack

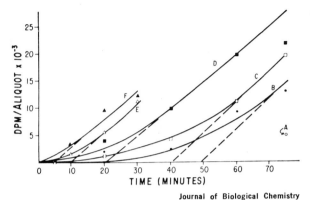

Figure 12. Variation in lag time for tyrosine hydroxylation by hamster melanoma tyrosinase. (A) no DOPA; (B) 4×10^{-6}M; (C) 8×10^{-6}M; (D) 1.6 $\times 10^{-5}$M; (E) 4×10^{-5}M; (F) 8×10^{-5}M (117).

of an ESR signal in the holoenzyme. The lack of visible absorption by native tyrosinase, as opposed to oxyhemocyanin, indicates a slightly different state for the copper in the two native proteins. If analogies are to be made, it would seem that native tyrosinase corresponds most closely to deoxyhemocyanin while the H_2O_2 derivative corresponds to oxyhemocyanin. If the latter is believed to have coupled Cu(II) pairs, perhaps the colorless proteins are more likely to contain Cu(I) pairs although the state of the copper in the presence of substrate may be different.

Catechol as an Activator of Tyrosinase. The phenolase activity of tyrosinase has been studied less completely than the catecholase activity, partly because of the lack of a satisfactory assay procedure. The phenolase reaction, however, is characterized by a "lag time" which can be abolished by adding dihydroxyphenylalanine (DOPA), the immediate product of the hydroxylation reaction (29–34, 102, 117). This phenomenon has been described by several investigators (29–34) and is illustrated in Figure 12, from Pomerantz and Warner (117), using the enzyme from Hamster melanoma. The same phenomenon has been analyzed by Duckworth and Coleman (102) for the mushroom enzyme. In the absence of DOPA, maximum velocity of the hydroxylase reaction is not reached for several minutes. Pomerantz and Warner (117) devised a convenient assay for the phenolase reaction by determining the radioactivity of an aliquot of ³HOH produced from incubations of L-tyrosine-3,5-³H and tyrosinase. The obvious conclusion from the data in Figure 12 is that DOPA activates the phenolase reaction, and maximum velocity is reached as soon as some small catalytic amount of DOPA is produced by the enzyme. This is supported by the fact that $1/t_{lag}$ is directly proportional to enzyme concentration (*102*).

The overall hydroxylase (phenolase) reaction would then be described by Equation 2 (*117*).

$$\text{L-Tyrosine-3,5-}^3\text{H} + \text{L-DOPA} + O_2 \rightarrow \tag{2}$$
$$\text{L-DOPA-5-}^3\text{H} + \text{L-DOPA quinone} + {}^3\text{HOH}$$

Thus a reduced catechol substrate would be required for the hydroxylation reaction, and the overall reaction simultaneously hydroxylates a phenol and oxidizes a catechol. It has been shown that DOPA is present in melanoma tissue (*117*).

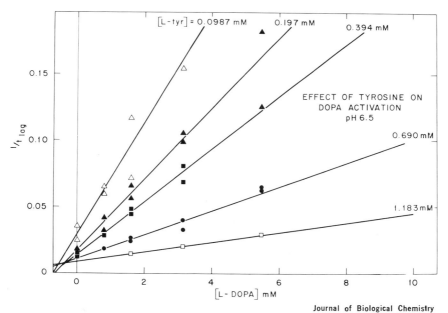

Journal of Biological Chemistry

Figure 13.   Plots of reciprocals of t_{lag} as a function of L-DOPA concentration at different concentrations of L-tyrosinase (102)

Such a mechanism accounts for the overall stoichiometry of four electrons passed in pairs to molecular oxygen. This stoichiometry must also be preserved in the catecholase reaction (*see* below). The mechanism also requires a second cosubstrate binding site for the activating catechol. A plot of $1/t_{lag}$ *vs.* DOPA concentration (Figure 13) shows that DOPA saturates a site responsible for the abolition of the lag time with a K_m of $0.47 \times 10^{-6}M$ in the case of the mushroom enzyme (*102*). This is more than two orders of magnitude smaller than K_m for DOPA in the catecholase reaction (Table IV). L-Tyrosine competes for the activator site since higher concentrations of tyrosine increase the concentration of DOPA required for equal shortening of the lag time (Figure 13) (*102*).

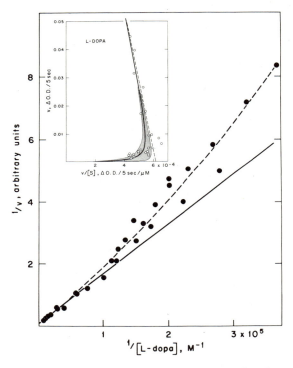

Figure 14.   Lineweaver-Burk plot and Eadie plot (inset) of the oxidation of L-DOPA by mushroom tyrosinase

Since one molecule of oxygen ultimately oxidizes two molecules of reducing substrate, the complete mechanism for the catecholase reaction must also involve two catechol molecules even if they are only sequentially oxidized. However, it might be that the activator site is also involved in the catecholase reaction and that the oxidations by tyrosinase involve a complex between two substrate molecules, copper, and oxygen, to allow for a simultaneous transfer of two pairs or four electrons.

Data on the possible participation of a second catechol are difficult to obtain since the kinetics (Figure 12) show that deviation from

Michaelis-Menten saturation kinetics should occur only at low substrate concentrations, near $10^{-6}M$. Using a spectrophotometric assay it is possible to observe statistically significant deviations as shown by the Lineweaver-Burke and Eadie plots in Figure 14 using DOPA as the substrate in the catecholase reaction.

Lastly there has been one attempt at a practical application of mushroom tyrosinase. Since L-DOPA is useful in treating Parkinson's disease, interest in its production from L-tyrosine has increased. The mushroom enzyme has been prepared in an immobilized form by attaching it to DEAE-cellulose by the crosslinking reagent, 2,4-dichloro-6-amino-*s*-triazine (*118*). The column made from this immobilized enzyme produced adequate amounts of L-DOPA from L-tyrosine put on the column (*118*). Further oxidation of DOPA was prevented by adding ascorbate. The drawback is that the immobilized enzyme, like the free enzyme, undergoes reaction inactivation presumably because of the nucleophilic addition of the quinone to the enzyme. If this could be prevented, this might be a practical method for synthesizing L-DOPA.

While many of these copper oxidases are essential to life as we know it on this planet, and some, like tyrosinase, are essential to our own health, their mechanism of action has remained relatively obscure, certainly in relation to our knowledge of some of the proteolytic enzymes. This paper has attempted to outline the considerable progress that has been made over the past several years in understanding these difficult enzymes. Much of the new information has been developed because of the presence of copper as a reporter group at the active site.

Literature Cited

1. Nakamura, T., *Biochim. Biophys. Acta* (1958) **30**, 44.
2. Omura, T., *J. Biochem.* (Tokyo) (1961) **50**, 264.
3. Malmstrom, B. G., Vänngard, T., *J. Mol. Biol.* (1960) **2**, 118.
4. Mosbach, R., *Biochim. Biophys. Acta* (1963) **73**, 204.
5. Fahraeus, G., Reinhammer, B., *Biochim. Biophys. Acta* (1967) **21**, 2367.
6. Nakamura, T., Ogura, Y., in "The Biochemistry of Copper," J. Peisach, P. Aisen, W. E. Blumberg, Eds., p. 389, Academic, New York, 1966.
7. Dawson, C., in "The Biochemistry of Copper," J. Peisach, P. Aisen, W. E. Blumberg, Eds., p. 305, Academic, New York, 1966.
8. Bouchilloux, S., McMahill, P., Mason, H. S., *J. Biol. Chem.* (1963) **238**, 1699.
9. Fling, M., Horowitz, N. H., Heinemann, S. F., *J. Biol. Chem.* (1963) **238**, 2045.
10. Amaral, D., Berstein, L., Morse, D., Horecker, B. L., *J. Biol. Chem.* (1963) **238**, 2281.
11. Hill, J. M., Mann, P. J. G., *Biochem. J.* (1964) **91**, 171.
12. Ambler, R. P., *Biochem. J.* (1963) **89**, 341.
13. Ambler, R. P., Brown, L. H., *J. Mol. Biol.* (1964) **9**, 825.
14. Suzuki, H., Iwasaki, H., *J. Biochem.* (Tokyo) (1962) **52**, 193.

15. Brill, A. S., Bryce, G. F., Maria, H., *Biochim. Biophys. Acta* (1968) **154**, 342.
16. Broman, L., Malmstrom, B. G., Aasa, R., Vänngard, T., *Biochem. Biophys. Acta* (1963) **75**, 365.
17. Gould, D. C., Mason, H. S., *Biochemistry* (1967) **6**, 801.
18. Omura, T., *J. Biochem.* (1961) **50**, 394.
19. Blumberg, W. E., Levine, W. G., Margolis, S., Peisach, J., *Biochem. Biophys. Res. Commun.* (1964) **15**, 277.
20. Peisach, J., Levine, W. G., Blumberg, W. E., *J. Biol. Chem.* (1967) **242**, 2847.
21. Katoh, S., Shiratori, I., Takamiya, A., *J. Biochem.* (Tokyo) (1962) **51**, 32.
22. Blumberg, W. E., Peisach, J., *Biochim. Biophys. Acta* (1966) **126**, 269.
23. Van Holde, K. E., *Biochemistry* (1967) **6**, 93.
24. Van Holde, K. E., Cohen, L. B., *Biochemistry* (1964) **3**, 1803.
25. Cohen, L. B., Van Holde, K. E., *Biochemistry* (1964) **3**, 1809.
26. Van Holde, K. E., Van Bruggen, E. F. J., in "Biological Macromolecules Series," S. N. Timasheff, G. Fasman, Eds., p. 1-53, Marcel Dekker, New York, 1971.
27. Thomson, L. C. C., Hines, M., Mason, H. S., *Arch. Biochem. Biophys.* (1959) **83**, 88.
28. Lontie, R., Witters, R., in "The Biochemistry of Copper," J. Peisach, P. Aisen, W. E. Blumberg, Eds., p. 455-462, Academic, New York, 1966.
29. Lerner, A. B., Fitzpatrick, T. B., Calkins, E., Summerson, W. H., *J. Biol. Chem* (1949) **178**, 185.
30. *Ibid.* (1950) **187**, 793.
31. *Ibid.* (1951) **191**, 799.
32. Pomerantz, S. H., *J. Biol. Chem.* (1963) **238**, 2351.
33. Pomerantz, S. H., Warner, M. C., *J. Biol. Chem.* (1967) **242**, 5308.
34. Burnett, J. B., *J. Biol. Chem.* (1971) **246**, 3079.
35. Ashida, M., *Arch. Biochem. Biophys.* (1971) **144**, 749.
36. Beinert, H., in "The Biochemistry of Copper," J. Peisach, P. Aisen, W. E. Blumberg, Eds., p. 213-234, Academic, New York, 1966.
37. Wharton, D. C., Gibson, Q. H., in "The Biochemistry of Copper," J. Peisach, P. Aisen, W. E. Blumberg, Eds., p. 235-244, Academic, New York, 1966.
38. Van Gelder, B. F., Slater, E. C., in "The Biochemistry of Copper," J. Peisach, P. Aisen, W. E. Blumberg, Eds., p. 245-265, Academic, New York, 1966.
39. Ehrenberg, A., Malmstrom, B. G., Broman, L., Mosbach, R., *J. Mol. Biol.* (1962) **5**, 450.
40. Broman, L., Malmstrom, B. G., Aasa, R., Vänngard, T., *J. Mol. Biol.* (1962) **5**, 301.
41. Vänngard, T., in "Magnetic Resonance in Biological Systems," A. Ehrenberg, B. G. Malmstrom, T. Vänngard, Eds., p. 213, Pergamon, Oxford, 1967.
42. Andreasson, L. E., Vänngard, T., *Biochim. Biophys. Acta* (1970) **200**, 247.
43. Poillon, W. N., Bearn, A. G., *Biochim. Biophys. Acta* (1966) **127**, 407.
44. Magdoff-Fairchild, B., Lovell, F. M., Low, B. W., *J. Biol. Chem.* (1969) **244**, 3497.
45. Ryden, L., *Eurp. J. Biochem.* (1972) **26**, 380.
46. *Ibid.* **28**, 46.
47. Yamada, H., Yasunobu, K. T., *J. Biol. Chem.* (1962) **237**, 1511, 3077.
48. Nara, S., Yasunobu, K. T., in "The Biochemistry of Copper," J. Peisach, P. Aisen, W. E. Blumberg, Eds., p. 423, Academic, New York, 1966.
49. Mahler, H. R., in "Trace Elements," C. A. Lamb, O. G. Bentley, J. M. Beattie, Eds., p. 311, Academic, New York, 1958.
50. McCord, J. M., Fridovich, I., *J. Biol. Chem.* (1969) **244**, 6056.

51. Mann, T., Keilin, D., *Proc. Roy. Soc., London, Biol. Sci.* (1939) **126**, 303.
52. Carrico, R. J., Deutsch, H. F., *J. Biol. Chem.* (1970) **245**, 723.
53. Sawada, Y., Okyama, T., Yamazaki, I., *Biochim. Biophys. Acta* (1972) **268**, 305.
54. Schoot Uiterkamp, A. J. M., Mason, H. S., *Proc. Nat. Acad. Sci. U.S.* (1973) **70**, 993.
55. Fridovich, I., *Accts. Chem. Res.* (1972) **5**, 321.
56. Carrico, R. J., Deutsch, H. F., *J. Biol. Chem.* (1969) **244**, 6087.
57. Hartz, J. W., Deutsch, H. F., *J. Biol. Chem.* (1972) **247**, 7043.
58. Weisiger, R. A., Fridovich, I., *J. Biol. Chem.* (1973) **248**, 4793.
59. "The Biochemistry of Copper," J. Peisach, P. Aisen, W. E. Blumberg, Eds., Academic, New York, 1966.
60. Malkin, R., Malmstrom, B. G., *Adv. Enzymol. Molecular Biol.* (1970) **33**, 177.
61. Vänngard, T., in "Biological Applications of Electron Spin Resonance," H. M. Swartz, J. R. Bolton, D. C. Borg, Eds., p. 411, Wiley-Interscience, New York, 1972.
62. Mason, H. S., *Annu. Rev. Biochem.* (1965) **34**, 595.
63. Nakamura, T., Makino, N., Ogura, Y., *J. Biochem.* (Tokyo) (1968) **64**, 189.
64. Blumberg, W. E., Horecker, B. L., Kelly-Falcoz, F., Peisach, J., *Biochim. Biophys. Acta* (1965) **96**, 336.
65. Yamada, H., Yasunobu, K., Yamano, T., Mason, H. S., *Nature* (1963) **198**, 1092.
66. Mondovi, B., Rotilio, G., Costa, M. T., Finazzi-Agro, A., Chiancone, E., Hansen, R. E., Beinert, H., *J. Biol. Chem.* (1967) **242**, 1160.
67. Rotilio, G., Finazzi-Agro, A., Calabrese, L., Bossa, F., Guerrieri, P., Mondovi, B., *Biochem.* (1971) **10**, 616.
68. Malmstrom, B. G., Vänngard, T., *J. Mol. Biol.* (1960) **2**, 118.
69. Taylor, J. S., Coleman, J. E., *J. Biol. Chem.* (1973) **248**, 749.
70. Taylor, J. S., Coleman, J. E., *Proc. Nat. Acad. Sci. U.S.* (1972) **69**, 859.
71. Kertesz, D., in "The Biochemistry of Copper," J. Peisach, P. Aisen, W. E. Blumberg, Eds., p. 359, Academic, New York, 1966.
72. Jolley, R. L., Evans, L. H., Mason, H. S., *Biochem. Biophys. Res. Commun.* (1972) **46**, 878.
73. Schoot Uiterkamp, A. J. M., Thesis, Rijksuniversiteit Groningen, The Netherlands, 1973.
74. Schoot Uiterkamp, A. J. M., Mason, H. S., *Proc. Nat. Acad. Sci. U.S.* (1973) **70**, 993.
75. Schoot Uiterkamp, A. J. M., *F.E.B.S. Letters* (1972) **20**, 93.
76. Tang, S-P.W., Coleman, J. E., Myer, Y. P., *J. Biol. Chem.* (1968) **243**, 4286.
77. Blumberg, W. E., in "The Biochemistry of Copper," J. Peisach, P. Aisen, W. E. Blumberg, Eds., p. 49, Academic, New York, 1966.
78. Brill, A. S., Bryce, G. F., *J. Chem. Phys.* (1968) **48**, 4398.
79. Nakamura, T., Ogura, Y., in "Magnetic Resonance in Biological Systems," A. Ehrenberg, B. G. Malmstrom, T. Vänngard, Eds., p. 205, Pergamon, New York, 1967.
80. Broman, L., Malmstrom, B. G., Aasa, R., Vänngard, T., *Biochim. Biophys. Acta* (1963) **75**, 365.
81. Katoh, S., San Pietro, A., in "The Biochemistry of Copper," J. Peisach, P. Aisen, W. E. Blumberg, Eds., p. 407, Academic, New York, 1966.
82. Nieman, R. H., Nakamura, H., Vennesland, B., *Plant Physiol.* (1959) **34**, 262.
83. Bishop, N. I., *Nature* (1964) **204**, 401.
84. de Kouchkovsky, Y., Fork, D. C., *Proc. Nat. Acad. Sci. U.S.* (1964) **52**, 232.

85. Rotilio, G., Morpurgo, L., Giovagnolli, C., Calabrese, L., Mondovi, B., *Biochem.* (1972) **11**, 2187.
86. Blumberg, W. E., in "The Biochemistry of Copper," J. Peisach, P. Aisen, W. E. Blumberg, Eds., p. 399, Academic, New York, 1966.
87. Coleman, J. E., Coleman, R. V., *J. Biol. Chem.* (1972) **247**, 4718.
88. Denning, R. G., Spencer, J. A., *Faraday Soc. Sym.* (1969) **3**, 84.
89. Rotilio, G., Calabrese, L., Coleman, J. E., *J. Biol. Chem.* (1973) **248**, 3855.
90. Williams, R. J. P., in "The Biochemistry of Copper," J. Peisach, P. Aisen, W. E. Blumberg, Eds., p. 131, Academic, New York, 1966.
91. Coleman, J. E., *Progr. Biorg. Chem.* (1972) **1**, 159.
92. Haffner, P., Coleman, J. E., *J. Biol. Chem.* (1973) **248**, 6626.
93. Liljas, A., Kannan, K. K., Bergsten, P.-C., Waara, I., Fridborg, K., Strandberg, C., Carbom, U., Järup, L., Lövgren, S., Petef, M., *Nature New Biol.* (1972) **235**, 131.
94. Malmstrom, B. G., Reinhammar, B., Vänngard, T., *Biochim. Biophys. Acta* (1968) **156**, 67.
95. Malkin, R., Malmstrom, B. G., Vänngard, T., *Europ. J. Biochem.* (1969) **7**, 253.
96. Malkin, R., Malmstrom, B. G., Vänngard, T., *FEBS Letters* (1968) **1**, 50.
97. Ehrenberg, A., Malmstrom, B. G., Broman, L., Mosbach, R., *J. Mol. Biol.* (1962) **5**, 450.
98. Aisen, P., Koenig, S. H., Lilienthal, H. R., *J. Mol. Biol.* (1967) **28**, 225.
99. Beinert, H., Griffiths, D. E., Wharton, D. C., Sands, R. H., *J. Biol. Chem.* (1962) **237**, 2337.
100. Kertesz, D., Zito, R., in "Oxygenases," O. Hayaishi, Ed., p. 307, Academic, New York, 1962.
101. Dawson, C. R., Magee, R. J., *Methods Enzymol.* (1955) **2**, 831.
102. Duckworth, H. W., Coleman, J. E., *J. Biol. Chem.* (1970) **245**, 1613.
103. "Physiology and Biochemistry of Hemocyanins," F. Ghiretti, Ed., Academic, New York, 1968.
104. Vanneste, W., Mason, H. S., in "The Biochemistry of Copper," J. Peisach, P. Aisen, W. E. Blumberg, Eds., p. 465, Academic, New York, 1966.
105. Williams, R. J. P., in "The Biochemistry of Copper," J. Peisach, P. Aisen, W. E. Blumberg, Eds., p. 471, Academic, New York, 1966.
106. Mallette, M. F., in "Copper Metabolism," W. D. McElroy, B. Glass, Eds., p. 48, The Johns Hopkins Press, Baltimore, 1950.
107. Levine, W. G., in "The Biochemistry of Copper," J. Peisach, P. Aisen, W. E. Blumberg, Eds., p. 371, Academic, New York, 1966.
108. Yoshida, H., *J. Chem. Soc.* (1883) **43**, 472.
109. Mason, H. S., Fowlks, W. L., Peterson, E., *J. Amer. Chem. Soc.* (1955) **77**, 2914.
110. Broman, L., Malmstrom, B. G., Aasa, R., Vänngard, T., *Biochtm. Biophys. Acta* (1963) **75**, 365.
111. Mason, H. S., Spencer, E., Yamazaki, I., *Biochem. Biophys. Res. Commun.* (1961) **4**, 236.
112. Mason, H. S., in "The Biochemistry of Copper," J. Peisach, P. Aisen, W. E. Blumberg, Eds., p. 339, Academic, New York, 1966.
113. Brooks, D. W., Dawson, C. R., in "The Biochemistry of Copper," J. Peisach, P. Aisen, W. E. Blumberg, Eds., p. 343, Academic, New York, 1966.
114. Wood, B. J. B., Ingraham, L. L., *Nature* (1965) **205**, 291.
115. Ingraham, L. L., *J. Amer. Chem. Soc.* (1957) **79**, 666.
116. Kertesz, D., Rotilio, G., Brunori, M., Zito, R., Antonini, E., *Biochem. Biophys. Res. Commun.* (1972) **49**, 1208.
117. Pomerantz, S. H., Warner, M. C., *J. Biol. Chem.* (1967) **242**, 5308.
118. Wykes, J. R., Dunnill, P., Lilly, M. D., *Nature* (1971) **230**, 187.

RECEIVED September 17, 1973. Work supported by National Institutes of Health grant AM 09070-09 and National Science Foundation grant BO-13344.

Flavoprotein Oxidases

HAROLD J. BRIGHT

Department of Biochemistry, School of Medicine, University of Pennsylvania, Philadelphia, Penn. 19174

Current knowledge of three flavoprotein oxidases—glucose oxidase, D-amino acid oxidase, and L-amino acid oxidase—is reviewed with respect to purification, molecular properties, kinetic mechanism, chemical mechanism, and applications. Kinetically, the three enzymes represent special cases of a single general mechanism which has been pieced together from the results of stopped-flow spectrophotometry. The chemical mechanism of flavoprotein catalysis has only recently begun to unfold from the results of derivatized and model substrates such as β-halogenated α-carboxylic acids and nitroalkanes. α-Proton abstraction appears likely to precede the redox process, and direct evidence for covalent attachment of the resulting carbanion to N^5 of the flavin nucleus has been obtained with the nitroalkanes. Applications of the flavo-protein oxidases in analysis and industrial processing are discussed.

The three flavoprotein oxidases reviewed here—glucose oxidase, D-amino acid oxidase, and L-amino acid oxidase—are termed "simple" flavoprotein oxidases because the only known redox cofactor is the flavin nucleus, which is present as FAD in all cases. They are curiously anomolous because none of them performs a function which is obviously crucial to the biology of their respective organisms and yet, between them, they have probably been investigated more than all other flavoproteins combined. In part, this results simply from the challenge, in terms of structure and function, which the ubiquitous flavin poses to biochemists and chemists alike.

This chapter is not comprehensive since it reflects the interest of the author in the kinetic and chemical mechanism of flavoprotein oxidase reactions. It is meant to complement the special interests of others who have written excellent reviews from their respective view points. The reader is referred especially to Refs. *1, 2, 3, 4* for a broader perspective. The chemical point of view, particularly in terms of model reactions, is covered only briefly even though much has been accomplished recently in this area (*see*, for example, Refs. *5, 6, 7*).

The Flavin Coenzyme

The flavin coenzyme occurs in each of the oxidases discussed here as flavin adenine dinucleotide (FAD). The R group attached to N^{10} is adenosyldiphosphoribityl. The flavin nucleus can exist in three redox states, each of which can adopt three ionization states (*8, 9*). Only two redox states—fully oxidized and fully reduced (*see* Equation 1)—are kinetically important in the simple flavoprotein oxidases under discussion.

$$\text{FAD} \qquad\qquad\qquad \text{FADH}_2 \tag{1}$$

$$[\text{E-FAD} = \text{E}_0] \qquad\qquad [\text{E-FADH}_2 = \text{E}_r]$$

The ionizations of possible biological consequence concern N^3–H in FAD and N^1–H in FADH$_2$. The optical characteristics of a large number of flavoproteins have been tabulated recently (*10*). The species E_o (λ_{max} at 452–460 nm and at 380–383 nm, with $\epsilon = 11.3$–14.1 $\times$ $10^3 M^{-1}$ cm^{-1} and 9.7–13.3 $\times$ $10^3 M^{-1}$ cm^{-1}, respectively) is yellow, whereas E_r is almost colorless and absorbs much less than E_o at wavelengths greater than 330 nm. The 450–460 nm wavelength region is excellent for monitoring flavoprotein oxidase kinetics, *e.g.*, glucose oxidase, by rapid reaction techniques because $\Delta\epsilon \approx 10^4 M^{-1}$ cm^{-1}. The complex of E_r with P ($E_r \cdots P$) in the case of the amino acid oxidases also has a unique absorption spectrum with sufficiently strong absorbance in the 550 nm region (where E_o and E_r do not absorb) to allow uncomplicated monitoring of its transient kinetics.

Purification and Physicochemical Properties

The most common source for D-amino acid oxidase is hog kidney. The enzyme is often prepared from this by the method of Brumby and Massey (*11*) which features the inclusion of benzoate, a tightly binding substrate–competitive inhibitor. The binary holoenzyme–benzoate complex has enhanced thermal stability and can be dissociated by the addition of D-alanine. Recently, two groups (*12, 13*) concerned with structural studies of the enzyme have obtained an electrophoretically homogenous enzyme through slight modifications of the procedure of Brumby and Massey. The molecular weight values for the monomeric enzyme in these highly purified preparations varied from 37,600 to 39,600 as determined by several procedures. Both the monomer and the dimer are catalytically active (*14, 15*). Each monomer contains a single FAD and a single active site. D-Amino acid oxidase is unique among the simple flavoprotein oxidases in that the FAD is appreciably dissociable at holoenzyme concentrations commonly used in conventional kinetic experiments. The dissociation constant for E-FAD is the range $1–6 \times 10^{-7}M$ (*11*). It is therefore necessary to add at least $5 \times 10^{-6}M$ free FAD in kinetic experiments where total enzyme concentration is $10^{-6}M$ or less. Nondestructive resolution of the holoenzyme is accomplished by a procedure involving dialysis against $1M$ KBr (*11*).

L-Amino acid oxidase occurs widely, but is most frequently obtained from the lyophilized venom of rattlesnakes, particularly *Crotalus adamanteus,* by the method of Wellner and Meister (*16*). The enzyme has a molecular weight of 128,000–153,000 and can be dissociated into two different polypeptide chains (*17*). There are two tightly, but noncovalently, bound FAD molecules in the holoenzyme, and kinetic evidence indicates that these reside at two separate and non-interacting catalytic sites (*18*).

Glucose oxidase is obtained from molds such as *Aspergillus niger* and species of *Penicillium.* In its gross molecular properties at least, the enzyme from *A. niger* resembles L- rather than D-amino acid oxidase (*19*) since it is a dimer of 186,000 MW with two tightly, but noncovalently, bound FAD molecules per dimer. These FAD moieties are kinetically independent and probably, therefore, are present at an active site on each subunit. The enzyme is remarkably stable over a wide range of pH and other variables, which makes it ideal for analytical and bulk conversion applications.

Nature of the Product Released by the Flavoprotein Oxidases

Bentley and Neuberger (*20*) provided early definitive evidence that glucose oxidase released D-glucono-δ-lactone as the oxidation product of D-glucose. By analogy, therefore, it was considered likely that the first product (P) of the amino acid oxidases was the α-imino acid which, through fairly rapid nonenzymatic hydration and ammonia elimination, yielded the α-keto acid which is actually observed in experiments with a time resolution no greater than several minutes. Yagi *et al.* (*21*) interpreted transient pH changes during D-amino acid oxidase turnover as being caused by the accumulation of α-imino acid, and Hafner and Wellner (*22*) used borohydride reduction during L-amino acid oxidase turnover to obtain racemic α-amino acid as more direct evidence for the realease of α-imino acid from the enzyme. However, Hafner and Wellner argued that the major free product species accumulating transiently was the carbinolamine, rather than α-amino acid. This is unlikely on chemical grounds (*23*), and Porter and Bright (*24*) were able, in fact, to show quantitatively that the major product species accumulating transiently after release from the enzyme is the α-imino acid.

As would be expected, therefore, the oxidized substrate (P) is released in all cases as the species $>\!\!C\!=\!\!X$ (*see* Equation 1), and further reactions of this product (to give D-gluconic acid or α-keto acid) are entirely solvent-catalyzed.

Kinetics—General Considerations

The initial steady state velocity data from flavoprotein oxidase reactions, when plotted in double reciprocal form, almost invariably generate families of parallel and usually straight lines (*see* Figures 1a and 1b).

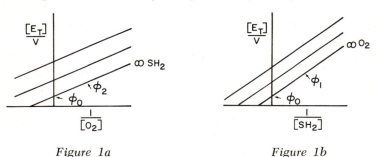

Figure 1a   Figure 1b

Such patterns are described by a three-term steady state rate equation (Equation 2)

$$\frac{E_T}{v} = \phi_0 + \frac{\phi_1}{[SH_2]} + \frac{\phi_2}{[O_2]} \tag{2}$$

A major objective of a kinetic study is to deduce the simplest kinetic mechanism which is capable of satisfying all the kinetic measurements. It then becomes possible to state the rate constant expression which represents each of the steady state coefficients (the ϕ's).

In general, a parallel line steady state pattern will be obtained when one or more of the steps interconnecting the two enzyme species which separately bind the two substrates during the catalytic cycle is an irreversible process under the conditions of the kinetic experiment. Product release is always an irreversible step in an initial velocity measurement. If the kinetic mechanism of a flavoprotein oxidase reaction is unbranched and the enzyme obligatorily oscillates between the free, fully oxidized and free, fully reduced, states, then the release of P and H_2O_2 must occur, respectively, in the reductive and oxidative half-reactions. This is the familiar ping pong mechanism, one version of which is the following (Equation 3):

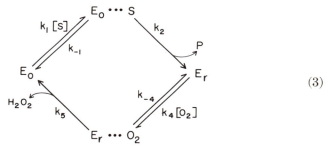

(3)

The steady state rate equation for this ping pong mechanism is Equation 4.

$$\frac{E_T}{v} = \frac{k_2+k_5}{k_2k_5} + \frac{k_{-1}+k_2}{k_1k_2[S]} + \frac{k_{-4}+k_5}{k_4k_5[O_2]}$$

(4)

Equation 4 clearly shows how, for this specific mechanism, the rate constants are related to the steady state coefficients (ϕ) in Equation 2. The glucose oxidase mechanism (*see* below) is a special case of equations 3 and 4.

In the case of the amino acid oxidase reactions, however, there is good evidence (to be discussed) that both SH_2 and O_2 combine with the enzyme in the major catalytic pathway before any product has been released. This represents a sequential mechanism from which, in general, the steady state patterns should consist of families of intersecting, rather than parallel, lines. The reason that intersecting patterns are not found in the flavoprotein oxidase reactions is that several highly irreversible steps, in addition to those involving product release, occur in the overall mechanism. These, of course, reflect the thermodynamics of the total reaction, which strongly favors product formation. In particular, both the binding of the first substrate, SH_2, by E_o and the subsequent monomolecular redox reaction, producing $E_r \cdots P$, are often highly irreversible under most experimental conditions. Since both of these processes precede the binding of O_2, the conditions for parallel line patterns are fulfilled regardless of whether the binding of O_2 precedes or follows the irreversible release of P. A major pathway for the amino acid oxidases, for example, is shown in Equation 5:

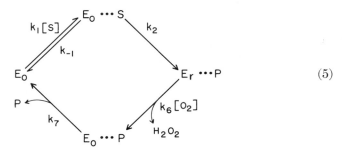

$$(5)$$

for which the steady state rate equation is Equation 6.

$$\frac{E_T}{v} = \frac{k_2 + k_7}{k_2 k_7} + \frac{k_{-1} + k_2}{k_1 k_2 [S]} + \frac{1}{k_6 [O_2]} \qquad (6)$$

This is a sequential mechanism which differs from the ping pong mechanism of Equation 3 in several important respects. Yet it is clear, by comparing Equations 4 and 6, which would be obtained from the rate data in coefficient form, that steady state kinetic analysis alone can not distinguish between the mechanisms of Equations 3 and 5. Because of the thermodynamics of the reaction, kinetic studies of the reverse reaction are not possible. Similarly, product inhibition measurements, which are often of great diagnostic value in steady state studies, see Ref. 25, are singularly unsuccessful, either because the true product is rapidly hydrolyzed nonenzymatically, e.g., the $\propto$-imino acid in the amino acid oxidase reactions, or because there is very little affinity of the product, e.g., H_2O_2 in all cases and the gluconolactone in the glucose oxidase reaction. Steady state methods, by themselves, are therefore of little value in deducing the kinetic mechanisms of flavoprotein oxidase reactions.

Fortunately, the characteristic absorbance of certain stable and transient enzyme species and, in some instances, of products, together with the fact that the two half-reactions can be studied separately, permits informative rapid kinetic measurements of the overall and partial reactions of flavoprotein oxidases. Stopped-flow spectrophotometric methods (26) have been particularly useful (the irreversibility of the partial and overall reactions rules out relaxation methods) because the measured rate constants often correspond in part or whole to the reciprocals of the steady state coefficients. This is the major reason for using the ϕ formulation

(27) rather than other conventions (25). Most importantly, the stopped-flow measurements are capable of distinguishing between most of the homeomorphic, *i.e.*, indistinguishable, mechanisms in terms of steady state measurements. Examples of such cases are given later.

Assay and Specificity

Early steady state kinetic studies established techniques for monitoring the overall reaction and for determining substrate specificity. The most generally applicable method for determining steady state rates of the oxidases is O_2 consumption. Oxygen electrode techniques (28) have now superseded earlier manometric methods. The enzyme preparations must either be completely free of catalase activity, as a result of high enzyme purity or addition of cyanide, or catalase must be added in amounts sufficient to prevent transient H_2O_2 accumulation.

$$2O_2 \xrightarrow{\quad 2S \quad 2P \quad} 2H_2O_2 \xrightarrow{\text{catalase}} 2H_2O + O_2 \qquad (7)$$

$$\text{Net:} \quad 2S + O_2 \longrightarrow 2P + 2H_2O \qquad (8)$$

When catalase activity is absent, H_2O_2 can be monitored directly spectrophotometrically (29) or by coupling to a peroxidase system (30). The weak $n \rightarrow \pi^*$ transition of the carbonyl group is present in the $\propto$-keto acid products of the amino acid oxidase reactions and can also be utilized under special conditions (31). However, all these methods present far more serious technical problems than the O_2 electrode technique and are used only in special circumstances.

The substrate specificity patterns of the oxidases are now well documented experimentally although their molecular interpretation is not entirely clear. Keilin and Hartree (32, 33) and others (34, 35) showed that the activity of glucose oxidase with β-D-glucose is at least 100 times greater than with any other sugar of biological origin under all conditions likely to be met in practice. This high specificity is the basis for the use of glucose oxidase for glucose analysis, which is discussed later. In contrast, the amino acid oxidases are not so specific and can not be used to estimate single amino acids in biological and other fluids. D-Amino acid oxidase oxidizes a variety of hydrophobic D-amino acids as well as D-arginine and D-lactate (3). Dicarboxylic amino acids are not substrates. L-Amino acid oxidase similarly prefers hydrophobic substrates, is less active with basic amino acids, and will not oxidize dicarboxylic amino acids (36).

As we have already emphasized, steady state kinetic methods by themselves are not capable of establishing the mechanism of flavoproptein oxidase catalysis. Therefore, the interpretation of steady state measurements together with the results of rapid kinetic studies will be discussed in the following section.

Kinetic Mechanism

Although there is not a total consensus regarding all details of the kinetic mechanism, the following scheme (Equation 9) appears to explain the kinetic behavior of glucose oxidase and both the amino acid oxidases.

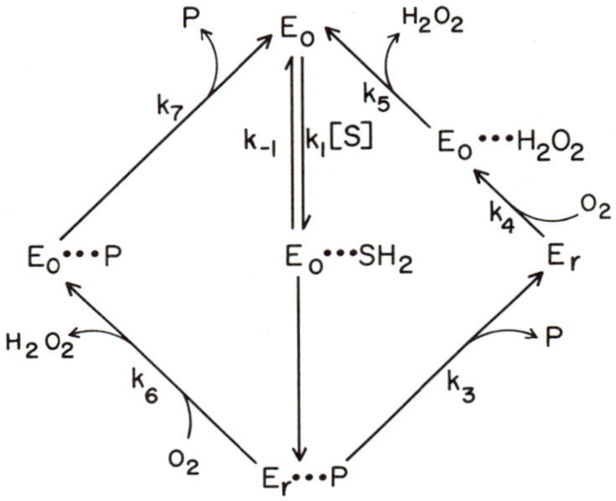

$$(9)$$

The most important evidence for this scheme is discussed for each enzyme in turn.

Glucose Oxidase. The first incisive kinetic evidence was presented by Gibson, Swoboda, and Massey (37) using the Aspergillus niger enzyme and by Nakamura and Ogura (38) using the *Penicillium amagasakiense* enzyme. Both groups agreed that, since ϕ_2^{-1} from steady state measurements (*see* Figure 1a and equation 2) was in excellent agreement with k_4 measured directly in stopped-flow oxidative half-reaction experiments $(E_r + O_2)$, the reaction of O_2 with free reduced enzyme (E_r) is an obligatory step in turnover. If E_rP reacts with O_2, then either $k_4 = k_6$, which is an unlikely coincidence in view of the amino acid oxidase results discussed later, or, more likely, k_3 is so much larger than $k_6[O_2]$ that flux through the left hand loop is negligible under attainable conditions. They also agreed that, by spectrophotometric criteria, the bound FAD oscillates between its fully oxidized and fully reduced states and that no free radical intermediates are kinetically competent. However, the groups disagreed as to the mechanism of the reductive half-reaction $(E_o + SH_2)$ with glucose. Gibson *et al.* (37) could not detect k_2, which could only be clearly resolved in the case of 2-deoxyglucose, and therefore concluded that $k_{-1} < k_2 > k_1[SH_2]$. If this is so, the reduction of E_o appears as a bimolecular reaction (Equation 10).

$$E_o + SH_2 \xrightarrow{k_1} E_r \cdots P \xrightarrow{k_3} E_r + P \qquad (10)$$

This behavior appears rather typical of flavoprotein oxidases, as discussed later. Nakamura and Ogura, on the other hand, with both the *Penicillium* (*38*) and *Aspergillus* (*39*) enzymes, concluded that k_2, which controls flavin reduction, was kinetically important and represented the maximum turnover number of the enzyme. These differences in interpretation may be more apparent than real, however, because $t_{1/2}$ for flavin reduction is clearly of the order of the dead time of the flow apparatus used. Consequently, limitations in time resolution will result in an underestimation of k_2. At 25°C, Gibson *et al.* showed (*37*), from stopped-flow measurements of the turnover number of the oxidized fraction of the enzyme, that k_3 becomes very large as the temperature is raised to 25°C and that the maximum turnover number at 25°C is entirely k_5. Subsequently, Bright and Gibson (*40*) showed that k_2 is clearly resolved in stopped-flow measurements when 1-^{2}H-glucose is used. The large kinetic isotope effect on flavin reduction, estimated to be 10–15-fold, dramatically shifts the rate-limiting first-order process in turnover from k_5 to k_2. Weibel and Bright (*41*) found that k_2, through a pH effect discussed later, could also be decreased sufficiently to be rate determining in turnover at pH values less than 5 through halide inhibition.

In summary, then, the kinetic mechanism of the glucose oxidase reaction is represented by the right hand loop of Equation 9. The rate-limiting processes under most conditions are $k_1[SH_2]$, $k_4[O_2]$, and k_5, although k_2 can be kinetically important under certain conditions.

L-Amino Acid Oxidase. In this reaction, there is good evidence that both the left- and right-hand loops of Equation 9 are kinetically significant. This conclusion can be reached much more easily here than in the case of glucose oxidase because $E_r \cdots P$, in both the L- and D-amino acid oxidase reactions, is purple and therefore has unique long-wavelength absorbance. Anaerobic reductive half-reactions ($E_o + SH_2$) measured in the stopped-flow apparatus at 550 nm, for example (*42*), show a rapid increase in absorbance, corresponding to the sequence $E_o \rightarrow E_o \cdots SH_2 \rightarrow E_r \cdots P$, followed by a first order decrease in absorbance, corresponding to $E_r \cdots P \xrightarrow{k_3} E_r + P$. With most substrates and at most pH values, k_3 is significantly smaller than the maximum turnover number. This clearly shows that the release of P from $E_r \cdots P$, and hence the right-hand loop of Equation 9, is not an obligatory turnover process and that O_2 must react with $E_r \cdots P$. This was first proposed by Massey and Curtis (42) and subsequently confirmed quantitatively by Porter (18, 43).

The most convincing demonstration (18) of the reactivity of $E_r \cdots P$ with O_2 comes from observation of biphasic plots of E_T/v versus $1/[O_2]$. At low O_2 concentrations, the steady state flux is through the right-hand loop of Equation 9 because $k_3 >> k_6[O_2]$. Under these conditions, the slope of the double reciprocal plot, with O_2 as variable substrate, is identical to the reciprocal of the value of k_4 obtained directly from stopped-flow oxidative half-reactions ($E_r + O_2$). At higher O_2 concentrations, $k_6[O_2] >> k_3$, and the slope of the double reciprocal plot increases about five-fold, corresponding to the value of k_6^{-1}. Product release, controlled by k_7, is rapid and is rate limiting only at very high O_2 pressures corresponding to 20mM dissolved O_2 or greater.

A well known property of L-amino acid oxidase is its susceptibility to substrate inhibition at high values of the ratio $[SH_2]/[O_2]$. Wellner and Meister (44) devised a reaction mechanism centered on this effect which required 2 FAD per active site. They assumed that $E_r \ldots P$ was $E \diagup \overset{(FADH \cdot)_2}{\underset{..P}{}}$ with coupled spins and that this species either reacts rapidly with O_2 or oxidizes a second SH_2 to form $E \diagup \overset{(FADH_2)_2}{\underset{..P}{}}$. If the latter species is less reactive with O_2 than the former, inhibition by SH_2 would result. Massey and Curti (42) pointed to several deficiencies in this mechanism, including the fact that the semiquinoid state of the enzyme is unreactive with SH_2. Additionally, a comparison of steady state and rapid kinetic data shows (18) that each active site requires one, not two, functional FAD. As DeSa and Gibson (45) first pointed out, inhibition by SH_2 could result if the species $E_r \cdots SH_2$ were less reactive with O_2 than is E_r. Curti et al. (46) obtained evidence for $E_r \cdots SH_2$ from freezing-inactivation data. Subsequently, Porter (18) showed quantitatively that the formation of $E_r \cdots SH_2$ from E_r in Equation 9 entirely accounts for substrate inhibition. In view of the preceding discussion of glucose oxidase, it is of interest that both $E_r \cdots P$ and $E_r \cdots SH_2$ are appreciably (about 10- and 4-fold, respectively) less reactive than E_r with O_2, which indicates that the presence of either SH_2 or P partially hinders the oxidation of E-FADH$_2$ by O_2.

D-Amino Acid Oxidase. The kinetic mechanism of this reaction is qualitatively, and quantitatively to some extent, very similar to that for the L-amino acid oxidase reaction. The long-wavelength intermediate ($E_r \cdots P$) is generally much more obvious in this reaction because of the remarkable slowness of P liberation. Yagi (3) has crystallized this intermediate under anaerobic conditions. Since k_3 is often, with different substrates, many orders of magnitude smaller than the maximum turnover

number, both Nakamura *et al.* (*47*) and Massey and Gibson (*48*) concluded that $E_r \cdots P \xrightarrow{k_3} E_r + P$ can not be an obligatory step in turnover
and that reoxidation must be initiated through the reaction $E_r \cdots P +$
$O_2 \xrightarrow{k_6} E_o \cdots P$. For this enzyme, k_6 tends to be slightly larger than k_4
(*49*). If the reactivity of $E_r \cdots S$ with O_2 were likewise greater than
that of E_r, this would explain why substrate inhibition of this enzyme
has never been observed. Yagi (*3*) has reviewed the enzymology of
D-amino acid oxidase recently in some detail.

Chemical Mechanism

As we have seen, the kinetic mechanisms of the oxidase reactions
differ only in detail, and the turnover cycle in each case is comprised of
a reductive half-reaction followed by an oxidative half-reaction. In this
section, current understanding of the chemical pathways of these half-
reactions is summarized, and the role of the apoenzyme in these processes
is assessed.

Reductive Half-Reaction. The process under discussion is the following (Equation 11):

$$E_o + HX{-}\overset{|}{\underset{|}{C}}{-}H \underset{k_{-1}}{\overset{k_1}{\rightleftharpoons}} E_o \cdots HX{-}\overset{|}{\underset{|}{C}}{-}H \xrightarrow{k_2} E_r \cdots X{=}C\diagdown \qquad (11)$$

There is now good reason to believe that homolytic (free radical) reactions are not involved. Therefore, the chemical mechanism underlying
the kinetic process controlled by k_2 must fall into one of the following
three categories. All the intermediates shown are enzyme bound.

$$HX{-}\overset{|}{\underset{|}{C}}{-}H \xrightarrow[H^+]{} HX{-}\overset{|}{\underset{|}{C^-}} \xrightarrow[\text{FAD,H}^+]{\text{FAD}} [HX{-}\overset{|}{\underset{|}{C}}{-}FADH] \longrightarrow \qquad (12)$$
$$\cdots\cdots\cdots\cdots\cdot{\blacktriangleright}FADH_2 + \diagup\!\!\!C{=}X$$

$$H{-}\overset{|}{\underset{|}{C}}{-}XH \xrightarrow[H^+]{\text{FAD,H}^+} [H{-}\overset{|}{\underset{|}{C}}{-}X{-}FADH] \longrightarrow$$
$$FADH_2 + \diagup\!\!\!C{=}X \qquad (13)$$

$$HX{-}\overset{|}{\underset{|}{C}}{-}H \xrightarrow{\text{FAD}} FADH_2 + \diagup\!\!\!C{=}X \qquad (14)$$

Both Equations 12 and 13 involve adduct formation between substrate and flavin, and both require removal of carbon-bound hydrogen as a proton. They differ substantially, however, in that adduct formation in Equation 12 occurs through the substrate carbon atom, whereas in equation 13 adduct formation occurs through nucleophilic attack of —XH on the flavin nucleus (50). Equation 14 involves no intermediate and can be visualized as a hydride transfer reaction.

Physiological substrates have been of little help in distinguishing between these mechanisms. There are two major reasons for this. First, no unique spectral species lying between $E_o \cdots HX-\overset{|}{\underset{|}{C}}-H$ and $E_r \cdots$

$X{=}C\overset{\diagup}{\diagdown}$ and having spectral properties corresponding to a flavin adduct has been detected by rapid reaction techniques. This means either that no such adduct exists or that it accumulates rapidly, and to only a very small extent, in a highly unfavorable equilibrium preceeding the rate-determining step, or that it is formed after the rate-limiting step. The fact that flavin reduction by $HX-\overset{|}{\underset{|}{C}}-{}^2H$ is often associated with a large kinetic isotope effect is certainly consistent with all three interpretations. The spectrum of the rate-limiting species in those cases (3) where the kinetic isotope effect is not observed (e.g., alanine with D-amino acid oxidase) has not been measured. It would obviously be of some interest to know whether this species has the oxidized benzoate type spectrum (3) or whether its spectrum resembles that expected for a flavin adduct. Since benzoate itself binds in a two-step mechanism (51), the former might well be the case. Second, even if an adduct occurred as an obligatory, non-rate-limiting intermediate, the technical difficulties in trapping it chemically in a process competitive with its rapid conversion to $E_r \cdots P$ would be formidable.

For these reasons, progress has been obtained with model, rather than physiological, substrates. In particular, recent studies of the reaction of the amino acid oxidases with β-halogenated-α-amino acids and of D-amino acid oxidase and glucose oxidase with nitroalkanes and their carbanions have begun to clarify the chemical mechanism of these reactions. The results and interpretations of these studies are discussed briefly below.

Walsh et al. (52), as well as ourselves (53), argued that if α-proton removal were an obligatory step in flavin reduction, then substitution of a good leaving group X, such as chloride, at the β-carbon of an amino acid oxidase substrate might reveal non-oxidative α-β elimination of HX

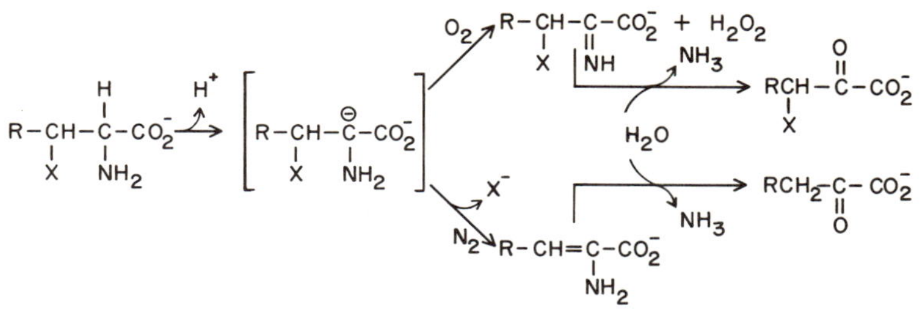

$$R-\underset{X}{\underset{|}{CH}}-\underset{NH_2}{\underset{|}{C}}-CO_2^- \xrightarrow{H^+} \left[R-\underset{X}{\underset{|}{CH}}-\overset{\ominus}{\underset{NH_2}{\underset{|}{C}}}-CO_2^- \right]$$

$$R-\underset{X}{\underset{|}{CH}}-\underset{NH_3}{\underset{|}{C}}=C-CO_2^- + H_2O_2$$

$$\xrightarrow{H_2O} RCH-\underset{X}{\underset{|}{C}}-CO_2^- \quad (15)$$

$$\xrightarrow{X^-} \quad \xrightarrow{NH_3} RCH_2-\underset{O}{\underset{||}{C}}-CO_2^-$$

$$\underset{N_2}{} R-CH=C-CO_2^- \atop NH_2$$

as a process competitive with the normal oxidative pathway (Equation 15).

The expected result was obtained since D-amino acid oxidase converted β-chloroalanine to pyruvate under anaerobic conditions, to chloropyruvate at high O_2 concentrations, and to mixtures of these at intermediate O_2 concentrations. Under steady state conditions, the reaction behaved as if cleavage of the α C-H bond were the rate-limiting process in turnover, although stopped-flow spectrophotometric measurements showed that this interpretation can not be entirely correct in this case (53) or in the case of β-choloro-α-aminobutyrate (54). α-β-Elimination has now been observed in three flavoprotein oxidase reactions (54) and can be considered strong circumstantial evidence for α-proton removal from compounds which closely resemble the physiological substrates.

Studies of nitroalkane oxidation by D-amino acid oxidase (55) and glucose oxidase (49, 56) have provided strong evidence both for intermediate substrate carbanions and for subsequent covalent adduct formation between these and the N^5 position of the flavin nucleus. The rationale for using nitroalkanes can be seen in the following reaction stoichiometries for D-amino acid oxidase (55):

$$H-\underset{|}{\overset{|}{C}}-NO_2 + O_2 + OH^- \longrightarrow \diagdown C=O + NO_2^- + H_2O_2 \quad (16)$$

$$H-\underset{|}{\overset{|}{C}}-NH_3^+ + O_2 + OH^- \longrightarrow \diagdown C=O + NH_3 + H_2O_2 \quad (17)$$

The unique advantage of the nitroalkanes is not only that they are carbon acids with pK_a values ranging from about 6 to 10 depending on the substituents, but that their conjugate acids and bases interconvert very slowly over the pH range and time scales of interest experimentally. Consequently, they provide the unusual opportunity of studying both the carbanion and its parent carbon acid as substrates. Thus, the carbanion of nitromethane is the most active reducing substrate known for glucose oxidase (56), being 10 times more reactive than D-glucose and 10^6 times more reactive than neutral nitromethane. Significantly, the latter substrate, in contrast to the carbanion, shows the same general-base dependence for flavin reduction as does D-glucose and other physiological substrates (56). These results suggest that the nitroalkanes are handled in much the same way as physiological substrates and that proton removal from substrate is a fundamental aspect of flavoprotein oxidase catalysis.

The case for obligatory substrate–flavin adduct formation during nitromethane–flavoprotein turnover by D-amino acid oxidase was established by stopped-flow analysis of the inhibition of this reaction by cyanide (55). Cyanide, as well as other nucleophiles, inhibits the reaction efficiently to yield an inactive enzyme-bound flavin–substrate adduct, EI, with spectral properties diagnostic of adduct formation at the N^5 position of the flavin nucleus. The evidence establishing that cyanide traps a kinetically competent intermediate on the catalytic pathway was obtained by comparing the flavin reduction rate at saturating substrate with the enzyme inhibition rate at saturating substrate and cyanide. Since these rates were identical, cyanide must interact with an intermediate EX which is formed rapidly from $E_o S$ at a rate controlled by flavin reduction. The exact locus of cyanide action was established by noting that O_2 was not required for inhibition and that O_2 did not compete with cyanide for $E_r P$. Thus, EX must reside between $E_o S$ and $E_r P$ (Equation 18).

$$E_o + S \underset{k_{-1}}{\overset{k_1}{\rightleftharpoons}} E_o \cdots S \overset{k_2}{\longrightarrow} EX \overset{fast}{\longrightarrow} E_r \cdots P \qquad (18)$$
$$CN^- \Big\downarrow fast$$
$$EI$$

The structures of EX and EI were deduced by resolution of EI into apoenzyme and free flavin–substrate adduct. The structure of this adduct was determined as 5-cyanoethyl-1,5-dihydro FAD and that of EX was deduced to be a cationic imine resulting from elimination of NO_2^- from the initial 5-nitroethyl-1,5-dihydro FAD adduct formed in the process controlled by k_2 by nucleophilic attack of nitroethane carbanion on the N^5 position of oxidized flavin. The chemistry of flavin reduction by nitroethane carbanion at the active site of D-amino acid oxidase is given by the following scheme (Equation 19) in which the kinetically important

$$(E_o \cdots S) \hspace{5cm} (EX)$$

$$E-FAD \cdots \overset{|}{\underset{|}{\ominus C}} - NO_2 \overset{k_2}{\longrightarrow} E-HFAD - \overset{|}{\underset{|}{C}} - NO_2 \overset{fast}{\underset{NO_2^-}{\searrow}} E-HFAD = \overset{\oplus}{C} \overset{/ fast}{\underset{\backslash H_2O}{\longrightarrow}} E-HFAD - \overset{|}{\underset{|}{C}} - OH$$
$$CN^- \Big\downarrow fast \hspace{3cm} \Big\downarrow fast \qquad (19)$$
$$E-HFAD - \overset{|}{\underset{|}{C}} - CN \hspace{2cm} E_r \cdots P$$
$$(EI)$$

intermediates are those given in Equation 18. It should be noted that the carbinolamine resulting from hydration of EX is formally analogous to the adduct (*see* Equation 12) which would result directly if physiological substrates were to react through carbon with N^5 of the flavin nucleus.

Our understanding of the chemical pathway of flavin reduction in flavoprotein oxidase catalysis may be summarized as follows: both the β-chloro-α-amino acid and nitroalkane results, taken separately and together, point strongly to the conclusion that the carbon-bound hydrogen is removed as a proton. Furthermore, the nitroalkane studies unequivocally implicate N^5 flavin–substrate covalent adducts as obligatory kinetic intermediates. Thus, although the gap dividing these substrate analogs with the behavior of physiological substrates has not been bridged, Equation 12 represents the only mechanism for which any substantial experimental evidence is available. It seems unlikely, in view of the very low acidity of substrates like glucose and the limited basicity of amino acid residues available to the apoprotein, that carbanions of demonstrable lifetime exist prior to flavin reduction by physiological substrates. Rather, as is indicated by the dashed arrow in Equation 12, proton removal and covalent adduct formation may be synchronous process, or, as indicated by the dotted arrow in Equation 12, the carbanion may be oxidized directly by the transfer of two electrons into N^5 of the flavin nucleus.

Although, as we have emphasized, none of the physiological substrates provides evidence which is both necessary and sufficient to support the mechanism of Equation 12 and to rule out Equations 13 and 14, two sets of evidence are clearly consistent with Equation 12, and there is no evidence contrary to the scheme of Equation 12. First, flavin reduction in glucose oxidase (40) and the amino acid oxidases (3, 57, 58) is usually associated with a large kinetic isotope effect when the deuterated substrate is used. Although some substrates of D-amino acid oxidase show no such isotope effect (3), the fact that substrate analogs such as benzoate bind by a two-step mechanism (51) and clearly do not form a covalent adduct with the oxidized flavin strongly suggests that an enzyme structural change must follow the initial binding event. Such a rate-limiting structural change, rather than covalent adduct formation without loss of substrate hydrogen (Equation 14) would readily explain, therefore, the behavior of a substrate such as alanine with D-amino acid oxidase (3). Second, flavin reduction in the glucose oxidase (41, 56) and amino acid oxidase reactions (3, 18) are clearly controlled by an ionizable residue in its basic state in the case of physiological substrates and neutral nitroalkanes, but not in the case of nitroalkane carbanions. Clearly, this behavior is consistent with general-base assisted proton removal from substrate preceding, or synchronous with, flavin reduction (Equation 12).

Oxidative Half-Reaction. The free, fully reduced flavoprotein oxidases react with O_2 at rates varying between about $2 \times 10^6 M^{-1}$ sec^{-1} (37) for glucose oxidase and about $2 \times 10^4 M^{-1}$ sec^{-1} (49) for D-amino acid oxidase. There is no kinetic evidence for the accumulation of intermediates corresponding to adducts between reduced flavin and O_2, although such presumably short-lived adducts are thought likely (59). Furthermore, the only reduction product of O_2 appears to be H_2O_2 in the case of simple flavoprotein oxidases. Thus $\overline{O}_2$ has not been detected in these reactions by superoxide dismutase, although metalloflavoproteins such as xanthine oxidase do produce this radical (60).

In sharp contrast to the reductive half-reaction, where the free oxidized flavin is totally inert in the presence of physiological substrates, reduced model flavins are appreciably reactive (nonenzymatically) with O_2 and other electron acceptors. However, the O_2 reactivity of reduced flavin is complicated for two perhaps related reasons (61). First, the reaction is autocatalytic owing to the formation of $2F\cdot$ (from F and FH_2) which in its anionic state is extremely reactive with O_2. Second, the superoxide radical is an important kinetic intermediate in O_2 reduction (59). Neither of these features is observed with the reduced flavoprotein oxidases.

Because of these clear differences of chemical pathway in the enzymatic and model reactions, it is difficult to assess the actual factor by which the apoenzyme enhances the reactivity of reduced flavin with O_2. It is fairly clear, however, that the environment of the apoenzyme suppresses homolytic pathways and, presumably by general-acid–base and steric mechanisms, allows only heterolytic bond cleavages.

The intermediate E_rP, which is the major species of reduced enzyme with which O_2 reacts in the amino acid oxidase reaction, is more reactive with O_2 than E_r in one case (49) (D-amino acid oxidase) but less reactive in the other (18) (L-amino acid oxidase). The reasons for such seemingly inconsistent behavior, as well as the virtual lack of reactivity of reduced flavins with O_2 in systems such as succinic dehydrogenase, will only become clear when the molecular details of the oxidation mechanism of reduced flavin are elucidated.

Flavoprotein Oxidase Applications

Glucose oxidase is widely used for quantitatively estimating glucose. It is ideally suited for this because of its stability, high specificity for β-D-glucose, high resistance to any kind of inhibition, virtual irreversibility of the reaction, large turnover number, and the ease with which pO_2 may be measured with the O_2 electrode. Semiquantitative analysis of glucose in diabetic and pregnancy urines, in blood, and in corn syrup can be achieved by the various dipsticks, such as Clinistix and Dextrostix from Miles-Ames and Tes-Tape from Eli Lilly, which are cellulose impregnated with glucose oxidase, peroxidase, and a reduced redox dye. The peroxidase-catalyzed reduction of H_2O_2 (equivalent to the amount of glucose oxidized) by the dye causes a color change in the latter which is related to the amount of glucose in the sample. This method can also be used in solution (*30*). Soluble glucose oxidase is used in the Beckman glucose analyzer, where the detection method is the O_2 depletion rate as measured by an O_2 electrode.

Recently, an apparatus was described which incorporates a small immobilized enzyme reactor followed by a vortex mixer and O_2 electrode and which is suitable for any oxidase-based analysis (*62*). For glucose analysis, glucose oxidase was covalently bound to porous glass particles by diazotization. The kinetic behavior of this immobilized enzyme had previously been determined by spectrophotometric and O_2-monitoring techniques (*63*). The accuracy, reproducibility, and longevity of the apparatus, in both kinetic and end point modes of analysis, were highly encouraging. Through the expedient of adding immobilized catalase to the enzyme reactor, it was later shown that this apparatus, used in end point mode of operation, was capable of reliable and rapid whole-blood glucose determinations (*64*). It is clear that the principles involved in such "reagentless" analysis will be extrapolated to more sophisticated applications in medicine and elsewhere. Similarly, we may expect the upgrading of the auto-analyzer technique through the adoption of immobilized flavoprotein oxidases and other enzyme systems.

While the use of glucose oxidase in analytical devices is likely to increase steadily, its importance in the food industry as a glucose and/or oxygen scavenger is less certain (*65*). It is currently used mainly to remove glucose from eggs prior to drying and, to a small extent, to remove oxygen from certain carbonated beverages, beer, prepared meats, salad dressings, and other foods with high fat contents. Its use as an antioxidant might increase, depending on the legislative fate of current chemical antioxidants and future trends in the cost of the enzyme. The continuous production of gluconic acid, possibly through the use of immobilized glucose oxidase, may also prove feasible.

The amino acid oxidases, mainly because of their lack of substrate specificity, are not likely to enjoy widespread analytical applications outside of the research laboratory. It remains to be established whether they will be used in industrial processing.

Literature Cited

1. Niems, A. H., Hellerman, L., *Ann. Rev. Biochem.* (1970) **39**, 740.
2. Palmer, G., Massey, V., "Biological Oxidations," T. P. Singer, Ed., p. 263, John Wiley and Sons, New York, 1968.

3. Yagi, K., *Advan. Enzymol.* (1971) **34**, 41.
4. Wellner, D., *Ann. Rev. Biochem.* (1967) **36**, 669.
5. Rynd, J. A., Gibian, M. J., *Biochem. Biophys. Res. Commun.* (1970) **41**, 1097.
6. Bruice, T. C., Hevesi, L., Shinkai, S., *Biochemistry* (1973) **12**, 2083.
7. Hemmerich, P., Jorns, M. S., *Proc. 8th F.E.B.S. Meet., Amsterdam* (1972) **29**, 95.
8. Beinert, H., *Enzymes* (2nd ed.) (1960) **2**, 339.
9. Hemmerich, P., Veeger, C., Wood, H. C. S., *Angew. Chem. Internat. Ed. Engl.* (1965) **4**, 671.
10. Ghisla, S., Massey, V., Lhoste, J.-M., Mayhew, S. G., *Biochemistry* (1974) **13**, 589.
11. Brumby, P. E., Massey, V., *Biochem. Prep.* (1968) **12**, 29.
12. Tu, S.-C., Edelstein, S. J., McCormick, D. B., *Arch. Biochem. Biophys.* (1973) **159**, 889.
13. Curti, B., Ronchi, S., Branzoli, U., Ferri, G., Williams, C. H., *Biochim. Biophys. Acta* (1973) **327**, 266.
14. Henn, S. W., Ackers, G. K., *Biochemistry* (1969) **8**, 3829.
15. Shiga, K., Shiga, T., *Biochim. Biophys. Acta* (1972) **263**, 294.
16. Wellner, D., Meister, A., *J. Biol. Chem.* (1960) **235**, 2013.
17. De Kok, A., Rawitch, A. B., *Biochemistry* (1969) **8**, 1405.
18. Porter, D. J. T., Ph.D. Dissertation, Univ. of Pennsylvania, 1972.
19. Swoboda, B. E. P., Massey, V., *J. Biol. Chem.* (1965) **240**, 2209.
20. Bentley, R., Neuberger, A., *Biochem. J.* (1949) **45**, 584.
21. Yagi, K., Nishikimi, M., Ohishi, N., Takai, A., *Biochim. Biophys. Acta* (1970) **212**, 243.
22. Hafner, E. W., Wellner, D., *Proc. Nat. Acad. Sci. USA* (1971) **68**, 687.
23. Jencks, W. P., "Catalysis in Chemistry and Enzymology," McGraw-Hill, p. 490, 1969.
24. Porter, D. J. T., Bright, H. J., *Biochem. Biophys. Res. Commun.* (1972) **46**, 571.
25. Cleland, W. W., *Biochim. Biophys. Acta* (1963) **67**, 104, 173, 188.
26. Gibson, Q. H., *Methods Enzymol.* (1969) **16**, 187.
27. Dalziel, K., *Acta Chem. Scand.* (1957) **11**, 1706.
28. Estabrook, R. W., *Methods Enzymol.* (1967) **10**, 41.
29. Bright, H. J., Appleby, M., *J. Biol. Chem.* (1969) **244**, 3625.
30. Bergmeyer, H.-U., Bernt, E., in "Methods of Enzymatic Analysis," H.-U. Bergmeyer, Ed., p. 123, Academic, 1965.
31. Pitts, J. N., Wan, J. K. S., *Chem. Carbonyl Group* (1966), 823.
32. Keilin, D., Hartree, E. F., *Biochem. J.* (1948) **42**, 221.
33. *Ibid.* (1952) **50**, 331.
34. Blakley, E. R., Boyer, P. D., *Biochim. Biophys. Acta* (1955) **16**, 576.
35. Sols, A., Fuente, G., *Biochim. Biophys. Acta* (1957) **24**, 206.
36. Meister, A., Wellner, D., *Enzymes* (2nd ed.) (1963) **7**, 609.
37. Gibson, Q. H., Swoboda, B. E. P., Massey, V., *J. Biol. Chem.* (1964) **239**, 3927.
38. Nakamura, T., Ogura, Y., *J. Biochem.* (1962) **52**, 214.
39. *Ibid.* (1968) **63**, 308.
40. Bright, H. J., Gibson, Q. H., *J. Biol. Chem.* (1967) **242**, 994.
41. Weibel, M. K., Bright, H. J., *J. Biol. Chem.* (1971) **246**, 2734.
42. Massey, V., Curti, B., *J. Biol. Chem.* (1967) **242**, 1259.
43. Porter, D. J. T., Bright, H. J., *Biochem. Biophys. Res. Commun.* (1972) **46**, 564.
44. Wellner, D., Meister, A., *J. Biol. Chem.* (1961) **236**, 2357.
45. DeSa, R. J., Gibson, Q. H., *Fed. Proc. Fed. Amer. Soc. Exp. Biol.* (1966) **25**, 649.
46. Curti, B., Massey, V., Zmudka, M., *J. Biol. Chem.* (1968) **243**, 2306.

47. Nakamura, T., Nakamura, S., Ogura, Y., *J. Biochem.* (1963) **54**, 512.
48. Massey, V., Gibson, Q. H., *Fed. Proc. Fed. Amer. Soc. Exp. Biol.* (1964) **23**, 18.
49. Porter, D. J. T., Bright, H. J., unpublished data.
50. Brown, L. E., Hamilton, G. A., *J. Amer. Chem. Soc.* (1970) **92**, 7225.
51. Nishikimi, M., Osamura, M., Yagi, K., *J. Biochem.* (1971) **70**, 457.
52. Walsh, C. T., Schonbrunn, A., Abeles, R. H., *J. Biol. Chem.* (1971) **246**, 6855.
53. Voet, J. G., Porter, D. J. T., Bright, H. J., Z. *Naturforsch.* (1972) **27b**, 1054.
54. Walsh, C. T., Krodel, E., Massey, V., Abeles, R. H., *J. Biol. Chem.* (1973) **248**, 1946.
55. Porter, D. J. T., Voet, J. G., Bright, H. J., *J. Biol. Chem.* (1973) **248**, 4400.
56. Porter, D. J. T., Voet, J. G., Bright, H. J., Z. *Naturforsch.* (1972) **27b**, 1052.
57. Porter, D. J. T., Bright, H. J., *Biochem. Biophys. Res. Commun.* (1969) **36**, 209.
58. Page, D. S., Van Etten, R. L., *Biochim. Biophys. Acta* (1971) **227**, 16.
59. Massey, V., Palmer, G., Ballou, D., *Oxidases Related Redox Syst.* (1973) **1**, 25.
60. Massey, V., Strickland, S., Mayhew, S. G., Howell, L. G., Engel, P. C., Matthews, R. G., Schuman, M., Sullivan, P. A., *Biochem. Biophys. Res. Commun.* (1969) **36**, 891.
61. Gibson, Q. H., Hastings, J. W., *Biochem. J.* (1962) **68**, 368.
62. Weibel, M. K., Dritschilo, W., Bright, H. J., Humphrey, A. E., *Anal. Biochem.* (1973) **52**, 402.
63. Weibel, M. K., Bright, H. J., *Biochem. J.* (1971) **124**, 801.
64. Dritschilo, W., Weibel, M. K., *Biochem. Med.* (1974) **9**, 32.
65. "The Present and Future Technological Status of Enzymes," B. Wolnak and Associates, Report **NSF-C752**, 1972.

RECEIVED March 11, 1974.

13

Lipoxygenases

BERNARD AXELROD

Department of Biochemistry, Purdue University, West Lafayette, Ind. 47907

Lipoxygenase, a dioxygenase that is distributed widely among plants, catalyzes the hydroperoxidation of cis-cis *pentadienes. The position of hydroperoxidation may vary with different lipoxygenases. Some lipoxygenases catalyze a secondary reaction involving substrate and hydroperoxide which results in chain scission, carbonyl formation, dimerization, and destruction of certain dyes. Various lipoxygenases are classified with respect to the secondary reaction. The reaction can: proceed aerobically or anaerobically, proceed anaerobically only, or not occur. Despite a long-held belief to the contrary, lipoxygenase contains a prosthetic group. Highly purified isoenzymes contain one atom of Fe per molecule. The metal can be removed directly by a Fe^{3+}-specific chelator or by a Fe^{2+}-specific chelator after reduction. The occurrence, role in food, and mechanism of action of lipoxygenase are discussed.*

Discovery of the unsaturated fatty acid oxidase, now called lipoxygenase, followed an indirect pathway. (Although a number of excellent reviews on lipoxygenase have appeared, Refs. *1, 2,* and *3* are recent ones.) Haas and Bohn (*4*) discovered in 1928 that the inclusion of small quantities of soybean flour in wheat flour dough decreased the normal yellow color of the wheat flour. Although they attributed the color loss to carotene destruction, it has since become obvious that the pigments are mainly xanthophylls. Their discovery resulted in a number of patents on bread dough bleaching (*5*). From the patent descriptions it is clear that they recognized the need for oxygen in the destruction of the pigments.

In the same period Andre and Hou (*6*) in France noted that fat in soybean curd made from unboiled soybean extract differed from that obtained from heated extract. Although they did not measure peroxide values, they concluded on the basis of other chemical changes that the fat in the unheated extract was oxidized by a heat-labile enzyme. They coined the term "lipoxydase" ("lipoxidase" in English), which was the accepted name for this enzyme until displaced by the more descriptive "lipoxygenase." These investigators also found "lipoxydase" activity in the common bean, *Phaseolus vulgaris,* and in the seed of another food legume, *Dolichos lablab* (7).

In 1929 Navez (8), working in the laboratory of Craig, found that an aqueous suspension of pulverized seeds of *Pisum, Phaseolus,* and *Lupinus* consumed oxygen, as measured in the Warburg respirometer. Craig (9) pursued this observation and found that suspensions of *Lupinus albus* consumed large quantities of oxygen compared with the carbon dioxide evolved. He further demonstrated that oxygen consumption could occur when a lipid and a crude protein fraction of the seed obtained by ammonium sulfate precipitation were mixed. He called this activity "unsaturated fat oxidase."

In 1940 J. B. Sumner and R. J. Sumner (10) and Tauber (11) recognized that the "fat oxidase" or "lipoxidase" and "carotene oxidase" were identical. From their work it became clear that the enzyme catalyzed the peroxidation of unsaturated fats as the primary reaction and that carotene destruction occurred only during the active enzymatic attack on the fat in a coupled reaction.

During this period much attention was directed toward carotene loss in curing alfalfa, especially by Hauge and his co-workers. As early as 1931 Hauge and Aitkenhead (12) postulated that alfalfa contained a carotene-destroying enzyme. Hauge (13) provided confirmation of this view in 1935, but no connection between the pigment-bleaching activity of soybean observed by Haas and Bohn and this enzyme was made. It remained for R. J. Sumner (14) to demonstrate lipoxygenase in alfalfa in 1943.

Considerable purification of lipoxygenase from soybeans was achieved by Balls *et al.* (15). They used conventional protein purification methods and the carotene oxidase reaction to follow the enzyme activity. Theorell *et al.* (16), in 1947, crystallized the enzyme, or more correctly, one of the several isoenzymes in soybean. We believe this was lipoxygenase-1 as described later.

Sources of Lipoxygenase

A nonending stream of papers report or rediscover new sources of lipoxygenase. In the early days of this field it was felt that the enzyme was largely restricted to legumes and had only a limited significance to general plant physiology. Now it has been established that lipoxygenase is widely distributed. When the enzyme is not found in a new species, one must ask whether it was truly absent or whether the detection method was sufficiently sensitive. Lipoxygenase can no longer be treated as a limited curiosity, and one cannot consider its physiological role trivial. In animal, particularly vertebrate, sources of the enzyme, Tappel (3) established that the lipid peroxidation reactions and associated rancidity ordinarily observed in meats and fish as well as the pathological effects arising from lipid peroxidation in living tissue cannot be attributed to lipoxygenase. On the other hand, prostaglandin synthetases are present in most mammalian tissues and attack, presumably by initial peroxidation, certain unsaturated fatty acids in a manner similar to lipoxygenase (17).

Lipoxygenase has been reported in many plants (Table I). Because the information is scattered this list is not comprehensive, and priorities of recognition are not necessarily indicated. While legumes are frequently reported to be lipoxygenase plants, such plants may be found throughout many other phyla.

Table I.   Some Plants Containing Lipoxygenase

Apple (fruit) (18)
Alfalfa (leaf) (12, 13, 14, 19)
Articum lappa (leaf) (20)
Barley (seed) (21, 22)
Beans (Phaseolus, various; seed) (7, 23, 24, 25, 26)
Cauliflower (florets) (27)
Centaurea spp. (leaf) (20)
Chlorella (whole cell) (28)
Cirsium arvense (leaf) (20)
Digitalis purpurea (leaf) (20)
Dimorphotheca sinuata (seed) (29)
Dornicium plantagineum (leaf) (20)
Egg plant (fruit) (30, 31)
Fava (seed) (32, 33)
Flax (seed) (34)
Garbanzo (seed) (33)
Locust, honey (leaf, pod, seed) (35)
Lupine (seed) (9)
Maize (germ) (36, 37)

Marijuana (seed) (38)
Olive (seed) (39)
Peas (seed) (19, 23, 26, 40, 41)
Peanut (seed) (26, 42)
Potatoes (tuber) (43)
Pumpkin (seed) (38)
Rape (seed) (38)
Robinia pseudoacacia (seed) (33)
Sinapis alba (seedlings) (44)
Solanum dulcamara (leaf) (20)
Squash (seedling) (45)
Soybean (seed) (4)
Sunflower (seed) (38)
Tomato (fruit) (46)
Trefoil (seed) (33)
Urd (seed) (26)
Urtica dioica (leaf) (20)
Vetch, purple (19)
Wheat (seed, leaf) (21, 23, 25, 26, 47)

Application to Food and Agricultural Science

The presence of lipoxygenase in plants may affect their storage and processing since it promotes the peroxidation of the polyunsaturated fatty acids (which are nutritionally essential) and can affect taste, odor, and color. Moreover lipoxygenase may influence ripening and abscission. In fact, it has been used to modify fatty acids, to bleach wheat flour, and to improve the rheological properties of wheat dough. An excellent discussion of lipoxygenase in the food applications of soybeans appears in the monograph by Smith and Circle (48). A fine review, in German, concerning lipoxygenase and its action in foods has been published recently by Grosch (1).

Flavor Alteration by Lipoxygenase. In normal preparation soybean meal is heated after oil extraction so that residual lipoxygenase activity is of no concern (48). Meal must be heated before it can be eaten to inactivate the trypsin inhibitor and hemogglutinins. The intact seed is stable, although cracking or bruising promotes lipoxygenase action. When dealing with soybean meal, especially full-fat meal intended for human consumption or meal to be processed in contact with aqueous medium, failure to inactivate lipoxygenase leads to instability and flavor deterioration (49). A correlation between intensity of bitterness and lipoxygenase activity was found as the soybeans developed and matured (50).

The connection between off-flavors and lipoxygenase in green peas has been studied by many investigators. Siddiqi and Tappel (41) implicated lipoxygenase in the off-odor of green peas. Wagenknecht and Lee (51) showed that endogenous lipase acting on unblanched frozen peas released fatty acids which served as substrate for the lipoxygenase. Whitfield and Shipton (52) found that raw green peas produced a variety of aromatic aldehydes. Murray *et al.* (53) identified alcohols as secondary products of lipoxygenase action on polyunsaturated fatty acid in unblanched, stored peas. Eriksson (54) substantiated the role of lipoxygenase in the development of off-odor compounds and showed that the enzyme was in the inner part of the cotyledons. He noted the presence of the substrates, linoleic and linolenic acids. He also demonstrated that an alcohol dehydrogenase in peas could convert the hexanal, derived from linoleic acid by the lipoxygenase-initiated reaction, to hexanol (55). Grosch (56) proposed that the action of lipoxygenase on linoleic acid followed by autoxidations produced decen-2-al, octanal, undecen-2-al. and nonenal.

Wagenknecht (57) attributed off-flavors in unbalanced sweet corn to lipoxygenase. Graveland *et al.* (22) suggested that 2-*trans*-nonenal, which contributes to the cardboard flavor associated with barley preparations, arises from polyunsaturated fatty acids *via* lipoxygenase action. They also found that lipoxygenase was associated with the glutenin. According to Koch *et al.* (58) in cottonseed oil the off-odor that sometimes develops is caused by the action of lipoxygenase on the glycerides of the polyunsaturated fatty acids.

Since gas chromatography is widely used to study the volatile products produced by plant materials, many aldehydes are being noticed, and there is much evidence that lipoxygenase is responsible for the aldehyde formation. Thus St. Angelo *et al.* (*59*) demonstrated hexanal in peanut volatiles. Saijya and Takeo (*60*) added ^{14}C-labeled linoleic and linolenic acid to fermenting tea and identified radioactive hexanal and *trans*-2-hexenal among the products. Tomatoes, too, were found by Jadhav *et al.* (*46*) to generate hexanal from unsaturated fatty acids. The possibility that lipoxygenase and linolenic acid can serve as the source of ethylene emanation in the apple is discussed below.

Pigment Destruction by Lipoxygenase. The first reaction attributed to lipoxygenase was the bleaching of yellow pigments in flour. The destruction of many pigments and other substances in the "flame" of the lipoxygenase fatty acid reaction has been observed—*e.g.*, xanthophylls (*4*), carotene (*10, 11, 12*), bixin (*61*), chlorophyll (*19*), cholesterol (*62*), thyroxine (*63*), dyes (*19, 64*), vitamin A (*65*), and phytofluoene (*66*). Although pigment destruction in wheat flour is desirable, the loss of carotene and its vitamin A potentiality is not wanted. The biological value of both α- and β-carotene and of vitamin A is lost in the lipoxygenase reaction, as shown by Frey *et al.* (*66*). In processing spaghetti, where it is desirable to retain the pigment, the lipoxygenase and polyunsaturated fat in the wheat flour are problems (*67*). For instance, the loss of chlorophyll in stored unblanched peas results in an unattractive product. Buckle and Edwards (*68*) believe that chlorophyll loss in peas requires a protein factor in addition to lipoxygenase. The loss of carotene from harvested alfalfa by endogenous lipoxygenase is a potentially serious problem but is readily controlled by heat inactivation of the enzyme (*13*).

Lipoxygenase in Baking. Lipoxygenase and pigment removal or retention in flour have been discussed. The action of lipoxygenase in wheat flour on the endogenous lipids is reported by Irvine (*69*) to influence baking quality. Presumably the beneficial action of lipoxygenase occurs through indirect oxidation of sulfhydryl groups in the protein to produce interpeptide disulfide bridges and thus to improve the rheological and baking properties of the doughs. Many workers are investigating this matter—*e.g.*, Auerman and co-workers (*70, 71*), Drapron and Beaux (*72*), and Frazier *et al.* (*73*). Mecham (*74*) has reviewed lipoxygenase in the baking industry.

Graveland (*75*) has investigated extensively the products from the enzymatic oxidation of the unsaturated lipids. Some of considerable interest to the mechanism of lipoxygenase activity (*see* below) include:

> 13-hydroperoxy-9-*cis*,11-*trans*-octadecadienoic acid
> 9-hydroperoxy-10-*trans*,12 *cis* octadecadienoic acid
> 9-hydroxy-10-*trans*,12,13 *cis*-epoxyoctadecadienoic acid
> 9,10-*cis*-epoxy-11-*trans*-13 hydroxyoctadecadienoic acid
> 9,12,13-trihydroxy-10-*trans*-octadecadienoic acid
> 9,10,13-trihydroxy-11-*trans*-octadecadienoic acid

Graveland also (76) observed that the lipoxygenase is reversibly absorbed on glutenin to form a complex which produces hydroxy-epoxy acids from linoleic acid in the presence of oxygen.

The importance of lipoxygenase increases when soybean flour is added to wheat flour to improve baking quality or when soybean flour is admixed in larger quantities to improve the protein nutritional quality.

Physiological Aspects of Lipoxygenase. Although no convincing general role for lipoxygenase in the physiology of the plant has been proposed, several significant observations have been made. Apples, as they approach the climacteric stage during ripening, show a marked increase in respiration accompanied by ethylene production. Lieberman and Mapson (77) suggested that ethylene was produced from linolenic acid. Wooltorton *et al.* (18) found that an increase in lipoxygenase activity paralleled the rise in both respiration and ethylene in the climacteric apple. Later Meigh and Hulme (78) noted that an increase in lipids, including linolenic and linoleic acids, occurred in the climacteric period. Galliard *et al.* (79) investigated the lipoxygenase reaction which initiated the conversion of fatty acid to ethylene and found an absolute requirement for linolenic acid and ascorbic acid. Lonna and Raffi (80) observed that abscisin-like substances are formed by degradation of carotenoids in coupled reactions with lipoxygenase and polyunsaturated fatty acids. They reported success with α- and β-retinene, xanthophyll, and to some degree with α- and β-cartotene. Firn and Friend (81) found that violaxanthin, in a coupled reaction with lipoxygenase and linoleic acid, formed xanthoxin, which is a plant growth inhibitor.

Oelze-Karow *et al.* (44) assert that lipoxygenase production in the seedlings of mustard, *Sinapis alba,* is under phytochrome control. When the far-red form of the pigment reaches a threshold level, lipoxygenase synthesis is rapidly repressed. When the concentration of the far-red form drops below this value, synthesis begins immediately.

Holman (82) found that soybean seedlings undergo a sharp loss in lipoxygenase content beginning on the second day. The loss parallels the decline in substrates. The change in lipoxygenase does not follow the rise and fall in catalase in this period, and it is therefore improbable that lipoxygenase is associated with the glyoxysomes.

With the prospect of increased use of soybean in human foods an effort is being made to breed varieties having a lower lipoxygenase level. Hammond and his co-workers have had encouraging results in their first efforts (83) and have noted that the lipoxygenase content is influenced by the genotype of the maternal parent.

Fritz *et al.* (84) established that lipoxygenase is a true dioxygenase. Using $^{18}O_2$ they proved that all of the oxygen acquired by the substrate in undergoing oxidation originated from molecular O_2. The possibility that lipoxygenase has a significant role as a terminal oxidase was proposed by Craig (9) when he discovered unsaturated fat oxidase. However, Fritz et al. (84) showed that plant tissue in an atmosphere labeled with $^{18}O_2$ incorporated amounts too small to sustain Craig's suggestion.

Properties and Enzymology of Lipoxygenase

Chemical and Physical Properties. Too little is known about the physical properties of pure lipoxygenases to permit their comparison. The separation of isoenzymes during their isolation and their differential behavior in electrophoresis indicate the existence of different enzymes. However characterization of the enzymes through their enzymatic properties shows that the lipoxygenases comprise a large group of diverse enzymes.

The original soybean lipoxygenase, crystallized by Theorell et al. (16), is a globulin (MW = ca. 100,000) and was reported to possess no components other than amino acids. No evidence for a prosthetic group or metal was noted. The amino acid composition was given (85) on the basis of the then imprecise methods. Stevens et al. (86) worked with apparently the same isoenzyme and reported that it contains eight half cystines—four present as cystine and four as cysteine. The free sulfhydryl groups are not accessible to ordinary sulfhydryl reagents when the protein is in the native state. Pistorius and Axelrod (35) have confirmed the difficult accessibility of the cysteines in the crystalline enzyme, using several denaturants, but have found about 12 half cysteines.

According to Stevens et al. (86) soybean lipoxygenase can be dissociated into two subunits of equal size. They assign a molecular weight of 108,000 to the intact enzyme. They observed dissociation of the enzyme with sodium dodecyl sulfate and with guanidinium chloride after reduction. The former appears to cause only partial dissociation while the latter seems to be substantially more effective although dissociation still appears less than complete. Christopher et al. (87) have failed to achieve any dissociation with sodium dodecyl sulfate, either with or without mercaptoethanol, using the procedure of Osborne and Weber (88) under conditions in which proteins having multiple subunits dissociate. On the other hand Grosch et al. (89) have reported that lipoxygenase, when first reduced and treated with iodoacetate, in the presence of sodium dodecyl sulfate yields a family of smaller species ranging down to an estimated molecular weight of 15,000, as judged on polyacrylamide gel electrophoresis in sodium dodecyl sulfate.

Isoenzymes. Kies (90), who questioned the identity of lipoxygenase and carotene oxidase, thought that the lipoxygenase activity of soybeans was caused by more than one enzyme. She discovered that a partially purified preparation of soybean lipoxygenase lost its carotene bleaching activity [assayed by the method of Balls *et al.* (*15*)] on two minutes exposure to 70°C without a concomitant loss of the peroxide-forming activity [assayed by the method of Theorell *et al.* (*91*)]. The differential heat stabilities of the two activities were verified by Haining (*92*). With crystalline Theorell-type lipoxygenase prepared by Schroeder (*93*), it was found that heating the enzyme for six minutes at 68°C produced little change in the amounts or ratios of carotene oxidase units to peroxide-forming units. The original ratio was 3.4×10^{-5}; the final 4.3×10^{-5}. On the other hand crude soybean extract which had an initial ratio of 1.5×10^{-2}, which was approximately 300 times greater in favor of carotene oxidase when compared with the crystalline enzyme, lost all of its carotene bleaching activity and none of its peroxidizing activity on heating (*94*). Thus the Theorell enzyme could not be the lipoxygenase species in soybeans which is mainly responsible for the carotene oxidase activity.

Koch *et al.* (*95*) had claimed earlier, on the basis of measurements of peroxide-forming ability, that soybeans contained two lipoxygenases— one with a preference for triglyceride and the other with a preference for free acids. Christopher *et al.* (*96*) purified to homogeneity a second isoenzyme from soybeans which differed from the first in pH optimum and in ester-acid preference. The original Theorell enzyme was designated lipoxygenase-1; the new isoenzyme, lipoxygenase-2. Yamamato *et al.* (*97*) also isolated an isoenzyme which they designated "b" and which may be identical with lipoxygenase-2. The guaiacol-linoleic hydroperoxide peroxidase of Grosch *et al.* (*89*) may also be identical with this isoenzyme.

Christopher *et al.* (*98*) isolated a third isoenzyme, lipoxygenase-3, from soybeans. It was distinct from the other two isoenzymes in: elution profile from DEAE-Sephadex, isoelectric point, pH-activity profile, and effect of Ca^{2+} on its activity. Although this preparation gave a single sharp peak on isoelectric focusing, it could be resolved into two fractions on elution from columns of hydroxyapatite or CM-Sephadex (*99*). These enzymes are so similar in their enzymatic properties that they are indistinguishable except for their behavior in the elutions mentioned. The more rapidly eluting component is lipoxygenase-3, the slower component, lipoxygenase-4.

Guss *et al.* (*25*), performing gel electrophoresis on acrylamide gels containing starch, were able to locate lipoxygenase by immersing the gels in linoleic acid and then in acidic KI to produce a blue band at the enzyme sites. They found multiple lipoxygenases in soybeans and wheat. The same group (*23*) showed that peas and beans also possessed multiple forms of the enzyme. Grosch *et al.* (*100*) also report the purification of an enzyme from soybean meal in addition to lipoxygenase-1.

Hurt (*101*) has partially separated two isoenzymes from dry bean seeds (*Phaseolus vulgaris,* variety Golden Wax)—one with a pH optimum near neutrality, the second at pH 9.0. Corn germ apparently contains two isoenzymes, one active near neutrality, described by Gardner and Weisleder (*36*), the other active at pH 9, demonstrated by Veldink *et al.* (*102*). As more sensitive methods are applied, isoenzymes of lipoxygenase will become more commonly recognized.

Assay. One of the older procedures for lipoxygenase assay is the coupled reaction using unsaturated fat or esters plus carotene (*15, 103*) and observing the bleaching rate. The carotenoid, bixin, was prepared by Sumner and Smith (*104*) to replace carotene because of its better compatibility with aqueous solutions and its greater molar absorption. This method was advantageous when ultraviolet spectrophotometers were not generally available. The carotene procedure is not recommended for routine assays because of the limited range over which activity is proportional to enzyme concentration, the instability of the substrate mixture, and the great differences between the carotene oxidase activities of different lipoxygenases compared with their peroxidizing activities. Despite the vagaries and difficulties of the coupled assay, phytoene has the advantage of high fluorescence and detection at low concentrations.

An assay which is based on the increase in absorption at 234 nm depends on the fact that the linoleic acid hydroperoxide formed by the action of the enzyme possesses two double bonds in a conjugated system. When arachidonic acid is used as substrate, the maximum is at 238 nm (*105*). The diene conjugation method, introduced by Theorell *et al.* (*91*), is sensitive and convenient for kinetic studies, especially when used with a recording spectrophotometer. When preparations of low specific activity are used, interference from protein absorption can be a problem. In this case the lipid products can be extracted with hexane. For routine work it is convenient to use Surrey's substrate mixture (*106*).

Activity can also be followed conveniently by measuring O_2 utilization with a recording O_2 electrode. This procedure is as sensitive as the spectrophotometric method, as simple to perform, and can be used with solutions that would be optically unsatisfactory in the spectrophotometer. Measurement of O_2 consumption by Warburg manometry should be avoided. Not only is it cumbersome and less precise, but lipoxygenase, especially when somewhat purified, is highly sensitive to surface denaturation on shaking.

The enzyme can also be assayed by measuring the peroxide formed. The commonly used, highly sensitive procedure depends on the conversion of $Fe(CNS)_2$ to the colored $Fe(CNS)_3$. Unfortunately the assay does not permit direct continuous measurement, and the color is unstable. Moreover there is no simple stoichiometric relation between the amount of color and the peroxide—not even a linear one. Nonetheless reasonable approximations may be obtained. The diene conjugation method and the O_2 electrode method are the best choices, although when secondary reactions occur, the two methods may not agree. Holman noted, in 1947 (*107*), that the diene yield falls short of theoretical yield, based on oxygen uptake.

Linoleic acid is probably the best substrate for assay. It is readily dispersed at alkaline pH, and at acid or neutral pH it is no more difficult to suspend than the methyl ester or the glyceride. Although some species of lipoxygenase show more favorable action with esters than others, we know of no instance where such enzymes do not also act well on linoleic acid. The poor water solubility of linoleic acid and its changing dissociation in the pH range of interest are a disadvantage and usually require emulsifiers. The sulfate ester of linoleyl alcohol, which does not possess this disadvantage, has been proposed by Allen (*108*) and appears to be a welcome improvement.

Substrate Specificity and Specificity of Attack. In characterizing the specificity of lipoxygenase, both its selectivity for certain substrates and for certain positions in the substrate must be considered. Substrates for lipoxygenase may be straight chain fatty acids, esters, alcohols, hydroxamates (*92*), or even halides (*109*) which contain two cis double bonds separated by a methylene group. This 1,4-pentadiene structure is found in the naturally occurring, nutritionally essential polyunsaturated fatty acids such as linoleic, linolenic, and arachidonic acid as well as in the families derived by further desaturation and chain elongation. Holman *et al.* (*110*) have thoroughly studied C_{18} dienoic acids in which the *cis-cis*-1,4-pentadiene structure appears in all possible positions. Enzymatic activities measured by absorption at 234 nm were compared, using a highly purified preparation of lipoxygenase-1. Based on rate, the 9,12-isomer (natural linoleic acid) was the best substrate while the 13,16-isomer was half as effective. However activity was found with all isomers from 5,8 to 13,16. Ca^{2+} had little effect on the rate with natural linoleic acid, but it caused some improvement with the other isomers. Many polyunsaturated C_{18} to C_{20} acids containing one or more *cis-cis*-1,4-pentadiene groups were also tested. Good rates were seen with compounds possessing ω-6 and ω-9 double bonds.

Hamberg and Samuelsson (*111*) had tried a number of polyunsaturated fatty acids earlier and concluded that the structural requirement for lipoxygenase attack was a *cis,cis*-1,4-pentadiene group whose methylene carbon was at position ω-8. In other words the requirement was a distal double bond in position ω-6. These authors also examined the modified products of the reaction by mass spectroscopy and found that in all cases O_2 was inserted at position ω-6 except for linoleic acid. In this case some O_2 was found at position ω-10. Although a commercial soybean preparation was used in these experiments, the reactions were carried out at pH 9, and one may assume that lipoxygenase-1 was the enzyme involved.

The generalities concerning the structural requirements for lipoxygenase noted here are developed from studies with soybean lipoxygenase and are not necessarily applicable to enzymes from other sources. Thus Hamberg (112) found that while linoleic, α-linolenic, γ-linolenic, and arachidonic acids are suitable for soybean lipoxygenase, the corn germ isoenzyme of Gardner and Weisleder (36) uses only the first two substrates.

Specificity for the site of O_2 insertion into the substrate is also a property of the enzyme although it can be influenced by experimental conditions. In autoxidation of linoleic acid it was known that a mixture of 9- and 13-hydroperoxides was formed (113). Values for the positional specificity for soybean lipoxygenase indicate a wide variation. Dolev et al. (114) found exclusive formation of 13-hydroperoxide. Other values for the ratio of 13- to 9-hydroperoxide are 95:5 by Eriksson and Leu (115), 92:8 by Hamberg and Samuelsson (111), 90:10 by Veldink et al. (116), 80:20 by Zimmerman and Vick (34), and 70:30 by Chang et al. (117), Veldink et al. (118), and Hamberg and Samuelsson (119). In many instances it is likely that no recognition was given to the possibility of isoenzymes of differing specificities being present. However, Galliard and Phillips (120) found that soybeans at pH 5.5 gave a ratio of 46:54 while at pH 9.2 the value was 74:26. They suggested that these results reflected the presence of two types of isoenzymes in soybeans. Christopher and Axelrod (121), using pure lipoxygenase-1 and lipoxygenase-2, obtained ratios of 13- to 9-isomer of 95:5 and 50:50, respectively. In a more extensive investigation, the effect of O_2 tension, pH, temperature, substrate, and the presence or absence of Ca^{2+} were studied. The results are shown in Tables II, III, and IV. Christopher (122) has also found that lipoxygenase-3 and -4 resemble lipoxygenase-2 in position specificity for O_2 insertion. Gardner and Weisleder (36) discovered that the corn germ isoenzyme, which is active at neutrality but inactive at pH 9.0, produced the 9-isomer almost exclusively. In these respects it resembles the highly purified potato enzyme of Galliard and Phillips (43).

Gardner et al. (29) have shown that Dimorphotheca sinuata (cape marigold) seeds contain a lipoxygenase that resembles soybean lipoxygenase-1 in its positional specificity but has a pH optimum at 6.0 and exhibits no activity at pH 9.0. Alfalfa lipoxygenase has an equal preference for 9 and 13 (117).

Privett et al. (123) made an important discovery in elucidating the structure of the products of the lipoxygenase reaction. They established the existence of trans-cis conjugated bonds in linoleic hydroperoxide and showed that the hydroxy compound from the hydroperoxide reduction was optically active. This latter finding dispelled the contention that lipoxygenase promoted the autoxidation of the lipid.

Table II.   Content of Hydroperoxide Isomers[a]

		Conditions			
		Solutions Gassed with Oxygen at 0°C		*Solutions Open to Atmosphere at 25°C*	
		L−1[b]	*L−2*[b]	*L−1*[b]	*L−2*[b]
pH 7	%9-isomer	24	60	52	70
	%13-isomer	76	40	48	30
pH 8	%9-isomer	15	N.D.[c]	46	67
	%13-isomer	85	N.D.[c]	54	33
pH 9	%9-isomer	10	N.D.[c]	52	55
	%13-isomer	90	N.D.[c]	48	45

[a] In incubations of each isoenzyme with linoleic acid dispersed in Tween 20 as substrate.
[b] L-1: lipoxygenase-1; L-2: lipoxygenase-2.
[c] Isomers were not detected.

An exploratory survey of crude extracts from a number of species for positional specificity of O_2 insertion has been reported by Vioque and Maza (*33*).

Hamberg and Samuelsson (*111*) showed that the hydroxy group was of the L configuration not only for the product of linoleic acid but also for that of 8,11,14-eicosatrienoic acid. Gardner and Weisleder (*36*) showed that corn germ lipoxygenase formed 9-D-hydroperoxy-10,12-octadecadienoic acid. Galliard and Phillips (*43*) identified the product obtained with the potato enzyme as the same compound. Veldink *et al.* (*116*) obtained a ratio of 13- to 9-isomers of 70:30 with soybean lipoxygenase at pH 9.0, as cited above. They also established that the 9-isomer was optically active (of the D-configuration), which proved that the product was of enzymatic origin.

Linoleic acid is readily autoxidized, and there is always the danger that the ratios of 13- to 9-isomers may not result solely from enzyme action. Hamberg (*112*) developed an ingenious technique for steric analysis of the reaction products. It is based on obtaining diasteroisomeric derivatives of the products which are then readily separated and identified by gas chromatography.

Hamberg found that the corn enzyme generated mainly 9-D-hydroperoxy-10,12-octadecadienoic acid from linoleic acid. The small amount of 13-hydroperoxy-10,12-octadecadienoic acid which was also formed

Table III.   Content of

<table>
<tr><td></td><td colspan="4">pH 7</td></tr>
<tr><td></td><td colspan="2">% 9-isomer</td><td colspan="2">% 13-isomer</td></tr>
<tr><td></td><td>+Ca</td><td>−Ca</td><td>+Ca</td><td>−Ca</td></tr>
<tr><td>Lipoxygenase-1</td><td>32</td><td>28</td><td>68</td><td>72</td></tr>
<tr><td>Lipoxygenase-2</td><td>48</td><td>38</td><td>52</td><td>62</td></tr>
</table>

[a] In incubations of each isoenzyme with linoleic acid dispersed in ethanol as substrate. Solutions were gassed with oxygen at 0°C. Calcium ion was either present (0.55mM) or absent.

consisted of equal amounts of 13-D and 13-L isomer. A comparable amount of 9-L isomer was also found. These compounds resulted from autoxidation. In a separate experiment, autoxidation of linoleic acid was shown to generate equal amounts of all four isomers.

In the formation of the conjugated diene from the cis-cis-1,4-penta-

diene the methylene carbon loses a H. Here too the question of positional specificity has been considered. Hamberg and Samuelsson (111), using stereospecifically ^{3}H-labeled cis-8,11,14-eicosatrienoic acid (labeled at position 13), found that with soybean lipoxygenase at pH 9, the L H (S in the Cahn-Ingold-Prelog designation) was removed. Egmond et al. (124) confirmed this for lipoxygenase acting on linoleic acid at pH 9.0. Corn germ lipoxygenase has a positional specificity for O$_2$ insertion at C-9 and removed the D H from the methylene group at pH 6.6. These authors conclude that if the pentadiene portion of the substrate is oriented on the enzyme in a planar form, O$_2$ insertion and H extraction must occur

Table IV.   Content of Hydroperoxide Isomers[a]

<table>
<tr><td></td><td colspan="2">pH 7</td><td colspan="2">pH 9</td></tr>
<tr><td></td><td>% 9-isomer</td><td>% 13-isomer</td><td>% 9-isomer</td><td>% 13-isomer</td></tr>
<tr><td>Lipoxygenase-1</td><td>24</td><td>76</td><td>19</td><td>81</td></tr>
<tr><td>Lipoxygenase-2</td><td>42</td><td>58</td><td>57</td><td>43</td></tr>
</table>

[a] In incubations of each isoenzyme with methyl linoleate dispersed in ethanol as substrate. Solutions were gassed with oxygen at 0°C.

Hydroperoxide Isomers[a]

pH 8				pH 9			
% 9-isomer		% 13-isomer		% 9-isomer		% 13-isomer	
+Ca	−Ca	+Ca	−Ca	+Ca	−Ca	+Ca	−Ca
18	10	82	90	11	8	89	92
48	59	52	41	53	60	47	40

on opposite sides of the plane. Veldink *et al.* (*125*) had previously proposed such an orientation of substrate on the molecule. An interesting consequence of this proposal is that enzymes with apparently opposite behavior with respect to 9 and 13 insertion and D or L loss of the methylene hydrogen can have stereochemically identical active sites. The differences can arise from the orientation of the substrate to the enzyme.

A strong isotope effect was observed when the methylene H was removed, indicating that this may be the rate-controlling step in the reaction sequence. These results confirm the earlier findings of Hamberg and Samuelsson (*111*) with stereospecifically-labeled H at C-15 of 8,11,14-eicosatrienoic acid.

Induction. The phenomena of induction must have some bearing on the reaction mechanism. Haining and Axelrod (*126*) found that when highly pure linoleic acid was used with soybean lipoxygenase, a lag occurred before the maximum rate was achieved. This lag could be abolished by the inclusion of a small amount of enzymatically oxidized or autoxidized linoleate. Linolenate or methyl linoleate that had been enzymatically oxidized also abolished the lag while autoxidation products of oleate or hydroperoxides of cumene, *tert*-butyl alcohol, and undecylenic acid could not replace the hydroperoxides of the lipoxygenase substrates. This specificity for product hydroperoxide in abolishing lag periods has been confirmed by Smith and Lands (*127*). These results imply that product hydroperoxide is essential for the reaction. This is supported by the observation of these workers that glutathione peroxidase plus reduced glutathione—a combination known to destroy lipid hydroperoxide (*128*)—inhibits lipoxygenase when included in the reaction mixture.

Smith and Lands have also shown that the lipoxygenase reaction is accompanied by reaction inactivation of the enzyme. The reaction rate is characteristic for each substrate and increases with the extent of unsaturation. They postulate that the enzyme has two sites—one for the substrate and one for the product hydroperoxide. They proposed a highly speculative mechanism in which the hydroperoxide combines with O_2 to form a planar tetroxide where the O_2 reaches the equivalent of a singlet state oxygen. The tetroxide molecule then attacks the substrate molecule to generate a perepoxide intermediate which rearranges to form a new molecule of product hydroperoxide. The tetroxide meanwhile reverts to the hydroperoxide *via* a hydroxy radical. These authors suggest that this radical, the tetroxide, the perepoxide, or possibly alkoxy radicals could account for the lipoxygenase self-destruction. It is customary to postulate radical intermediates in the lipoxygenase reaction. However without further experimental work, it is not possible to adduce which, if any, of these hypothetical intermediates is responsible for the destruction.

Mechanism of Action. It has been fairly well established that the mechanism of enzymatic action does not involve a free radical chain process as in autoxidation. The chain process is ruled out on kinetic grounds (*129*), and moreover this process also seems unlikely in view of the sharp stereospecificity of the reaction. A common representation of the reaction as described by Tappel (*2*) is:

$$
\begin{array}{cccc}
 & \text{c} & \text{H} & \text{c} \\
\text{R—CH}=\text{CH—C—CH}=\text{CH—R}_1 + O_2 \\
 & & \text{H} &
\end{array}
$$

$$
\begin{array}{cccc}
 & \text{c} & \text{H} & \text{c} \\
\text{R—CH}=\text{CH—C—CH}=\text{CH—R}_1 + \cdot\text{OOH}
\end{array}
$$

$$
\begin{array}{cc}
\text{c} & \text{t} \\
\text{R—CH}=\text{CH—CH}=\text{CH—CH—R}_1 + \cdot\text{OOH}
\end{array}
$$

$$
\begin{array}{cc}
 & \text{OOH} \\
\text{c} & \text{t} \\
\text{R—CH}=\text{CH—CH}=\text{CH—CH—R}_1
\end{array}
$$

Although the abstraction of the H as a free radical is shown to occur *via* O_2, it is possible that it is removed by the enzyme and held in a transitory binding, and the H radical or its equivalent is then placed on the O_2 after the O_2 is inserted. A concerted reaction could also occur in which the H radical is removed and the O_2 inserted.

Chan (*130*) has suggested that lipoxygenase behaves in some respects as if it generated singlet O_2 when it attacks linoleic acid. [His claim that photochemical oxidations of certain organic compounds, which are well known to occur *via* singlet O_2, can be mimicked by the co-oxidation of these compounds in the lipoxygenase-linoleic reaction has been withdrawn.] Finazzi and his co-workers (*131*) provided additional indirect support for this concept by observing that soybean lipoxygenase, while acting on linoleic acid, generates a chemiluminescent substance. Since singlet O_2 fluoresces, and since they found that the addition of superoxide dismutase abolishes the chemiluminescence, they concluded that singlet O_2 is produced. It is significant that superoxide dismutase does not influence the normal course of the lipoxygenase reaction. Thus, singlet O_2 does not seem to be on the main path of the normal hydroperoxidation. It is known that the co-oxidation of secondary substrates has a slight effect on the extent of the primary reaction.

Formation of Secondary Products and Lipohydroperoxide Destruction. As early as 1945 Holman and Burr (*132*) found that crude soybean lipoxygenase acting on a number of substrates produced carbonyl-containing material in addition to diene. Holman, as noted above (*107*), used his crystalline enzyme and found that it was difficult to establish a correspondence between O_2 consumption and diene conjugation. The diene concentration always tended to be too low. Privett *et al.* (*123*) found that the reaction products varied with enzyme concentration and method of addition. Vioque and Holman (*133*) identified 9-keto-11,13- and 13-keto-9,11-octadecadienoate with the usual hydroperoxides in a reaction carried out with linoleic acid and a relatively large amount of crude soybean lipoxygenase at pH 9.

Garssen *et al.* (*134*) observed that soybean lipoxygenase, presumably lipoxygenase-1, acting at pH 9.0 under anaerobic conditions and in the presence of linoleic acid and 13-hydroperoxy octadeca-*cis*-9-*trans*-11-dienoic acid (*i.e.*, the major product formed in the hydroperoxidation of linoleic acid by this enzyme) gives 13-keto-octadeca-9,11-dienoic acid and the split products, pentane and 13-keto-trideca-*cis*(*trans*)-9-*trans*-11-dienoic acid. The D-9-hydroperoxy compound cannot substitute for the L-13-hydroperoxide (*135*). Dimers are also formed under these conditions. These reactions do not occur under aerobic conditions. Two possible pathways for the anaerobic reaction suggested by these workers are shown in Figures 1 and 2.

The formation of 280 nm-absorbing material has been independently noted by my co-workers and me (*136*) while following the action of lipoxygenase-1 on linoleic acid in a recording spectrophotometer. When the absorption at 234 nm ceased to increase, the calculated amount of conjugated hydroperoxide was equivalent to the O_2 originally present in the reaction mixture. At the moment the absorption at 234 nm ceased, the absorption at 280 nm commenced. Introduction of oxygen caused the increase at 280 nm to stop while the absorption at 234 nm resumed its climb. We also noted that product peroxide alone was not affected by the presence of the enzyme and that the enzyme had no effect on linoleic acid alone under anaerobic conditions. However both product hydroperoxide and substrate were required together. We did not recognize that dimers were being formed or that cleavage of the chain was occurring. We did note a new parameter in lipoxygenase specificity. Lipoxy-

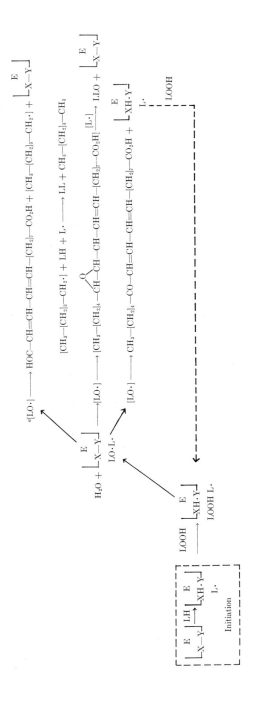

Biochemical Journal

Figure 1. Proposed pathway for the anaerobic reaction between soybean lipoxygenase, linoleic acid, and 13-hydro-peroxylinoleic acid (135)

^{a}[LO·] is $CH_3-[CH_2]_4-CH(O)-CH=CH-CH=CH-[CH_2]_7-CO_2H.$

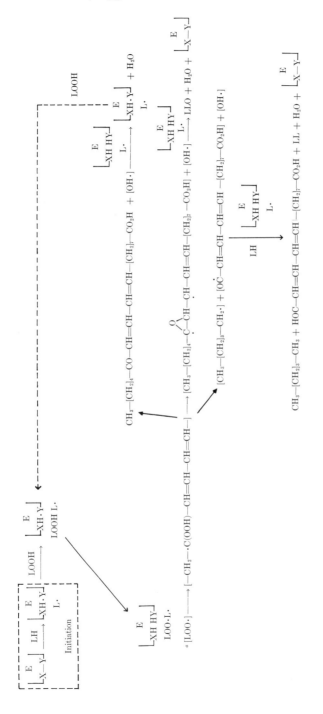

Figure 2.   Alternate proposed pathway for anaerobic reaction between soybean lipoxygenase, linoleic acid, and 13-hydroperoxy-linoleic acid (135)

genase-3 generated both the normal hydroperoxydienoic acid and the 280 nm-absorbing compounds in air. Lipoxygenase-2 resembled lipoxygenase-3 although it was less effective in generating the 280 nm-absorbing material. The coupled destruction of 2,6-dichlorophenolindophenol in the lipoxygenase reaction has been described by Fritz and Beevers (*64*). With lipoxygenase-1 we found that dye destruction was coupled to the reaction, but only in the anaerobic phase. On the other hand with lipoxygenase-3, dye destruction began immediately in the aerobic phase of the reaction.

Lipohydroperoxide Destruction and Secondary Products. Investigators have noted that during a lipoxygenase reaction the hydroperoxide concentration, as measured with $Fe(CNS)_2$, rises and then declines [*e.g.*, Balls *et al.* (*15*), Blain and Barr (*137*), Gini and Koch (*138*)]. This phenomenon has been attributed to a lipohydroperoxide-destroying enzyme, or lipohydroperoxidase, as well as to hematin compounds (*3, 143*). However Pistorius and Axelrod reproduced this phenomenon with crystalline lipoxygenase-1 (*139*). There is little doubt that the anaerobic destruction of lipohydroperoxide, as detailed by Garssen *et al.* (*134*), is responsible for this.

On the other hand an enzyme, discovered by Zimmerman and Vick (*140, 141*) in flax seed, converts 13-L-hydroperoxide of linoleic acid to 12-keto-13-hydroxyoctadeca-*cis*-enoic acid. This isomerization abolishes the conjugated diene and can be assayed by measuring the absorption loss at 234 nm. Veldink *et al.* (*116*) showed that the isomerase converts only the 13-hydroperoxide and not the 9-isomer. Gardner (*142*) found a similar enzyme in corn germ, which appears to attack both the 9- and 13-isomers.

Galliard and Phillips (*144*) have found that homogenates of potato tubers contain an enzyme which converts the primary lipoxygenase product of linoleic acid, the 9-D-hydroperoxyoctadecadienoic acid, into an 18-carbon ether (Figure 3). Here too, cleavage of the C_{18} chain must occur. The formation and role of possible cleavage products from linoleic and linolenic acids in aroma and physiology of plant tissues have already been mentioned.

Johns *et al.* (*145*) verified the results of Garssen *et al.* (*134*)—that soybean lipoxygenase-1 generates pentane from linoleic acid under anaerobic conditions. Lipoxygenase-2 did not produce pentane under either aerobic or anaerobic conditions.

Role of Iron in Lipoxygenase. Apart from its practical significance and the curiosity concerning its function, lipoxygenase has been intriguing. Since its purification and compositional analysis (*85*), it has had no place in the logical scheme of oxygenases. It appeared to be the exception that proved the rule that all oxygenases must possess a prosthetic group, cofactor, or metal. Its untypical behavior grew more serious as the list of well-defined oxidases, and particularly dioxygenases, grew. Numerous attempts were made to see if redox metal was present, principally through the use of chelators in the reaction mixture. All results appeared to support the conclusion that such metals were absent. The possibility of disulfide bonds which might provide transient configurations to handle an extra unpaired electron was considered. Walker (*146*) reported a signal in ESR studies that could be ascribed to ·S. Another solution to the dilemma of the absent prosthetic group was that the enzyme activated the substrate so that it was directly attacked by O_2, and that unlike the case in other dioxygenase reactions, the O_2 did not have to be activated.

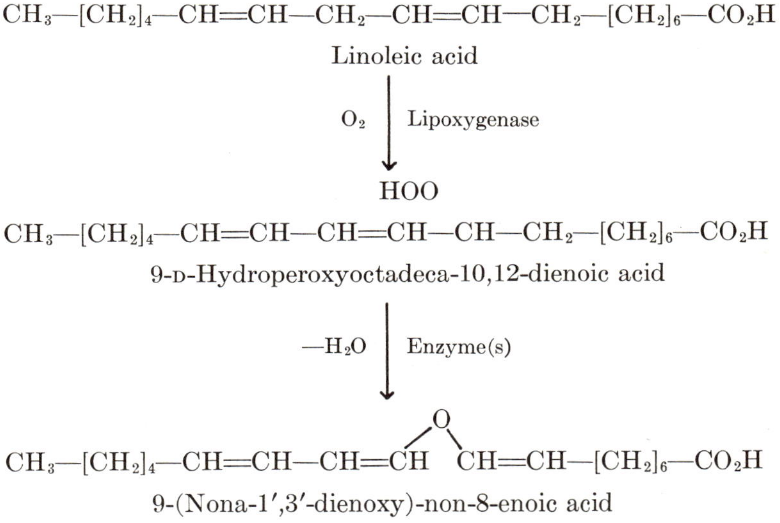

$$CH_3—[CH_2]_4—CH=CH—CH_2—CH=CH—CH_2—[CH_2]_6—CO_2H$$

Linoleic acid

$$O_2 \quad \Big| \quad \text{Lipoxygenase}$$

HOO

$$CH_3—[CH_2]_4—CH=CH—CH=CH—CH—CH_2—[CH_2]_6—CO_2H$$

9-D-Hydroperoxyoctadeca-10,12-dienoic acid

$$—H_2O \quad \Big| \quad \text{Enzyme(s)}$$

$$CH_3—[CH_2]_4—CH=CH—CH=CH \quad CH=CH—[CH_2]_6—CO_2H$$

9-(Nona-1',3'-dienoxy)-non-8-enoic acid

Biochemical Journal

Figure 3.   Proposed enzymic pathway for formation of nonadienoxy-nonenoic acid from linoleic acid (144)

At the 11th World Congress of the International Society for Fat Research in Gothenberg, in June 1972, H. W. S. Chan discussed dioxygenation of various organic compounds in the coupled reaction with lipoxygenase. He found Fe in purified commercial lipoxygenase (apparently the lipoxygenase-1) in stoichiometric quantities of one to two atoms per molecule. Moreover he reported that extensive dialysis of lipoxygenase against a variety of metal chelators caused loss of enzymic activity.

Elfriede Pistorius and I were not successful in repeating Chan's experiments on dialysis inactivation with pure lipoxygenase-1. Using atomic absorption analysis we did find, however, that all four isoenzymes

(L-1, -2, -3, and -4), when freshly prepared and highly active, contained close to one atom of Fe per molecule of lipoxygenase (*147, 148*). The Fe could not be removed with o-phenanthroline which is a strong chelator for Fe^{2+}. However if a low concentration of mercaptoethanol were added, itself insufficient to inactivate the enzyme, Fe^{2+} was readily extracted, with a corresponding decline in enzyme activity. Other reducing agents could be substituted for the cysteine. Tiron, a chelator with an excellent stability constant for Fe^{3+}, slowly inactivated the enzyme in the absence of cysteine. The denatured enzyme readily gave up its iron to Tiron (*147*). Lipoxygenase does not contain acid-labile sulfur. Chan has now reported that lipoxygenase-1 contains one atom of iron per molecule (*149*). Roza and Francke have obtained a similar result but could not verify Chan's results with the chelators (*150*).

Since the senses are sharpened by prior expectation, we could detect by eye a slight reddish-yellow coloration in concentrated solutions of the enzyme. On spectrophotometric examination a gradually ascending, unstructured spectrum was seen which rose from the red into the ultraviolet region where interference from the normal protein spectrum of the concentrated enzyme became serious. The molar extinction coefficient at 400 nm for lipoxygenase-1 is approximately 300 and varies with the age and quality of the preparation.

In collaboration with Graham Palmer of the University of Michigan, Pistorius and I made a preliminary examination of lipoxygenase-1 with ESR at about $10°K$, which revealed no significant signal. However when the reaction mixture obtained by adding insufficient linoleic acid (with respect to the O_2 present) was tested, a prominent signal at $g = 6$ appeared. In excess linoleic acid the reaction mixture forms hydroperoxide and quickly becomes anaerobic. Under these conditions the hydroperoxide formed is destroyed by the linoleic acid and enzyme present, and no signal is seen.

Experiments using the UV-visible spectrophotometer at $25°C$ gave results in accord with the ESR studies. When lipoxygenase was titrated with enzyme-generated 13-L hydroperoxide under anaerobic conditions, the absorption at 360 nm increased to a plateau. Addition of linoleic acid abolished the increase at 360 nm. These results indicate that the structure responsible for the signal at $g = 6$ is also responsible for the increase in absorption at 360 nm.

A related experiment, carried out with the cooperation of Karl G. Brandt on a stopped-flow apparatus, confirmed the results of the spectrophotometric experiment. It also provided an order of magnitude for the rate of the production and destruction of the 360 nm-absorbing structure under empirical conditions. In this experiment enzyme solution was rapidly mixed with substrate solution containing excess linoleic acid. As the hydroperoxide product was generated, the absorbance at 360 nm increased, reaching a peak in about 30 msec. Since the solution was now anaerobic and excess linoleic acid was still present, the increment of absorption was quickly wiped out.

We believe that these results indicate that the normal enzymatic reaction involves Fe. However further investigation is required to explain the changes in the visible spectrum and in the ESR signals in terms of ligand shift, structure of the presumed Fe complex, and valence change in Fe. The enzymes must be fresh and of high potency for an ideal exhibition of the behavior described.

Acknowledgments

I am indebted to Elfriede K. Pistorius and John P. Christopher for their skilled work, stimulation, kindness, and tolerance—all of which contributed to our laboratory results included in this review.

Literature Cited

1. Grosch, W., *Fette, Seife, Anstrichmittel* (1972) **74**, 375.
2. Tappel, A. L., *Enzymes* (1963) **8**, 235.
3. Tappel, A. L., *Autoxidation and Antioxidants* (1961) **1**, 325.
4. Bohn, R. M., Haas, L. W., in "Chemistry and Methods of Enzymes," J. B. Sumner and G. F. Somers, Eds., 3rd ed., Academic, New York, 1953.
5. Haas, L. W., Bohn, R. M., U.S. Patents **1,957,333-37** (1934).
6. Andre, E., Hou, K.-W., *Compt. Rend. Acad. Sci., Paris* (1932) **194**, 645.
7. Andre, E., Hou, K., *Compt. Rend. Acad. Sci., Paris* (1932) **195**, 172.
8. Navez, A. E., cited in *J. Biol. Chem.* (1936) **114**, 727.
9. Craig, F. N., *J. Biol. Chem.* (1936) **114**, 727.
10. Sumner, J. B., Sumner, R. J., *J. Biol. Chem.* (1940) **134**, 531.
11. Tauber, H., *J. Amer. Chem. Soc.* (1940) **62**, 2251.
12. Hauge, S. M., Aitkenhead, W., *J. Biol. Chem.* (1931) **93**, 657.
13. Hauge, S. M., *J. Biol. Chem.* (1935) **108**, 331.
14. Sumner, R. J., *Ind. Eng. Chem., Anal. Ed.* (1943) **15**, 14.
15. Balls, A. K., Axelrod, B., Kies, M. W., *J. Biol. Chem.* (1943) **149**, 491.
16. Theorell, H., Holman, R. T., Åkeson, A., *Arch. Biochem.* (1947) **14**, 250.
17. Hamberg, M., Samuelsson, B., *J. Biol. Chem.* (1967) **242**, 5536.
18. Wooltorton, L. S. C., Jones, J. D., Hulme, A. C., *Nature* (1965) **207**, 999.
19. Strain, H. H., *J. Amer. Chem. Soc.* (1941) **63**, 3542.
20. Holden, M., *Phytochemistry* (1970) **9**, 507.
21. Franke, W., Frehse, H., *Z. Physiol. Chem.* (1953) **295**, 333.
22. Graveland, A., Pesman, L., van Erde, P., *Tech. Quart.* (1972) **9**, 97.
23. Hale, A. S., Richardson, T., Von Elbe, J. H., Hagedorn, D. J., *Lipids* (1969) **4**, 209.
24. Koch, R. B., *Arch. Biochem. Biophys.* (1968) **125**, 303.
25. Guss, P. L., Richardson, T., Stahmann, M. A., *Cereal Chem.* (1967) **44**, 607.
26. Siddiqi, A. M., Tappel, A. L., *J. Amer. Chem. Soc.* (1957) **34**, 529.
27. O'Reilly, S., Prebble, J., West, S., *Phytochemistry* (1969) **8**, 1675.
28. Zimmerman, D. C., Vick, B. A., *Lipids* (1973) **8**, 264.
29. Gardner, H. W., Christianson, D. D., Kleiman, R., *Lipids* (1973) **8**, 271.

30. Süllmann, H., *Experentia* (1945) **1**, 322.
31. Grossman, S., Trop, M., Avtalion, R., Pinsky, A., *Lipids* (1972) **7**, 467.
32. Beaux, Y., Drapron, J. N., Caillat, J. M., *Biochimie* (1973) **55**, 253.
33. Vioque, E., Maza, M. P., *Grasas Aceites* (1971) **22**, 22.
34. Zimmerman, D. C., Vick, B. A., *Lipids* (1970) **5**, 392.
35. Pistorius, E., Axelrod, B., unpublished results.
36. Gardner, H. W., Weisleder, D., *Lipids* (1970) **8**, 678.
37. Fritz, G., Beevers, H., *Arch. Biochem. Biophys.* (1956) **55**, 436.
38. Frehse, H., Franke, W., *Fette, Seifen, Anstrichmittel* (1956) **58**, 403.
39. Cantarelli, C., *Olii Mineral., Grassi Saponi Colori Verniei* (1960) **37**, 2.
40. Erikson, C. E., Svensson, S. G., *Biochim. Biophys. Acta* (1970) **198**, 449.
41. Siddiqi, A. M., Tappel, A. L., *Arch. Biochem. Biophys.* (1956) **60**, 91.
42. Dillard, M. G., Henick, A. S., Koch, R. B., *J. Biol. Chem.* (1961) **236**, 37.
43. Galliard, T., Phillips, D. R., *Biochem. J.* (1971) **124**, 431.
44. Oelze-Karow, H., Schopfer, P., Mohr, H., *Proc. Nat. Acad. Sci. U.S.A.* (1970) **65**, 51.
45. Surrey, K., *Plant Physiol.* (1967) **42**, 421.
46. Jadhav, S., Singh, B., Salunkhe, D. K., *Plant Cell Physiol.* **13**, 449.
47. Irvine, G. N., Anderson, J. A., *Cereal Chem.* (1953) **30**, 334.
48. Smith, A. K., Circle, S. J., *Soybeans: Chem. Technol.* (1972), **1**.
49. Mustakas, G. C., Albrecht, W. J., McGhee, J. E., Black, L. T., Bookwalter, G. N., Griffin, E. L., *J. Amer. Oil Chem. Soc.* (1969) **46**, 623.
50. Rackis, J. J., Honig, D. H., Sessa, D. J., Moser, H. A., *Cereal Chem.* (1972) **49**, 586.
51. Wagenknecht, A. C., Lee, F. A., *Food Res.* (1956) **21**, 605.
52. Whitfield, F. B., Shipton, J., *J. Food Sci.* (1966) **31**, 328.
53. Murray, K. E., Shipton, J., Whitfield, F. B., Kennett, B. H., Stanley, G., *J. Food Sci.* (1968) **33**, 290.
54. Eriksson, C. E., *J. Food Sci.* (1967) **32**, 438.
55. Eriksson, C. E., *J. Food Sci.* (1968) **33**, 525.
56. Grosch, W., *Z. Lebensm. Unters. Forsch.* (1968) **139**, 1.
57. Wagenknecht, A. C., *Food Res.* (1959) **24**, 539.
58. Koch, R. B., Smuli, J. W., Henick, A. S., *J. Amer. Oil Chem. Soc.* (1959) **36**, 205.
59. St. Angelo, A. J., Dupuy, H. P., Ory, R. L., *Lipids* (1972) **7**, 793.
60. Saijyo, R., Takeo, T., *Plant Cell Physiol.* (1972) **13**, 991.
61. Sumner, J. B., Smith, G. N., *Arch. Biochem.* (1947) **14**, 87.
62. Teng, J. I., Smith, L. L., *Fed. Proc. Fed. Amer. Soc. Exp. Biol.* (1972) **31(2)**, 912.
63. Wynn, J., *J. Biol. Chem.* (1970) **245**, 3621.
64. Fritz, C., Beevers, H., *Plant Physiol.* (1955) **29**, 69.
65. Frey, C. N., Schultz, A. S., Light, R. F., *Ind. Eng. Chem.* (1936) **28**, 1254.
66. Axelrod, B., unpublished results.
67. Matsuo, R. R., Bradley, J. W., Irvine, G. N., *Cereal Chem.* (1970) **47**, 1.
68. Buckle, K. A., Edwards, R. A., *J. Sci. Food Agr.* (1970) **21**, 307.
69. Irvine, G. N., *J. Amer. Oil Chem. Soc.* (1955) **32**, 558.
70. Auerman, L. Ya, Popov, M. P., Dubtsov, G. G., Samsonov, M. M., *Prikl. Biokhim. Mikrobiol.* (1971) **7**, 678.
71. Auerman, L. Ya., Ponomareva, A. N., Polandova, R. D., Klimova, G. S., *Tr. Vses. Nauch. Issled. Inst. Khlebopek Prom.* (1971) **12**, 95, in *Chem. Abstr.* (1972) **77**, 356.
72. Drapron, R., Beaux, Y., *Comp. Rend. Acad. Sci. Paris, Ser. D* (1969) **268**, 2598.
73. Frazier, P. J., Leigh-Dugmore, F. A., Daniels, N. W. R., Eggitt, P. W. R., Coppock, J. B. M., *J. Sci. Food Agr.* (1973) **24**, 421.
74. Mecham, D. K., in "Wheat, Chemistry and Technology," I. Hlynka, Ed., p. 353, American Association of Cereal Chemists, St. Paul, 1964.

75. Graveland, A., *J. Amer. Oil Chem. Soc.* (1970) **47**, 352.
76. Graveland, A., *Biochem. Biophys. Res. Commun.* (1970) **41**, 427.
77. Lieberman, M., Mapson, L. W., *Nature* (1964) **204**, 343.
78. Meigh, D. F., Hulme, A. C., *Phytochemistry* (1965) **5**, 863.
79. Galliard, T., Hulme, A. C., Rhodes, M. J. C., Wooltorton, L. S. C., *FEBS Letters* (1968) **1**, 283.
80. Lona, F., Raffi, F., *Ateneo Parmense Sez. 2*, (1972) **8**, 45.
81. Firn, R. D., Friend, J., *Planta* (1972) **103**, 263.
82. Holman, R. T., *Arch. Biochem.* (1948) **17**, 459.
83. Hammond, E. G., Fehr, W. R., Snyder, H. E., *J. Amer. Oil Chem. Soc.* (1972) **49**, 33.
84. Fritz, G. J., Miller, W. G., Burris, R. H., Anderson, L., *Plant Physiol.* (1958) **33**, 159.
85. Holman, R. T., Panzer, F., Schweigert, B. S., Ames, S. R., *Arch. Biochem.* (1950) **26**, 199.
86. Stevens, F. C., Brown, D. M., Smith, E. L., *Arch. Biochem. Biophys.* (1970) **136**, 413.
87. Christopher, J. P., Pistorius, E. K., Axelrod, B., unpublished results.
88. Weber, K., Osborn, M., *J. Biol. Chem.* (1969) **244**, 4406.
89. Grosch, W., Höxer, B., Stan, H.-J., Schormüller, J., *Fette, Seifen Anstrichmittel* (1972) **74**, 16.
90. Kies, M. W., *Fed. Proc. Fed. Amer. Soc. Exp. Biol.* (1947) **6**, 287.
91. Theorell, H., Bergström, S., Akeson, A., *Pharm. Acta Helv.* (1946) **21**, 318.
92. Haining, J. L., "The Mechanism of Action of Lipoxidase," M.S. Thesis, Purdue University, 1957.
93. Schroeder, D. H., "Some Properties of Soybean Lipoxygenase," Ph.D. Thesis, Purdue University, 1968.
94. Kies, M. W., Haining, J. L., Pistorius, E., Schroeder, D. H., Axelrod, B., *Biochem. Biophys. Res. Commun.* (1969) **36**, 312.
95. Koch, R. B., Stern, B., Ferrari, C. G., *Arch. Bioch. Biophys.* (1958) **78**, 165.
96. Christopher, J., Pistorius, E., Axelrod, B., *Biochim. Biophys. Acta* (1970) **198**, 12.
97. Yamamoto, A., Yasumoto, K., Mitsuda, H., *Agr. Biol. Chem.* (1970) **34**, 1169.
98. Christopher, J. P., "Isoenzymes of Soybean Lipoxygenase: Isolation and Partial Characterization," Ph.D. Thesis, Purdue University, 1972, and Pistorius, E. K., unpublished results.
99. Christopher, J. P., Pistorius, E. K., Axelrod, B., *Biochem. Biophys. Acta* (1972) **284**, 54.
100. Schormüller, J., Neber, J., Höxer, B., Grosch, W., *Z. Lebensm. Unters. Forsch.* (1969) **139**, 357.
101. Hurt, G. B., unpublished results.
102. Veldink, G. A., Garssen, G. J., Vliegenthart, J. F. G., Boldingh, J., *Biochem. Biophys. Res. Commun.* (1972) **47**, 22.
103. Cosby, E. L., Sumner, J. B., *Arch. Biochem.* (1945) **8**, 259.
104. Sumner, J. B., Smith, G. N., *Arch. Biochem.* (1947) **14**, 87.
105. Miller, J. J., unpublished results.
106. Surrey, K., *Plant Physiol.* (1964) **39**, 65.
107. Holman, R. T., *Arch. Biochem.* (1947) **15**, 403.
108. Allen, J. C., *Chem. Commun.* (1969) **16**, 609.
109. Blain, J. A., Shearer, G., *J. Sci. Food Agr.* (1965) **16**, 373.
110. Holman, R. T., Egwin, P. O., Christie, W. W., *J. Biol. Chem.* (1969) **244**, 1149.
111. Hamberg, M., Samuelsson, B., *J. Biol. Chem.* (1967) **242**, 5329.
112. Hamberg, M., *Anal. Biochem.* (1971) **43**, 515.

113. Bergström, S., *Ark. Kemi* (1946) **21A**, 1.
114. Dolev, A., Rohwedder, W. K., Dutton, H. J., *Lipids* (1967) **2**, 28.
115. Eriksson, C. E., Leu, K., *Lipids* (1971) **6**, 144.
116. Veldink, G. A., Vliegenthart, J. F. G., Boldingh, J., *Biochem. J.* (1970) **120**, 55.
117. Chang, C. C., Esselman, W. J., Clagett, C. O., *Lipids* (1971) **6**, 100.
118. Veldink, G. A., Vliegenthart, J. F. G., Boldingh, J., *Biochem. Biophys. Acta* (1970) **202**, 198.
119. Hamberg, M., Samuelsson, B., *Biochem. Biophys. Res. Commun.* (1965) **21**, 531.
120. Galliard, T., Phillips, D. R., *Biochem. J.* (1971) **124**, 431.
121. Christopher, J., Axelrod, B., *Biochem. Biophys. Res. Commun.* (1971) **44**, 731.
122. Christopher, J., unpublished results.
123. Privett, O. S., Nickell, C., Lundberg, W. O., Boyer, P. D., *J. Amer. Assn. Oil Chem.* (1955) **32**, 505
124. Egmond, M. R., Vliegenthart, J. F. G., Boldingh, J., *Biochem. Biophys. Res. Commun.* (1972) **48**, 1055.
125. Veldink, G. A., Vliegenthart, J. F. G., Boldingh, J., *Biochim. Biophys. Acta* (1970) **202**, 198.
126. Haining, J. L., Axelrod, B., *J. Biol. Chem.* (1957) **232**, 1958.
127. Smith, W. L., Lands, W. E. M., *J. Biol. Chem.* (1972) **247**, 1038.
128. Chistophersen, B., *Biochim. Biophys. Acta* (1968) **164**, 34.
129. Tappel, A. L., Lundberg, W. O., Boyer, P. D., *Arch. Biochem. Biophys.* (1953) **42**, 293.
130. Chan, H. W. S., *J. Amer. Chem. Soc.* (1971) **93**, 2357.
131. Finazzi Agro, A., Giovagnoli, C., De Sole, P., Calabrese, L., Rotilio, G., Mondovi, B., *FEBS Letters* (1972) **21**, 183.
132. Holman, R. T., Burr, G. O., *Arch. Biochem.* (1945) **7**, 47.
133. Vioque, E., Holman, R. T., *Arch. Biochem. Biophys.* (1962) **99**, 522.
134. Garssen, G. J., Vliegenthart, J. F. G., Boldingh, J., *Biochem. J.* (1971) **122**, 327.
135. Garssen, G. J., Vliegenthart, J. F. G., Boldingh, J., *Biochem. J.* (1972) **130**, 435.
136. Axelrod, B., Pistorius, E. K., Christopher, J. P., *Proc. Intern. World Congress Soc. Fat Res., 11th,* 1972 (abstr.).
137. Blain, J. A., Barr, T., *Nature* (1961) **190**, 538.
138. Gini, B., Koch, R. B., *J. Food Sci.* (1961) **26**, 359.
139. Pistorius, E., Axelrod, B., unpublished results.
140. Zimmerman, D. C., *Biochem. Biophys. Res. Commun.* (1966) **23**, 398.
141. Zimmerman, D. C., Vick, B. A., *Plant Physiol.* (1970) **46**, 445.
142. Gardner, H. W., *J. Lipid Res.* (1970) **11**, 311.
143. Ben-Aziz, A., Grossman, S., Ascarelli, I., Budowski, P., *Anal. Biochem.* (1970) **34**, 88.
144. Galliard, T., Phillips, D. R., *Biochem. J.* (1972) **130**, 435.
145. Johns, E. B., Pattee, H. E., Singleton, J. A., *J. Agr. Food Chem.* (1973) **21**, 570.
146. Walker, G. C., *Biochem. Biophys. Res. Commun.* (1963) **13**, 431.
147. Pistorius, E. K., Axelrod, B., *Fed. Proc. Fed. Amer. Soc. Exp. Biol.* (1973) **32(3)**, 544.
148. Pistorius, E. K., Axelrod, B., *J. Biol. Chem.* (1974) (April).
149. Chan, H. W.-S., *Biochim. Biophys. Acta* (1973) **327**, 32.
150. Roza, M., Francke, A., *Biochim. Biophys. Acta* (1973) **327**, 24.

RECEIVED September 17, 1973. Purdue University Agricultural Experiment Station Journal Paper No. 5259.

INDEX

The text of this book is set in 10 point Caledonia with two points of leading. The chapter numerals are set in 30 point Garamond; the chapter titles are set in 18 point Garamond Bold.

The book is printed offset on Danforth 550 Machine Blue White text, 50-pound. The cover is Joanna Book Binding blue linen.

Jacket design by Norman Favin.
Editing and production by Virginia Orr.

The book was composed by the Mills-Frizell-Evans Co., Baltimore, Md., printed by The Maple Press Co., York, Pa., and bound by Complete Books Co., Philadelphia, Pa.